Franz Keller, und Andere

Explorations Made in the Valley of the River Madeira, from 1749 to

1868

Franz Keller, und Andere

Explorations Made in the Valley of the River Madeira, from 1749 to 1868

ISBN/EAN: 9783744791434

Printed in Europe, USA, Canada, Australia, Japan

Cover: Foto ©berggeist007 / pixelio.de

More available books at **www.hansebooks.com**

EXPLORATIONS

MADE IN THE

VALLEY OF THE RIVER

MADEIRA,

FROM

1749 to 1868.

PUBLISHED FOR THE NATIONAL BOLIVIAN NAVIGATION COMPANY.

1875.

INDEX.

— ◆ -

INTRODUCTION.

THE falls and rapids of the River Madeira, as the only barrier which exists between commercial nations and a direct access to the richest slope of the Andes, are worthy of the study of the geographer, the traveller, the lover of progress, and especially of those ambitious to develop commercial intercourse among nations.

Those interested in defeating the efforts of Bolivia to open a trade avenue to the Atlantic have sought for and accepted information from many sources relative to the country traversed by the river Madeira. Whether the informant was from Soudan, Sumatra, Chili, China, Patagonia or Peru, was immaterial, provided he possessed the primary qualification of never having been within a thousand miles of the district he professed to describe. With the hope of contributing to the knowledge of those who desire to impartially examine the geographical and engineering features of the problem Bolivia is endeavouring to solve, I have collected, translated, and caused to be translated the following works. These are the results of scientific and other voyages given to the world long previous to the time when the Government of Bolivia invited me to undertake the task of constructing a railway to connect the 3,000 miles of navigable upper waters of the Madeira with a point on the lower river called San Antonio, accessible to sea-going steamers.

The first efforts to penetrate the valley of the Madeira were from its head waters, near Cuzco, upon the accession of the tenth King, the Ynca Yupanqui, to the throne of the Empire of the Yncas. He invaded Musu, now called Moxos, with an army of 10,000 men in a vast fleet of canoes, built for the purpose. Upon the conquest of Perú by Spain, several expeditions were sent to make discoveries in the region drained by the Beni, the Mamoré, and other tributaries of the Madeira. The first was dispatched by Pizarro in 1539, and numbered 200 men, under Captain Candia. It penetrated to a populous town called Oputari, only 30 leagues to the east of Cuzco. The expedition, again reorganised under a new commander, Peranzures, explored much of the country towards the Beni and Moxos, penetrating 60 leagues from Cuzco over a well-cleared, Inca road.

In 1561, two different expeditions were organised by the Viceroy Count de Nieva, the only tangible result of which was the founding of the town of Apolobamba.

In 1562, Anton de Gastos, with a small party, entered Moxos from Cochabamba and discovered the great river Mamoré.

In 1563, under authority from the Viceroy, Diego Aleman reached a point 60 leagues north of Cochabamba.

In 1565, Luxan led a party of eight men in search of gold to a point 20 leagues north of Cochabamba, where they were all killed.

The great expedition of Juan Alvarez Maldonado, of Cuzco, invaded the Moxos district in 1567, and did not return until 1569, making many valuable discoveries, and tracing the "east and west" course of the great Mayu-tata branch of the Beni river for a long distance.

In 1569, a small expedition of 60 men performed a journey on their own account north of Cochabamba into the valley of the Mamoré, but were ordered back by the Audiencia de los Charcas.

Numerous other exploring and warlike parties continued to penetrate Moxos and Chiquitos during the settlement of the present Bolivia by the Spaniards.

The eastern part of Bolivia, drained by the head waters of the Madeira, was settled by Spanish expeditions from Paraguay, the first important point populated being Santa Cruz de la Sierra.

During the 16th and 17th centuries, the Jesuits and Franciscans made numerous reductions of the Indians of the Beni, Moxos and Chiquitos, establishing many flourishing towns.

The first expedition which ascended the Madeira into Bolivia of which I have been able to gain any knowledge was made by the Portuguese from the city of Parà in 1723, under Francisco Melho Palheta. No definite account of it has fallen into my hands. If any exists it is probably to be found in the Brazilian archives. The party reached the present Bolivian town of Exaltacion on the river Mamoré.

The first descent of the rapids of the Madeira of which I have found any account was made by a small party from Mato Grosso in 1742.

In 1749, by command of the King of Portugal, a large expedition ascended the Amazon, Madeira, and Itinez rivers to Mato Grosso in Western Brazil. A full history of the exploration is found herein. Its lack of mathematical and scientific data renders it of little real value; and it can only serve as a rough description of the country traversed. The sixteenth rapid which they found, and which is described on page 282, no longer exists. The river has cut away the small obstacles which caused the violent current then existing at that point.

Next came Señor José Augustin Palacios, a Bolivian engineer of much talent and painstaking observation, who, in 1844, was instructed by his Government to examine the Beni Department and explore also the rapids of the Madeira. He did this as well as the feeble means at his command would permit. The result is herein published, and his report is replete with valuable information. I found Señor Palacios still living at La Paz in 1872.

In 1853, the United States Government took a great interest in the problem of opening Bolivia to the commercial world. I have therefore thought it desirable to give extracts from the work of Lieut. Gibbon relative to his exploration of the falls of the Madeira.

The pamphlet of Señor Arauz reproduced gives a few valuable and practicable observations from an energetic Bolivian who has seen much of the country of which he speaks.

In 1861, the celebrated General Quintin Quevedo of Bolivia descended the rapids of the Madeira, and became so convinced that this was the natural outlet of this country, that he wrote the pamphlet found in these pages, describing the falls of the great river and the best method of surmounting them.

When the boundary line treaty was made in 1867-8 between Brazil and Bolivia, the former Government sent the Keller expedition to make a thorough examination of the rapids of the Madeira, and report upon the best method of avoiding these obstacles to commercial communication. The report of the Messrs. Keller will be found of scientific interest, and for accurate and valuable data it is of great importance.

My own journals kept in my descent of the Piray, the Mamoré, and Madeira I have never published.

The object of the present volume is to give in a compact form other evidence than my own relative to the river Madeira and the rich country which it drains, and to call greater attention to the vast system of natural canals which are destined to play a great rôle in the commercial and political history of South America.

Very truly,

GEORGE EARL CHURCH.

Exploration of the River Madeira.

REPORT

OF

JOSÉ AND FRANCISCO KELLER,

MADE TO THE

IMPERIAL GOVERNMENT OF BRAZIL,

AND

PUBLISHED IN THE GOVERNMENT "RELATORIO" OF 1870.

Translated from Portuguese by GEORGE EARL CHURCH.

LONDON:

WATERLOW & SONS, PRINTERS, GREAT WINCHESTER STREET

1873.

RELATION OF THE EXPLORATION OF THE RIVER MADEIRA:

OF THE PART INCLUDED BETWEEN THE RAPID OF SANTO-ANTONIO AND THE MOUTH OF THE MAMORÉ.

A.—Introduction.

B.—Short description of the voyage of the expedition.

C.—Climatic and geologic ideas.

D.—Results of hydrographic measurements.

E.—Projects for the improvement of the actual route, as much with reference to facilitating navigation as to the construction of a marginal road.

F.—Statistical data relative to commerce and the productions of the valley of the Madeira, Guaporé, Mamoré, and branches.

G.—Comparative calculation of freights by the different routes.

H.—Conclusion.

A.—INTRODUCTION.

By orders of the 10th October, 1867, we were charged by the Imperial Government to explore the River Madeira, in the part filled with rapids from Santo-Antonio to the mouth of the River Mamoré, and to elaborate the most appropriate projects for the improvement of this important line of communication with the province of Mato-Grosso and the Republic of Bolivia.

We embarked with this object the 15th day of November, 1867. on board the steamer *Paraná*, of the line of steam packets for the northern provinces, in which we arrived December 1st at Pará;

and afterwards, taking the river-steamer Belem, we landed at the port of Manãos the 10th December.

The then President of the Province of Amazonas, Dr. José Coelho da Gama e Abreu, who knows perfectly the climatic character of those regions, informed us that the season of the year was least favourable to undertake such an exploration, the ascending of the rapids and the study of the river in the time of floods being difficult and arduous.

That we might employ our time in a useful manner, His Excellency, the same President, in accord with His Excellency the Minister, instructed us to make a plan of a part of the course of the river Negro, as well as that of the city of Manãos. The latter work was executed with all exactitude by our assistant, Mr. José Manuel da Silva, but for the execution of the first it was not possible to find at Manãos or in its vicinity the necessary number of oarsmen to man two canoes, it remained in project, and in place of it we made a plan and estimate for the reconstruction of one of the bridges existing in the city which was in a ruined condition.

To our great regret, it was during this time Dr. Gama e Abru removed to the Province of Goyaz as President, and all our efforts to obtain of his successor the persons necessary to enable us to comply with the orders of His Excellency the Minister of Public Works relative to the exploration of the river Madeira were fruitless; not even could we obtain the number of guards, nor the small steamer *Jurupensen*, which, with so much economy for the public coffers, might have carried the *personnel* and train of the expedition to Santo-Antonio, although she was unemployed and at the disposal of the President himself in the port of Manãos.

With the firm resolution to accomplish the honourable mission with which we had been charged, despite all impediments and mischances, we applied to Mr. Ignacio Araus, Vice-Consul of the Republic of Bolivia, who at that time was at Manãos.

He having already received previous notice from his Government with respect to the projected exploration, showed himself disposed to aid us in every way possible. It was through him that we made the acquaintance of a Bolivian merchant, who had been to Pará, and from there was returning to Bolivia.

We agreed then, that, from the residence of the above Vice-Consul, the merchant should furnish a well-manned canoe for the use of the expedition, and that from Serpa to that place we were to go in our own canoe.

The said Vice-Consul gave us hopes that in Bolivia we might find, finally, through his Government, a sufficient number of oarsmen to return and execute the hydrographic studies.

In spite of having in this manner the disgust of seeing a commission under orders of the Imperial Government continue its voyage under the auspicies of a private individual, we were forced to accept the proposition, and after having bought a canoe of 300 arrobas burden in Serpa of Major Damaso de Souza Barriga, badly constructed and in a bad condition for want of another, we continued our voyage the 30th May, 1868, following the Bolivian merchant with only seven boatmen, when the canoe required at least twelve.

B.—DESCRIPTION OF THE VOYAGE.

The cargo of our craft consisted of supplies for four months, iron for construction and repair of the canoe, medical instruments, awnings and tents.

The Amazon being 55 palms or 12 mètres above low water, it was impossible for us to overtake the Bolivian merchant, who had the start, for want of a favourable wind, and against a strong current.

In the town of Borba, 25 leagues above the mouth of the Madeira, where we arrived the 9th of June with a still reduced crew, from the fact that two of our rowers ran away from us, it was impossible to obtain from the authorities the aid necessary to engage others.

After having passed a settlement of Muras Indians, called Sapucaia-oroca, the island of Araras and the mouth of the River Aripuaná, we had, on the night of the 14th-15th of June, to unload and repair our craft in all haste, for it made water so fast that it threatened to founder.

On the afternoon of the 15th we passed to the rocks of Uroá.* At this place the fall of the river augments considerably, and the main channel on the left side, between the island of the same name and the main land, has, in the season of low water, only a depth of $1\frac{1}{2}$ mètres of water over the layers of sandstone.

On the 18th of June we finally arrived at the establishment of Señor Ignacio Araus, where the Bolivian merchant awaited us with impatience.

After having transhipped the cargo into a more appropriate canoe, we returned the first to Major Damaso do Souza Barriga, in Serpa.

The party consisted in all of eight whites and seventy Bolivian Indians from the missions of the Mamoró, as rowers and pilots of the seven canoes, and thus we continued our voyage the 21st of June.

The waters of the river, which were yet six mètres above low water, were going down continually, and the banks, which crumbled all along, required great attention, principally in the selection of an anchorage.

On the 27th of June we left the Baêtas, at the mouth of the little affluent on the left bank of the Madeira, where, a few years ago, there was a *malocca* of Muras Indians, who dispersed on account of being persecuted by recruiters.

On the 30th of June we met a family of these same Muras, true nomads of the Amazon, who in ten small canoes were going up the river to be present on the banks of the Upper Madeira at the time the turtles go there to lay their eggs. For some days they followed our canoes, and thus we were able to buy from them some turtles, for the hunting of which they have, with an arrow termed *sararaca*, an extraordinary aptitude. The river was then at medium height, and was five mètres above low water mark, as we learned from some indiarubber collectors, and also by the elevation of the characteristic vegetation on the river margins.

Above the place called Tres-Casas, we visited the shed of some Bolivian rubber gatherers, who with a number of Mojos Indians, employed themselves in the extraction of this lucrative resin. They

---

* Note by Translator.—Much doubt exists about the depth of water at this point. It was passed, so I have been informed by good authority, by the steamer *Jenmary*, drawing 8 feet of water, in the driest of the dry season of 1872.

showed themselves exceedingly satisfied with the hope of seeing in a short time the waters of the lower Maderia at least ploughed by steamers, and filled with hope with regard to the results of the exploration with which we were entrusted.

On the 5th of July we arrived near Crato, now a cattle farm, for which the neighbouring fields, only in part known, and which in all probability extend as far as the Beni, are admirably adapted. At the date of our arrival the detachment of San Antonio was there, which, on account of the fever season that prevails every year at San Antonio at the beginning of the rainy period, had retired to a healthier place.

The commander of the detachment, for whom we brought orders from the Government, received us most cordially, and it was through him that we became acquainted with the actual proprietor of the place, Captain Tenorio.

The flourishing establishment, for the foundation of which the first cattle came from Bolivia from the estates of the distinguished Brazilian citizen, Antonio de Barros Cardozo, who was the first to have the nerve to transport cattle in launches or great canoes along the rapids of the Madeira. This establishment, we repeat, will one day be of immense advantage to the Province of Amazonas, which lacks in fresh meat.

The width of the river front of Crato is 900 mètres, while 1,000 mètres below, at the the widest point at the islands, it is 1,500 mètres.

The climate of Crato is excellent, despite the unfortunate fame which the outpost of the same name enjoyed in colonial times and the first years of independence. It would appear all the more extraordinary since the neighbouring fields always existed, and since no greater destruction has occurred, if it were not known that the old Crato is not located in the same place; but, beyond doubt, higher up at the mouth of the Jammary, a place which to-day is yet reputed as very unhealthy.

The river Madeira at this time had already lowered to 4 mètres above low water mark. The banks of sand on the convex sides showed themselves more and more.

We cannot fail to mention here a grave fact of which the rubber collectors of the Madeira generally complain, it is that the Indian boatmen of the Bolivian canoes, with the consent and approval of their masters, who often fail to bring with them sufficient supplies for such a long voyage, constantly rob the neighbouring farms on the river margins, even opposing force to the proprietor who wishes to defend the fruits of his work. To remedy this evil, at least to a certain extent, the authorities at the port of departure for the canoes, in Serpa, should require the owner of the boats to prove that his supplies are sufficient for his crew; considering that the length of the voyage is almost always the same. At the detachment in Crato or in San Antonio the same could be again examined, holding responsible the owners of the canoes, who are almost always the same, for the damages to the complaining rubber gatherers.

On the 10th of July we passed the Island of Abelhas, taking the channel of the left margin, that of the opposite margin being in part obstructed by rocks of ferruginous conglomerate which is so frequently found in the vast valley of the Amazon.

The mouth of the Pirapitinga being passed, we arrived at that of the Jammary which has two mouths, its width being 50 mètres. The fisheries at the mouth of this stream are extraordinary, and this would be an excellent point for a settlement were it not for the malignant fevers which infest it.

The 16th of July we passed the Island of Mutuns and the shore Tamanduá, where, in the month of September thousands of turtles, which go up the river, deposit their eggs in the sand. It is the same place where a great number of rubber gatherers. fishermen, &c., annually meet for the purpose of collecting eggs for the manufacture of butter, and to turn over and carry away a great number of turtles.

The destruction proceeding from these manufactures of butter can be estimated, say by an approximate calculation. There are 2,000 jars of butter made on the shore. Now, as for each jar 2,000 eggs more or less are required, as I am informed, it amounts to the enormous quantity of 4,000,000 of eggs, the number annually destroyed, to make an oil which hardly serves for a lamp. Yet, not contented with the destruction of eggs, and with the carrying off thousands of great turtles, they return again after a certain time to carry away canoe-loads of the small turtles, scarcely out of the shell. In consequence of such great destruction repeated every year, aside from hunting them, which is done at all times with arrows, these animals must diminish considerably, and if the Government does not take some means to restrict at least the manufacture of butter, which has an advantageous substitute in oils of vegetable origin, this important article of food for the inhabitants of the province of Amazonas will disappear little by little.

On the 16th of July we finally arrived at the rapid of San Antonio, at an elevation of 61ᵐ 6 above the level of the sea. On the left bank we caught sight of the huts of straw of the detachment, which, in consequence of the fevers we have already stated, had retired to Crato. At this first rapid it is necessary to unload the canoes and transport everything 450 mètres over the rocks on the left bank. Accompanying the Indian boatmen we discovered on different ledges of granite rock on the left bank open grooves in the rock, which showed straight lines crossing each other, now at acute angles, now at right angles, and presenting in the cavities, of 0ᵐ.01 of depth, a polished surface. Later we found the same signs in greater numbers on the rocks at the fall of Theotonio, and upon others higher up; yet at the rapid of Ribeirão, below the mouth of the small affluent that gives its name to the rapid, we found, in a ledge on the right bank, and almost at low-water level, clearly defined signs in the extremely hard rock, which have the character of letters, of which the eaten outlines indicate an extraordinary age.

The great and patient labour which was necessary to cut these grooves in stone of this nature, without any iron tools, and only by erosion with another stone, leads to the belief that these signs are not the labour of indolence, and that they have some signification, principally those of Ribeirão, which form an interesting parallel with the rough representations of celestial objects and of animals upon the rocks of the Orinoco, described by Humboldt.

MACACOS—THEOTONIO.

On the 18th July we arrived at the current of Macacos, from which is seen the smoke which rises from the fall of Theotonio. The height of the fall of a rapid depends upon the condition of the river, that is, of the height of the waters above low water. We found that of Theotonio three mètres above low water was eight mètres. The width of the river was 1,100 mètres, so that despite the enormous volume of water and the height above, the picturesque effect of the fall was not great.

The hills on each side of the fall are from thirty to fifty mètres high. The temperature of the bank of sand below the fall where we had rested was almost insupportable, and passed 32° R.

The canoes after being unloaded are carried over the land on rollers, over an elevation of fifteen mètres, where yet are found walls of the outpost, which at the close of the past century the Portuguese Government ordered to be established at this point.

After having transported the cargo a distance of 500 mètres above the fall and the canoes by land, one of these unfortunately, by a blow against a stone, stove in some holes; but the voyager on the Madeira carries all his iron and carpentry with him, and in less than a day the canoe was repaired. This delay, like analogous ones at other rapids, were always improved by making astronomical and hypsometrical observations, for which, at least, going up stream, the sky being clear, the best occasion offered.

The riches in fish (surubins, pintados, tambakis, etc.) in the still water below the fall is astounding, and at certain seasons of the year they can be taken with a harpoon at the moment they make efforts to overcome the small rapids below the principal fall.

MORRINHOS.

On the 22nd of July we continued our voyage, arriving in the afternoon at the rapid of Morrinhos, which has two distinct falls; the lower one of 1$^{m.}$ 50 the upper of 0$^{m.}$ 70 at the time in which the river is 5 m. above low-water. At the first it was necessary to unload and have the canoes dragged up by ropes, the second was ascended loaded.

On the 25th, between the Morrinhos and the Caldeirão do Inferno, about ten leagues of distance without any impediment, we found some canoes contructed of the bark of the Jatubá, of the Caripunas who inhabit the banks in this neighbourhood. The crew, men and women, to the number of ten or twelve, of whom the first were nude, invited us to go to their *malocca*, which we did, presenting them with knives, treasures, fish-hooks, etc. In exchange they gave us some mandioca roots and some corn in the ear. All comported themselves well, and we separated in perfect amity.

Some days after, we met above the Caldeirão do Inferno a second horde of Caripunas, of whom in our descent, we bought in exchange for some iron, part of the rich captures they had made in hunting the tapir and the wild-hog. All of them upon either of these occasions behaved themselves always with the same calmness, without showing fear or hostile intentions of any kind. Thus it appears to us since they had only a few years previously attacked several different parties, but now recognise it to be more advantageous to themselves to live in peace, that religious instruction would now find fertile ground among them. It would be sufficient for a missionary to appear among them furnished with some sulphate of quinine to alleviate their fevers, to which they are equally subject, and by this means for ever gain their affections. In a short time this poor tribe would be transformed into beings who would be of some use to the human community.

Some old navigators of the Madeira assured us that during the last 15 or 18 years the Caripunas have considerably diminished in number, but none of them could inform us if it was in consequence of illness or on account of their retiring further from the bank of the river and from contact with the white man, so often fatal to them.

The number of members of each one of the tribes or families which we found in the ascent or descent did not exceed fifty, the total one hundred persons.

Caldeirão do Inferno.

The 26th, we saw up stream in the morning a line of hills and at mid-day we arrived at the lower part of the dangerous rapid of Caldeirão do Inferno. All along the left bank of the river, and on one of the islands, rudely-made crosses give proof of the sad fame of this rapid. Not only the terrible whirl, another Charybdis, at the upper entrance of it, which has already engulphed several canoes, killing all their crews, but it also appears that the fevers of this part of the river are more perilous than in any other.

Among the persons who have lost their lives at this rapid we will only cite the Peruvian Colonel of Engineers, Maldonado, who, persecuted in his country for political motives some five or six years since, embarked upon the Madre do Dios, descended by this to the Beni, and by the Beni to the Madeira, dying at the rapid of Caldeirão do Inferno. He thus furnished material proof that the Madre do Dios is an affluent of the Beni and not of the Purús. It is to be regretted that all the papers of Maldonado were lost on this occasion. Two of the rowers alone barely saved themselves. They were ignorant men who were unable to give extended information upon the region traversed.

Our party also had to pay a tribute to this sad spot. We were obliged to bury one of our Indian rowers who died of inflammation of the intestines, probably caused by excessive eating and constipation by the cold at night.

The bed of the river of this rapid is very irregular and divided by seven large islands, into branches of different sizes.

At medium high water and floods the canoes take the channel of the left bank, but at low-water this channel becomes nearly dry, and voyagers find themselves almost obliged to take the centre canal where the slightest inattention in the management of the canoe carries it to the abyss.

The total height of the fall, which is distributed in two principal volumes, is four mètres at medium water.

On the 27th of July we saw the hills, which are at the foot of the fall of Girão, whose general direction is from north-east to south-west. The bed of the river narrows from 1,500 to 700 mètres and the total fall of the strong rapid, at the time at which the waters were two mètres above low-water mark, is eight mètres, the the entire length being 900 mètres. At this point, 98 mètres above the level of the sea, the thermometer at night descended to 15° R. it being during the day 32° R. And the temperature of the sand on the shores 48° R.

The transportation of the cargoes and the dragging over land the canoes for a distance 720ᵐ demanded a persevering work of about four days and only on the 31st of July were we able to continue our voyage.

GIRÃO.

The 1st of August we arrived at a point where the strata of argillaceous gritstone, which appear at the bend, are known to navigators under the denomination of Pedras de Amolar. There the strong currents require the employment of ropes, still continuing upwards for four currents more, created by some banks of ferruginous conglomerate, which are passed in the same manner. At flood time nothing can be seen of all these currents, the river becoming perfectly levelled.

TRES-IRMÃOS.

The same afternoon we arrived at the stronger current of Tres-Irmãos, which has a fall of 0ᵐ 80, and close by, above it, is found the mouth of the river of the same name, 25ᵐ wide. On the left bank some hills are seen. They attain the height of 50ᵐ to 60ᵐ above the level of the river. The bed of the river, divided by a large island into two parts, is strewed with granitic rocks, yet offering channels of sufficient depth for navigation.

PAREDÃO.

The 3rd and 4th of August we passed the *cachoeira* (rapid) of Paredão, the fall of which is divided into two parts. The lesser, one mètre high, is surmounted by ropes. The upper is divided by a promontory of huge blocks of granite, characterized by enormous feldspathic crystals ; advances for 70 mètres from the right bank, and requires all the canoes to be unloaded and their cargoes carried overland. The difference of level caused by the advanced rocks is 1^m 50 with the river 3^m above low water. The width of the river divided by two islands into two branches is 1,000^m.

PEDERNEIRA.

The following day we arrived at Cachoeira Pederneira, so called from the veins of quartz which appear in the seams of the metamorphic rocks. The total fall is 1^m 70, and the cargoes are carried overland. At this point the hills constantly became more distant from the margin, and the meagre and mixed vegetation gives with the general impression of the passage an immoderate feeling of sad monotony.

The part of the river just above Pederneira is from its slight fall the most appropriate for navigation. A little above Pederneira is found on the left bank the mouth of the river Abuna, 60^m wide.

ARARAS.

The next rapid is Araras, with a fall of 1^m 60 and an island, the river attaining at the same point the considerable width of 1,500^m. The united force of all our crews, more than seventy men, was necessary to haul up our canoes one after the other through the strong currents of the left bank.

PERIQUITOS.

The current following, that of Periquitos, is caused by the rocks, which obstruct the bed of the river. It has a difference of level of 1^m 5 at low water. It was overcome in the same way.

RIBEIRÃO.

On the 9th of August at midday we arrived at the lower part of the *cachoeira* of Ribeirão. It is formed by five distinct rapids, of which the last of 5^m is the most considerable. The entire fall is 13^m at low water. The total length of this long series of obstacles to navigation is nearly six kilomètres or one league. The passage of the canoes at the different points depends, as at all the others, upon the height of the water. When we went up, about about 1ᵐ 50 above low water, we passed the first by means of ropes without much difficulty, with reference to the second and third, it was necessary to unload the canoes. It was at this point where the unknown inscription above mentioned is found upon a block of stone. In the passage of the third a small canoe (*montaria* of the Amazon) slipped from the cable and went to the bottom, the bowman escaping by a miracle.

For the passage of the last fall, called Cabeça do Ribeirão, it is not only necessary to carry by land the cargoes, but also drag the canoes a distance of 300 metres.

It was near this last point that, at the side of the small Ribeirão, from which the rapid takes its name, there existed thirty years ago a military outpost, the commander of which was assassinated by the Caripuna Indians for violence practised against them.

The passage of the *Cachoeira* of Ribeirão took us six days: that is from the 9th to the 15th of August.

MISERICORDIA.

The current of Misericordia, a little above Ribeirão, comes from two advanced rocks on the right bank. It is one of the most difficult to pass in high water; at low water it offers scarcely any obstacle whatever. It is one of those interesting localities where, by the character of the banks, the declivity at floods is greater than at low water, the profile across which the floods have to pass being much smaller than that of the profile usually found at the rapids where the width is generally above the normal.

It is an error committed at times, even by engineers, in not observing that the difference of level of a rapid, or the fall of it is not constant, but varies with the height of water generally in the same direction, but at times opposite, as happens in the present instance.

MADEIRA.

In the afternoon we arrived in a heavy tempest at the rapid of Madeira, after passing with the cable a small fall of 0ᵐ 75, with the water two metres above low water. The fall of the principal rapid, the river being in the same condition, is 2_m 50.

At the rocks and points of the islands of this rapid are found enormous accumulations of wood, principally of cedar trunks from the Beni, the mouth of which is found a little above.

At the same place, a few months before our arrival, a Bolivian merchant was assassinated by his own Indian rowers, and some of our men found a trunk in the bushes with some of his goods and papers, which were afterwards sent to the Bolivian authorities.

The mouth of the river Beni, the waters of which are of a yellow colour, has a width of $1,002^m$ without the islands found at the same point.

Lages.

In the afternoon we arrived at the *Cachoeira* of Lages, where, up stream, is seen a line of hills. The rapid has 2^m 70 of fall, with 3^m of low water, and a length of 700 metres. We passed it the 18th of August, by ropes, with the canoes loaded. In flood time, when by the exceptional configuration of the bed, the incline augments, it is necessary to unload the canoes transporting the loads by land a distance of about 2,000 metres.

Pão Grande.

The rapid of Pão Grande, with a total fall of 2 m 10, which is distributed along 275 metres of length, is passed by towing along the left bank of the river with the canoes unloaded.

Above the rapid, on the left margin is found the mouth of the river Jata, 40^m wide.

The waters of the Madeira, above the mouth of the Beni, appeared to us clearer. The width of the river was considerably reduced, and the elevation of the land upon the banks was less, without, however, decending below the flood level, as the character of the vegetation clearly showed.

Bananeiras.

Before arriving at the fall of Bananeiras, we had to overcome two small currents. At this considerable fall of a total height of

6 ^m 35, with 3 ^m above low water, we arrived the 19th of August, after having sought in vain for a practicable channel in the labyrinth of rocks formed just below it.

The passage of the fall, which required the entire unloading of the canoes and the transportation of these over the rocks in the middle of the river, took us two days.

In the night, from the 21st to the 28th, we had, after an extraordinary drought, a heavy thunderstorm with rain, in consequence of which the thermometer of Réaumur descended at dawn to 11°.

A short distance up the river the *Serra* of Pacca-Nova was observable with its summits 100 ^m above the level of the river. The river is free and has a normal width of 275 mètres.

Guajará-Guassú.

On the 23rd of August, we arrived at the rapid of Guajará-Guassú, the next to the last of the long series of obstacles. It has a fall of 1 ^m 70 in a length of 200 mètres. The loads were carried overland and the canoes were towed up by water.

Guajará-Merim.

The following day we passed the last, Guajará-Merim, which with a fall of 1 ^m 25 in a distance of 500 ^m does not offer great difficulty.

It was at this point that we met a large party of ten Bolivian canoes, loaded with hides and tallow, and by the proprietor of them, who was from Santa-Cruz de la Sierra, we were enabled to send letters to Pará and Rio de Janeiro.

The same day we reached still farther, to the mouth of the stream of Pacca-Nova at the foot of the range of hills of the same name.

The character and features of the river above Guajará-Merim, change entirely. The inclination is perfectly regular and very small. The margins of alluvium with the *igapós** on the convex side, the steep banks on the concave, and the old beds of the river, almost closed and filled up appear like those of the lower Madeira.

Some rocks of ferruginous conglomerate which are occasionally

---

* Note by Translator.—See p. 31.

found here and there in mid-river offer no obstacle whatever to navigation, and it makes one sad to see so great a river extension, navigable for large steamers, separated from the equally navigable lower part by the impediments of the rapids.

The canoes of the expedition were all more or less in a defective condition owing to the passages over the rocks. They required indispensable repairs; consequently an appropriate place was selected to haul them out of the water and caulk them with oakum of the country, which the chestnut trees of the neighbouring forest gave in abundance.

THE MAMORÉ.

The 1st of September at 8 o'clock in the morning we arrived at the mouth of the river Mamoré, or better, the junction of the Mamoré with the Guaporé or Itenez. The width of the first is 300 mètres, and of the last, 600.

The waters of the Guaporé, the colour of which is of a bright green, differed much from the yellow waves of the Mamoré, despite its not having attained as yet its lowest stage. The temperature of the waters of the Mamoré was less by some degrees than those of the Guaporé.

At the junction of the two rivers, and on the lands immediately below and above, live a tribe of Indians, whose name itself is not known; although they have already acquired a sad fame from the frequent assassinations they have committed with unheard of temerity, as well among the crew of canoes as among the soldiers of the Fort of Principe da Beira, who go fishing in sight of the Fort.

The vegetation of these lands, in comparison to that of the region of the rapids, is of a notable poverty and scantiness, proceeding from the quality of the soil which already forms a transition for that of the fields.

At night, from the spots we occupied on the vast shores of the Mamoré, we could sometimes see the light caused by the burning prairies of the ancient missions.

The transition of the prairies was every day more visible, and already in many places the zone of the low shrubs upon the margins was narrower, allowing us to see the vast horizon, behind it.

A fresh north-east wind carried our canoe in a short time to the Cerrito,* the estate of St. Antonio de Barros Cardozo, on the bank of the Mamoré. This distinguished Brazilian citizen, who has lived in Bolivia for 15 years, was one of the first to ascend the Madeira in modern times with large craft of 500 arrobas tonnage. He received us with the greatest affability, offering us all in his power and his valuable aid. By lending us canoes, which were required, and providing us with necessary supplies, he gave signal service to the expedition as he had before done to former explorers, (Gibbon) and to merchants. Sr. Cardozo had the kindness to accompany us to Exaltacion, the first town found on the margin of Mamoré, at some leagues distance above the Cerrito. At Exaltacion he placed his house at our disposal, for our residence during the time we were forced to remain at said place to treat for the enlistment of rowers for our return.

The voyage from Serpa to Exaltacion where we arrived on the 10th of September, was 103 days. The total distance traversed was about 500 leagues. The *Corregedor* or Administrator of the town of Exaltacion de Santa Cruz, to whom we had to apply to engage the Indian rowers, informed us that, in anticipation, the Prefect of the department of the Beni had received orders from the Government of the Republic to render all aid to the expedition. He did not doubt that we should be able in a short time to find the necessary number of Indians.

At the same time he might have added, that in the town of Exaltacion, there were but fifteen available men. We judged it advisable to consult with the Prefect of the department, despite the distance of more than sixty leagues, which still separated us from the capital of Trinidad.

Francisco Keller was the member of the commission entrusted with this task. He was well received by the Bolivian authorities, and especially by the Prefect, A. Morant, and after an absence of eleven days, succeeded in bringing to Exaltacion specific orders for the *Corregedores* of Exaltacion and of S. Joaquim to furnish the

* NOTE BY TRANSLATOR.—Now the property of the National Bolivian Navigation Company.

balance of the rowers, he himself bringing eight at the same time from Trinidad.

In this interval of time the canoes, supplies, etc., were obtained, so that upon the arrival of the rest of the crew, on the 15th of October we were able to set out immediately for Cerrito, where the final preparations, repairs of canoes, etc., had to be made.

The convoy was formed of five craft of different sizes, with a total of 32 crew ; this being the minimum of men that can be taken for a certain tonnage of canoes, on account of the dragging by land.

The size and quality of the craft were as follows :—

1. A *galeota* lent by Sr. Cardozo ; burden 150 arrobas.
2. An *igarité* bought in Trinidad ; burden 200 arrobas.
3. An *ubá* lent by Sr. Cardozo ; burden 45 arrobas.
4. An *ubá* for measuring.
5. A *montaria* (small canoe) for measuring.

EXALTACION.

Before describing our return, allow us to give some data relative to the town of Exaltacion, the old Jesuit mission, in lat. S. 13° 18′ 20″, longitude West of Rio 22° 42′ 40″, and at an elevation above the sea of 159^m 20 (height of the level of the Mamoré at low water, in the lower port of the Mission.)

Exaltacion de la Santa Cruz, is situated in the midst of the prairie, in one of the immense bends of the Mamoré, on the lands of the Cayubabas Indians. The date of its foundation was probably at the end of the 17th century. The width of the river at low water is 200 mètres, and the elevation of the land of the margins above the same level is 12 mètres. The medium depth of the water in time of drought is sufficient the whole of the distance to Trinidad for craft that draw one mètre.

RISE OF RIVER.

At the end of October the river commences to rise, inundating a little the prairies for a considerable distance, and barely leaving the highest points like islands out of the water. In Exaltacion, for instance, the waters reach up to the church door.

The streets of the town are regularly traced, crossing each other at right angles, and leaving in the centre a spacious plaza, on the front of which are found located the church and the ancient college. This general arrangement is the same in all the missions founded by the Jesuits, as we can attest from the ruins of the missions of Nossa Senhora do Loreto and of Santo Ignacio, on the banks of the Paranapanema in the province of Paraná.

By the remainder of the houses on the prolongation of the streets the number of the old population may be readily calculated at 3,000, the present being scarcely 1,800. The houses of the Indians, which are contiguous, and along the front of which runs a veranda, are constructed of alcoves, and covered partly with tiles and partly with thatch. The church and college are built in the same manner, and it is very interesting to observe how, with such limited means, without stone and lime, they managed to build an edifice of a marked architectural character. Still, to-day, as nothing is done to preserve these edifices, they are falling little by little, and part of the college is already in ruins.

In the interior of the church are the altars and pulpit, first-class works of the carver's art. The sacristy holds great wealth in worked silver, which with greater advantage might be employed in the material repairs of the town, all the more so as, for many years, there has been no vicar. The digging of wells, for example, to supply the town with good water, which it lacks much, would be one of the most necessary improvements.

Among the causes which tend to contribute to the decadence of so flourishing a town we will cite the fevers which of late years appear to have taken an endemic character. The growth of the use of intoxicating drinks, the almost total destruction of the old riches in cattle, and the budding navigation of the River Madeira—in consequence of which part of the Indians have removed to Brazilian territory, while elsewhere they pay with their lives the fatigues and privations of long journeys—are other numerous reasons of decadence.

Their tillage is insignificant. They cultivate only as much corn, mandioca, and tobacco as they require for their own consumption.

The excessively low pay which they receive when they work for a white man and the innate indolence of this race make it extremely difficult to engage workmen there for any service. Only by means of the *cacique*, who maintains as yet a certain moral influence over them, can a large number of rowers be obtained.

The total indigenous population of the Department of the Beni is about 30,000 souls, they being distributed in fifteen missions or towns of the following names :—

1. Trinidad del Beni, capital.
2. Nuestra Señora de Loreto.
3. San Ignacio.
4. San Javier.

All these formed by the Mojos tribe, properly called.

5. San Pedro, by the *Canichanas* Indians.
6. San Ramon.
7. Santa Maria Magdalena.
8. San José de Guacaraje, by the Itonamas Indians.
9. Nuestra Señora de Concepcion de Baures.
10. Nuestra Señora del Carmen de Chapacora.
11. San Joaquim, by the Baures Indians.
12. Exaltacion de la Santa Cruz, by the Cayubabas.
13. Santa Ana, by the Mobimas.
14. Reyes.
15. San Borja, by the Maropas.

In the chapter which treats upon the commerce and agriculture of the Madeira and Mamoré we will resume the subject of the missions.

On the 19th of October, 1868, we left the estate of Cerrito to make, as decided, the detailed plan of the course of the river, and to make, moreover, some measurements of the cubic volume of the waters of the Mamoré, the Beni, and the Madeira. The astronomical and hypsometrical observations were concluded ascending.

We regretted our inability to continue the exploration of the Upper Guaporé, especially as near the fort of Principe da Beira there yet exists another rapid of which no explorer speaks. Moreover, the advanced season of the year and the detention which a voyage up the Guaporé would have occasioned, would have let slip the principal aim of the expedition, the examination of the rapids of the Madeira. It would have suffered considerably, especially as the waters of the rivers were already rising, and as the soundings offer but little interest if they are not taken when the river is low. We therefore gave up the idea.

The opening of a straight track the length of the rapids became impossible, as the President of the province of Amazonas did not furnish us the necessary men for whom we officially asked several times, and the Bolivian Indians are with difficulty engaged at great cost for a voyage by water. Such an opening would only be valuable to study better the forests between the Madeira and the Tapajos, for the track of a cart road must perforce follow the course of the river more or less, on account of the considerable elevation of the Serra of Pacca Nova, which is no other than a continuation of the Serra-Geral, which divides the waters of the affluents of the Amazon from the head waters of the Paraguay. The pretended shortening of the length of the road is only illusory, because the configuration of the land *in no case* permits a tracing of a direct line from one terminal point to the other; and, in all probability, the curves of the small affluents which it would be necessary to follow to reach the culminating point of the *serra* would be greater than those of the river. Still it is not alone the length of the line, but also its profile which influences its selection, and this last will be difficult so soon as the general slope of the valley of the great river is abandoned.

The American, Gibbon, who, not being a professional, first suggested the idea of a straight line of 40 to 50 leagues traversing these mountains, which could not have escaped his observation, after all only speaks of a common mule road. Such a road would be entirely insufficient in this case, and the actual navigation, imperfect as it is, is many times preferable.

As we have said above, there exists near the fort of the Principe da Beira, a rapid, which at low water requires the complete unloading of canoes, while at floods the same loads are hauled up the river with greater facility. From this rapid down there are only found light currents in the Guaporé, which not so much for their rapidity as from the lack of water, oppose difficulties to navigation. This want of sufficient water in dry times is still greater above Villa Bella ; and the junction of the head waters of the Guaporé, with one of those of the Jaurú could not be effected except by means of canalization with sluices or locks over a long distance down stream.

The 21st October we found ourselves at the mouth of the Mamoré, where we proceded to take an exact measurement of the cubic volume of the waters of the two rivers Mamoré and Guaporé, occupying ourselves two days at this work.

On the 24th of October we arrived at the island called the Cavallo Marinho, where we commenced detailed measurements with the micrometer.

GUAJARÁ-MERIM.—GUAJARÁ-GUASSÚ.

A strong tempest at night announced the approach of the rainy season of the Madeira, which having already reached its lowest stage, was now 0m. 85 above low-water mark. The explorations and soundings of Guajará-Merim were made the following day, despite the rain, and only at Guajará-Guassú was there an indispensable delay for the transportation of the cargoes by land.

BANANEIRAS.

The 27th of October, with the river risen 0^m 90 above low water, we passed the fall of Bananeiras. One of our craft struck against the shore, and such was the force and weight of the furious waves, that the most persevering work of the whole crew was necessary to save it.

Some of the Indian rowers and one of the Engineers of the Commission was attacked at the same time with intermittent fever, which was immediately treated with sulphate of quinine. The heat (35° R.) was excessive, and at night we had a heavy tempest.

PÃO-GRANDE.—LAGES.

On the 30th we arrived with the measurements at the rapid of Pão-Grande, which was passed by rowing through the main channel, as well as that of Lages.

The Beni.

The 2nd of November we reached the mouth of the Beni, where we entered to measure the volume of the waters. The Beni appeared to be at that time 2m. above low-water, and had a medium velocity of 1m. 50 per second. The margins are five and six mètres above low-water, and up the river, for a distance of two leagues, a rich vegetation and high banks indicate high land. For lack of time, we could not carry into effect the idea of ascending to the first rapid.

Madeira.—Misericordia.

At the rapid of Madeira the cargoes were carried overland, while the canoes passed by water. With greater ease Misericordia current was passed.

Ribeirão.

The passage of Ribeirão, which demanded the entire unloading of the canoes in two places, took us two days, so that only on the 5th did we find ourselves below it. The weather continued almost always rainy during the time, with the sky overcast.

Periquitos—Araras.

Periquitos as well as Araras we passed rowing. The 7th November, we continued, despite the rain, with our measurements, but we were finally forced to stop and erect *barracas* and tents.

Pederneira—Paredão.

The following day, with little rain, we were able to continue, passing Pederneira and arriving in the evening at Paredão, which we also passed, rowing in the middle of the angry waters of the main channel.

Tres-Irmãos.

The 11th of November, the river being already 3 mètres above low stage, we passed Tres-Irmãos. We regretted that we could not ascend one of the little hills on the left bank, from which might be enjoyed a most beautiful surrounding view, as the weather was cloudy and inappropriate. More and more we found ourselves more or less attacked by the fever, and debilitated by the privations inevitable in such a long journey.

GIRÃO.

Arriving at the fall of Girão, we found that two of the canoes required prompt repairs, and we therefore made them without delay.

CALDEIRÃO DO INFERNO.

On the 13th and 14th we passed the Caldeirão do Inferno, but not without one of the cargo canoes running great danger of being dashed to pieces on the rocks.

We found the Caripuna Indians below the rapid. We met them on the passage up, and, aside from this tribe, yet another of the same number more or less. All showed themselves satisfied to see us again, and we acquired the firm conviction that to catechise among them would be the easiest thing in the world.

MORRINHOS.

On the 16th we passed the mouth of the Jassiparaná, 80 mètres wide, arriving at and descending the same afternoon the rapid of Morrinhos.

THEOTONIO.

The following day we arrived at Theotonio after having passed with ease the little currents above them. The transportation of the loads and the dragging of the canoes was effected in the same way as the ascent, and the 18th of November we were able finally to leave the great beach below the fall.

MACACOS SAN ANTONIO.

Neither the passage of the current of Macacos, nor the rapid of San Antonio offered the least difficulty, from the fact that the river had already risen considerably, it being four mètres above low water. The waters were turbid, and we suppose that it was the daily rise of this water which changed the condition of our health.

The rainy season of the year was entering with all its force, and it is certain that with any further delay on the Mamoré or Guaporé it would have been impossible for us to finish the exploration of the Madeira.

Having concluded in this manner the expedition entrusted to us, we improved a part of the night below San Antonio to shorten our voyage, and only thus were we able to arrive the 21st of November, at Crato, the first point where we found any supplies.

We found, at this point, a Bolivian merchant, who having started some days before us, had suffered great delays by the sickness and death of a companion, he himself being so debilitated by fevers that he was in no condition to continue his voyage without recuperating himself.

We had the pleasure of finding letters and journals in the hands of the commander of the detachment, which through the attention of the worthy Colonel Leonardo Ferreira Marques, then Vice-President of the Province of Amazonas, were occasionally sent to Crato.

Before our departure we left in charge of the commander of the detachment, one of the *ubás* (canoes) to forward on the first occasion to Senor Cardozo, to whom it belonged.

Having passed the Manicoré, we arrived the 25th at the site of Sr. J. Araus, where we left one of our embarkations, and part of our crew, continuing the voyage with three canoes.

At the town of Muras, in Sapucaia-oroca, we made a measurement of the cubic volume of the waters of the Madeira, as much to have a comparison with the results of the anterior measurements as for a calculation of the increase in the lower course.

The following night we passed the town of Borba, and seeing the following day some banks of sand we concluded that the Amazon was still comparatively low. In effect, we afterwards ascertained that the Amazon was only two mètres above the level of low water.

On the afternoon of the 30th of November we finally arrived at Serpa, where we delivered our canoes to Major Damaso de Souza Barriga, giving immediately official notice of this to the president of the province. We afterwards went to Manáos, where we had to deposit in the arsenal some arms and other articles. **There we had occasion** to compliment the worthy president of the province, who perfectly understands that interesting part of the empire.

We arrived on the 14th of December at Pará, and on the 4th of January at Rio de Janeiro, after an absence of 14 months, and found that false news had reached there relative to the fate of the expedition.

Finally, we have great pleasure in recognising here the perfect application and zeal, of which our assistant Sr. José Manoel da Silva, more than once gave ample proofs, not only in the voyage and during the exploration, but also when the work of the office, the calculations, the designs of maps and profiles were required.

C.—IDEAS UPON THE GEOLOGICAL FOUNDATION AND CLIMATE OF THE LANDS TRAVERSED.

With reference to the geologic studies and observations of the immense basin of the Amazon, they are as yet entirely insufficient to enable me in forming an exact idea upon the geologic formation of an area of such immense extent. There exist some data which tend to prove that the beds of gritstone, and principally the calcareous, which is found in different places (River Maués), belong to the Silurian and Devonian formation. In the calcareous one of Maués, the following fossils are found :—

Productus antiquatus,
Spirifer trigonalis,
Terebratula porrecta,
Orthis orbicularis, and others,

all characteristic of the Silurian and Devonian formations.

The ferruginous conglomerate which is found on the surface of the earth is only covered with a bed of clay of from five to six metres of thickness, is a conglomerate of gritstone, little pieces of dolerite cemented with oxide of iron, full of openings and cavities, which give it the appearance of a sponge or scoria.

Its beds are generally horizontal, and are from four to five mètres thick. In the inferior beds the seams are smaller, at some points disappearing entirely, and forming then a more homogeneous mass of red gritstone, very argillaceous.

The extension over which this formation is found is enormous. We saw it at Manãos and on the margins of the Rio Negro, and afterwards along the whole length of the Lower Madeira, perforated and in part destroyed by the rapids, subsequently in the Upper Madeira, the Guaporé and the Mamoré, extending in only this direction more than 12 deg. of latitude.

In the lower Madeira there is, at the place called Matucaré, a bank of ferruginous conglomerate, which traverses the whole width of the river. . It gives origin to the only current of any importance in this river, in which, notwithstanding, there exists a good channel for navigation. Examining with some attention the formation of this interesting point, it is seen that the *pedra-canga* (ferruginous conglomerate) has more resistance than the underlying beds of argillaceous gritstone. The latter crumbles by the action of the water, the upper beds becoming in this way undermined. These last, from time to time, then break and fall in large pieces, which disappear at the bottom of the river. In the course of ages rapids and currents disappear in this manner, and the inclination of the river becomes more regular. On the river margins, at three distinct points below the Matucaré, may be seen the remains of rock banks which have been destroyed in this manner.

The waters, coming from the Cordillera, little by little cut their course through the beds of clay deposited in the vast basin of the Amazon, in all probability, by a similar erosion. In the meantime this basin was raised above the Silurian seas, the ice contributing in great part to produce a more powerful and efficacious action.

The occasional unequal resistance of the margins causes the course of the waters to become serpentine, abandoning the old course and creating new ones. It also forms banks of sand on the convex and gnaws the concave margin, until a great flood nearly perforates the isthmus formed by the serpentine—a work which continues, a true *perpetuum mobile* even to-day, its effects being visible at each step.

The lakes found on both banks of the Amazon, Solimões, Madeira, etc., and which in Bolivia are called *Madres*, have no other origin. They generally mark the course which the river has followed in former times, and which the floods are not yet able to fill up with detritus, sand and mud which they deposit. These continuous changes of duration in the course of the rivers in alluvial soil will not stop until the banks are fixed by art, although among the rapids, on solid rocky ground, they are less perceptible, but still they do not fail to exist.

The considerations we have stated are made, principally, for the purpose of correcting erroneous ideas with respect to the actual state of the affluents of the Amazon which exist even among professionals. We still state that the change of course of all the rivers in alluvial soil being general, neither the bed of one or the other among them can be considered still in the process of formation.

The direction of the general elevation of the lands south of the Amazon, so to speak, the first line commencing at the Serra-Geral, between the Paraguay and the southern affluents of the Amazon, runs in a north-east direction from the first rapid of the Madeira, passing more or less towards the first in the Tapajoz, Xingú and Tocantins. To the west of the Madeira it curves more to the west and south, so much so that the Purús nearly to its head waters has no rapid whatever.

In conformity with this the lands on the left bank of the Madeira are less mountainous than those of the right bank, where the ramifications of the Serra-Geral (mount of the Parecis) reaches to the margin of the river in different places.

With reference to the mineral wealth of the territory traversed, we must cite gold as the principal. It is undoubtedly found in the white quartz veins which are met with in many places among the rocks of the rapids; and what is notorious from the earliest times of the Conquest, it has been discovered among the head waters of the various branches of the right bank of the Amazon—among others, the Upper Guaporé. In modern times the mining of gold in this region has much diminished, principally on account of intermittent fevers, which appear to be the inseparable companions of miners. In Bolivia we learned that in the *serra* of the Guarajús, on the left bank of the Guaporé, near the abandoned mission of San Simon, there are found veins of auriferous quartz of great richness, which with regular mining would yield great returns, and which have as yet only been explored by mine hunters without capital.

Before describing the climate of these regions, it becomes necessary to throw light upon the terms *igapó*, *vargem* and *terra firma*, which are found in the mouths of the inhabitants of the Amazon, and which they employ at every instant.

The *igapó* is the most modern alluvium of the convex margins, whose elevation as yet is not more than four or five mètres above low-water, and consequently at medium floods it commences to be inundated.

The character of its vegetation is well marked, producing, aside from *capim*, called *canarana*, in the banks, and some low trees of white and soft wood, the *embaúba* (cecropia) and the *seringa* (ficus elastica).

The second, the *vargem*, includes the lands whose elevation is between medium water and floods, or, therefore, which are only inundated but a short time during the year. Its characteristic form consists in the different palm trees, principally the muru-murú, uru-cury, etc., the mulatto wood, the pacova-sororoca (Urania), the cacáo, etc. The culture of the sugar-cane is appropriate for the high vargem.

The third grade, the *terra-firma*, are the remains of the ancient bottom of the basin of the Amazon, in which the rivers have formed their courses. It, therefore, may have very different altitudes, but the class of land, a hard clay, of red colour, is always the same.

It is only on *terra-firma* that the good classes of wood are found. Among the characteristic trees we will only cite the gigantic chestnut (Bertholetis excelse).

With respect to the medium temperature of the region of the rapids, we only have the data of the months of June and October, which are not sufficient for a calculation of an annual medium. It appears probable, however, that an isothermal of 20° R. may be that of the lower rapids. The barometrical observations, which we principally made for our knowledge, to furnish us the elevation above the level of the sea, follow hereafter, along with the temperatures of the above months.

The principal direction of the wind is from south-west to north-east. The first, to which the Cordillera gives birth was that which brought the rain of the months of September, October and November, while that from the other direction was dry.

Despite its being impossible for us to make direct observations for the calculation of the cubic volume of rain which falls every year, we have by the floods, combining the cubic volume of their waters with the superficial area of the tributary lands, at least a comparative scale for the calculation of the rain which falls upon a given amount of tributary surface, less the loss by infiltration and evaporation. The result of the calculation is that the rivers Madeira, Beni, Mamoré, and Guaporé, give a drainage of 1.03 cubic mètres of water for each square league of superficial surface. The rainy season generally commences on the Madeira at the end of October and finishes in March.

For the medium velocity of the waters, it may be calculated that the wave of the first flood requires from 20 to 21 days to pass the distance from Exaltacion on the Mamoré to the mouth of the Madeira. The waters of the Mamoré and Beni begin to swell a little before reaching the Guaporé, this difference proceeding from the different climatic and hypsometric condition of the regions in which are found the principal head waters.

With respect to the influence of climate upon the sanitary condition of the inhabitants and upon travellers, it becomes us to note that in the dry season intermittent fevers are extremely rare. They commence to appear with the first wave of the flood, becoming then the real scourge of those regions.

Leaving to professional men to clear and explain the following contradictions, we cite only : —

1st. That the intermittent fevers are severer in the region of the rapids than above or below them, where there are still more fens.

2nd. That in the prairies of Bolivia, where, after the floods, there is much stagnant water, which is drank without much precaution, the fevers are comparatively rare.

It is certain that aside from the quality of potable water, there is yet another factor—the force and direction of the winds—not less important than the first; so much so, that some of the houses of the rubber collectors, in unhealthy spots, become healthier, to some extent, after the forests have been cut down in a certain direction.

D.—RESULTS OF HYDROGRAPHIC MEASUREMENTS.

I. Results of Astronomical Observations.

Names of Places.	Lat. S	Long. W. of Rio.
On the Lower Madeira—		
Murassutuba ...	5° 37′ 37″ 0	
Isle of Baêtas ...	6° 18′ 28″ 7	
Espíritu Santo	6° 43′ 20″ 5	
Crato	7° 31′ 3″ 4	
12 kil. above Abelhas	8° 15′ 33″ 1	
Domingos Leigue, rubber gatherer	8° 36′ 4″ 0	
Upper Madeira—		
San Antonio	8° 49′ 2″ 6	21° 29′ 8″
Theotonio	8° 52′ 41″ 6	21° 30′ 57″
Morrinhos	9° 1′ 45″ 3	21° 36′ 30″
Caldeirão	9° 15′ 48″ 7	21° 52′ 14″
Girão	9° 20′ 45″ 7	21° 54′ 22″
Pederneira	9° 32′ 81″ 3	22° 20′ 20″
Paredão ...	9° 36′ 37″ 7	22° 13′ 4″
Araras ...	9° 55′ 5″ 8	22° 15′ 20″
Ribeirão	10° 12′ 52″ 1	22° 8′ 30″
Mouth of Beni ...	10° 20′ 0″ 0	22° 12′ 20″
Guajará-Merim	10° 44′ 32″ 8	22° 3′ 42″

II. Distances between the different points of the Course, measured with the Micrometer.

Names of Places.	Distances in Metres.
1. Islands of Cavallo-Marinho to the rapid of Guaporó-Merim (N. 1—13)	21,239.00
2. From Guaporó-MerimtoGuajará-Guassú(13—21)	6,842.00
3. From Guajará-Guassú to Island (N. 30) ...	15,379.00
4. From point of Island N. 30 to beginning of the rapid of Bananeiras (N. 30—42)	13,153.00
5. From point of N. 42—53 to the principal fall of Bananeiras	494.75
6. From point N. 53 to the beginning of the Cachoeira of Pão-Grande, 162	8,266.10
7. From point N. 65—71 to the end of Pão-Grande	5,040.70
9. From point N. 71—87 to the mouth of the Rio Beni	9,408.00
10. From point of 87—102 to the rapid of Misericordia	17,614.00
11. From point N. 102—106 to the beginning of Ribeirão	4,120.00
12. From point N. 106-116 to the current of Periquitos	16,230.00
13. From point N. 116—124 to the rapid of Araras	15,667.00
14. From point N. 124—153 to the mouth of the river Abuna	44,942.00
15. From point N. 153—161 to the rapid of Pederneira	13,300.00
16. From point N. 161—177 to the rapid Paredão...	23,300.00
17. From point 177—197 to the current of Tres-Irmãos	30,700.00
18. From point N. 197—222 to the fall of Girão ...	37,600.00
19. From point N. 222—235 to the rapid of Calderão do Inferno...	10,330.00
20. From point N. 235—261 to the *Malocca* of the Caripunas...	28,700.00
21. From point N. 261—269 from the *Malocca* of the Caripunas to the point of island	9,400.00
22 From point N. 269—282, from the point of island to the rapid of Morrinhos ...	16,005.00

Mètres.

23. From point N. 282—300, from Morrinhos to the
fall of Theotonio 26,260.00
24. From point N. 300—314, from Theotonio to San
Antonio 10,600.00
25. From point 314—318, end of micrometrical
measurements 5,600.00
The total length of the course of the river in the part
filled with rapids between Guajará-Merim and
San Antonio is therefore ... *363,846.00
or 70.19 leagues of 18 to the degree.
Adding to this yet the distance from the islands of
Cavallo-Marinho to Guajará, about ⁄ 21,239.00
and that from San Antonio to the end ... 5,600.00
We have as the total length of micrometric measure- ————
ments ... ... †390,685.00
or 63.18 leagues.

Leagues.

26. The length of the course of the Madeira, between
San Antonio and its mouth is approximately ... 174
27. The length of the course of the Madeira, between
Guajará-Merim and the mouth of the Mamoré ... 32
28. The length of the course of the Mamoré from its
mouth to the town of Exaltacion 40.5
29. Length of the course of the Mamoré from Exaltacion to Trinidad 58.5
(Summing up, therefore, all these distances, it is
seen that the expedition traversed in canoes, a
distance of 749.2 leagues, going and coming,
upon the rivers Madeira and Mamoré.)
Some distances, which may be of some use when the
commerce of the Madeira and Bolivia is undertaken, are as follow :—

---

* (?) 369,246.00.
† NOTE BY TRANSLATOR.—Evidently a mistake. The 70.19 leagues above
represent less *mètres* than the 63.18 here.

		Leagues.
30. The length of the Mamoré from Trinidad to Vinchuta		57·0
31. The distance from Trinidad to Sucre, capital of Bolivia		135·0
32. Distance from Sucre to Cobija, the only Bolivian port on the Pacific		220·0
33. Distance from Trinidad to Santa Cruz ...		94·0

III.—RESULTS OF THE HYPSOMETRIC OBSERVATIONS.

The heights of the level of the low-water above the level of the sea at the different points, are as follow:—

Name of Place.		Elev. in Mètres.
Town of Serpa in the Amazonas ...	...	18.00
Mouth of the Madeira ...	... ...	21.00
Manicoró, on the Madeira	... ...	28.00
Baêtas, ,,	... ...	40.00
Tres-Casas, ,,	... ...	50.00
Isle of Salamão, ,,	... ...	53.00
Domingos Leigue, on the Madeira	... ...	54.00
Barra do Jamary	... ...	56.80
Rapid of San Antonio, lower part	... ...	61.60
Fall of Theotonio ,,	...	83.40
Rapid of Morrinhos ,,	... ...	87.70
Rapid of Caldeirão do Inferno, lower part ...	...	92.80
Fall of Girão ,, ...	...	102.00
Guajará-Merim, lower part, end of the rapids	...	144.60
Mouth of the Mamoré ,,	...	150.40
Town of Exaltacion, on the Mamoré	... ...	152.20

The vertical fall of the rapids together with the length of the same is found in the following table; water at medium height:—

Name of Rapid.				Height. Mètres.	Length. Mètres.
1. Guajará-Merim	...	...	...	1.2	1,500
2. Guajará-Guassú	...	...	...	1.7	450

			Mètres.	Mètres.
3.	Upper part ⎫		1.2	400
	Central part ⎬ Fall of the Bananeiras ...		6.0	500
	Lower part ⎭		1.5	1,500
4.	Pão-Grande		2.0	400
5.	Lages		2.5	750
6.	Madeira		2.5	900
7.	Misericordia		0.6	100
8.	*Riberião.*	Principal fall	4.1	400
		1st current below fall	1.5	275
		2nd ,, ,, ...	2.7	1,000
		3rd ,, ,, ...	0.9	250
		4th ,, ,, ...	1.5	900
9.	Periquitos		0.8	300
10.	Araras		1.4	700
	Araras, current below it as far as the river Abuna		0.5	250
11.	Pederneira		1.1	250
	Current below		0.9	350
12.	Paredão		1.7	550
	1st current below		1.5	750
	2nd ,, ...		1.2	700
	3rd ,, ...		0.5	250
13.	Tres-Irmãos		0.6	150
	,, Current below		0.3	70
	,, ,, ...		0.5	150
	,, ,, ...		0.7	200
14.	Girão, principal fall ...		8.0	700
	,, Current below ...		0.6	200
	,, ,, ...		0.6	200
15.	Caldeirão do Inferno		2.2	400
	1st current below		1.9	1,170
	2nd ,, ,, ...		0.7	250
	3rd ,, ,, ...		0.4	150
	4th ,, ,, ...		0.5	300
	5th ,, ,, ...		0.3	120

				Metrés	
16. Morrinhos	..	1. 1	450
1st current below		0. 3	120
2nd ,, ,,	...	0. 4	100
3rd ,, ,, ...		0. 6	250
4th ,, ,, ...		0. 3	60
5th ,, ,, just above Theotonio		0. 7	150
17. Fall of Theotonio		7. 5	300
Current below it		0. 3	300
,, ,, (Macacos)		0.45	150
18. San Antonio		12.0	300
Total fall and length of rapids ...		69. 6	20,169*

The fall existing in the distances between the rapids from San Antonio to Guajurá-Merim is found by deducting the height of the rapids from the total difference of level between the same points, being, 83m—69m, 60=13m.40.

IV.—WIDTH OF THE RIVER, DEPTH, INCLINE, AND VOLUME OF THE WATERS.

The first three are the factors for the calculation of the cubic volume of the water which a river carries in a certain space of time, and it being possible, from the plan as well as from the levels presented to deduce these data for any point of the course, we only cite the following inclines, etc.

Inclines.

1. The Mamoré between Exaltacion and its mouth at
   the Madeira 1 32,104
2. Madeira from mouth of Mamoré to Guajará-Merim   1 30,000
3. General incline of the Madeira from Guajará to San
   Antonio 1 5,303
4. Madeira from San Antonio to its mouth ...   1 26,490

* NOTE BY TRANSLATOR.—Ought to be 19,665m.

Depth of water (at the time of low-water) :   Mètres.

1. Maximum depth of the Mamoré at its mouth...   10
2. Minimum depth of the Mamoré below Exaltacion,
   caused by a reef of ferruginous conglomerate,
   taken in a straight line 0.75
3. Maximum depth of the Madeira between the mouth
   of the Mamoré and Guajará   15
4. Minimum depth of the Madeira at the same place
   taken in a direct line 1. 4
5. Depth of Beni at mouth 15. 0
6. Maximum depth between the rapids, immediately
   above Theotonio 37. 4
7. Maximum depth of the lower Madeira (near
   Sopucaia-oroca) 36. 8
8. Minimum depth on the line of *pego*, at Uroá   1. 3

1. Mamoré at mouth, low water   295
   ,,   ,,   high water   475
2. Guaporé at mouth, low water   500
   ,,   ,,   high water   700
3. Madeira, medium width above rapids   ...   435
4. Madeira, among rapids, minimum   350
   ,,   maximum   2,000
5. Lower Madeira (Sapucaia-oroca)   730
6. Beni   1,000

ELEVATION OF THE BANKS, DIFFERENCE OF LEVEL BETWEEN HIGH
AND LOW WATER, CUBIC VOLUMES OF THE SAME IN THE MAMORÉ,
GUAPORÉ, BENI AND THE MADEIRA.

At Exaltacion we found the elevation of the prairies near the
banks 13m. above low water, this elevation being the same as that
attained by the floods. At the mouth of the Mamoré, however,
this proportion is different, the banks having an average elevation
of eight mètres above low water, while the floods rise to nine
mètres, inundating the prairies in consequence.

The same difference between low water and floods is noted in the Guaporé and Madeira, at least as far as the first rapid of Guajará-Merim where the general declivity suddenly changes.

We found the margins of the Beni near its mouth to have an elevation of six mètres above low water, there appearing above the left bank some hills entirely covered with dense forests.

In the part containing the rapids, the elevation of the banks is very different. They are generally low immediately above the rapids, where also the difference between low water and floods is the least. In these places at times, the margins are not elevated more that 2m.50 above low water, the same taking place with reference to floods.

In the lower Madeira the normal elevation of the margins is seven mètres above low water, this elevation being little inferior to floods. In some places, however, as for example in Sapucaia-oroca this difference rises to 12m. the right bank with a base of ferruginous conglomerate, being 13m. while the opposite margin is only 10m.

CUBIC VOLUME OF THE WATERS OF THE MADEIRA AND ITS AFFLU-
ENTS AT DIFFERENT SEASONS OF THE YEAR CALCULATED IN CUBIC
MÈTRES AND PER SECOND :—

River.	Low Water.	Medium Water.	Floods.	Observations.
	M. C.	M. C.	M. C.	
Guaporé at its mouth ..	.663	1 . 879	5 . 120	Measurement made at low water.
Mamoré at its mouth ..	.835	2 . 530	7 . 024	Measurement made 2m. 5c. above low water.
Madeira between mouth of Mamore and Guajará ..	1 . 498	4 . 310	12 . 144	
Beni at mouth	1 . 383	4 . 344	13 . 109	Measurement about 2 m. 5c. above low water.
Lower Madeira at Sapucaia-oroca	4 . 142	14 . 642	39 . 106	Measurement about 5m. above low water

The tributary areas approximatively calculated for the above rivers are as follow : —

For the Guaporé ... ... 9,715 square leagues.
,, ., Mamoré... ... ... 9,982 ,,
., ,, Beni ... ... 7,068 .,
,, ,, Lower Madeira... 11,016 ,,
———
37,781

The tributary service of the Beni is probably greater than indicated above. With the aid of the new and complete map of the Empire now being made, this area can be obtained with greater exactitude.

The cubic volume of the waters of the Beni being equal to those of the Mamoré and Guaporé united proves the Beni to be the true main source of the Madeira. With respect to the waters of the Mamoré and Guaporé, it may appear singular that their cubic volumes do not preserve the same proportion of tributary surface, taking note, however, that the head-waters of the Mamoré are found in the upper Cordillera while those of the Guaporé come from inconsiderable elevations of the Serra-Geral, this apparent contradiction is solved.

E.—PROJECTS FOR THE IMPROVEMENTS OF THE LAND TRANSIT AS WELL AS OF THE RIVER ROUTE, THAT IS THE CONSTRUCTION OF A ROAD, WITH APPROXIMATIVE CALCULATIONS FOR EACH ONE.

For the past five or six years there has existed on the Madeira a small traffic by canoes between the town of Serpa in the Amazonas and the old Missions of the Mamoré. The influence of the same has extended on the one hand to the city of Santa Cruz de la Sierra, while on the other it has become a commerce of the city of Pará which furnishes the imported goods receiving those which come from Bolivia.

The number of canoes which annually descend from Bolivia being from fifty to sixty, with a medium freight of 350 *arrobas* each

(6¼ net tons each) the total weight of goods transported becomes about 700 tons.

The part of the river Madeira not only below San Antonio, but that above Guajará, and the Mamoré as far as Trinidad, are perfectly navigable, the river admitting in that part of its course steamers of sufficient draft.

The minimum depth of water in the dry season is 1 m. 80 in the portion of the river below San Antonio, it being one metre in the part above the rapids. * In no part does the velocity of the water during floods exceed 1 m. 50 per second, or three miles per hour.

There are but 70·67 leagues between San Antonio and Guajará well filled with rapids, among which are found insuperable impediments to steamers, obliging even the actual canoe traffic to make frequent landings and transits by land.

To diminish the dangers, the detentions and fatigues of the actual navigation, and to develope the commerce, which hardly born, already threatens to perish for want of an easier method of communication, three plans offer themselves :—

1st. The construction of inclined planes, by which the vessels may overcome the heavy inclines.

2nd. The opening of a canal on the right bank.

3rd. The construction of a railway.

1st.. On the inclined planes or *mortonas* the vessels with their loads are placed upon an iron carriage running on rails, which runs under the water to a necessary depth. This carriage, with the boat, is raised by means of a winch and brass cable to the height of the platform (more or less 15m. above low water), from which it again descends to the water, and leaves the carriage to continue its voyage—having in this manner overcome the rapid.

Such works were constructed some time ago in Morris, of the United States, and in Eastern Prussia, where they do excellent service.

Conjointly with the plans and levels of the Madeira we will present an outline of the mechanism of this construction.

---

* Note by Translator.—I believe both these depths are greatly understated, if I may judge from soundings made in person.

2nd. The opening of a canal for navigation on the right bank for a length of fifty leagues, more or less practicable for small screw towboats, finds in the strong general slope of this part of the river a considerable difficulty.

Accordingly, the general declivity being, as indicated by the levels 1 : 5265, it becomes indispensable to construct locks, for the velocity of the floods with the above slope would give a volume that might impede navigation, the preservation of the canal becoming at the same time very difficult.

Taking 3m. 50 as the maximum height overcome by a lock, and the total fall from San Antonio to Guajará, equal to 83m., it is easily found that the number of locks required is $83 \div 3.50 = 24$.

The width of the canal could not be less than 20m., and the depth of water 1m. $80 = 2m$.

3rd. The construction of a railway on the right bank, whose line would be approximately 50 leagues in length.

This line would not be straight between San Antonio and Guajará on account of the profile of a track in this direction being, per force, very defective and entirely improper for a wheel road on account of the ramifications of the Serra-Geral, which extends itself to the right bank of the river. Notwithstanding, it would not be necessary to follow at all points the curves of the river, it being possible to cut off different arcs among them.

Another consideration, not less important, which prevents the tracing of a straight line, even though it might be possible, technically speaking, is that the new line of communication should touch the bank of the Madeira opposite the mouth of the Beni, it being the rich lands on the margins of these rivers which will give some day an addition to the commerce of the Madeira.

As, in the description of the voyage, we have already had occasion to state, we could neither obtain the necessary crew, nor had we the time required to open, not alone the hypothetical straight line, but that near the right margin, and cutting off only the shorter curves by a definite tracing of the line.

We add that the opening of a track in the entire length between San Antonio and Guajará, for the length of fifty leagues, with plans, levels, definite location of the road or railway, would not require less than two years, although the trained staff was sufficiently numerous to work on different sections.

The approximate calculations which follow may serve, however, as a base for the definite selection of one or the other among the different projects presented.

1.—APPROXIMATE CALCULATION FOR THE CONSTRUCTION OF INCLINED PLANES AT THE DIFFERENT POINTS OF THE RIVER.

To overcome the rapids between San Antonio and Guajará 20 *mortonas* are required, with a total length of 2,160m., or in a round sum, 20 kilometres, this being the entire length of the rapids and currents to be overcome.

A. The construction of 20 kilomètres of iron track costs, in conformity with the detailed calculation, adopting the rail of the Ahlburg system, for each kilomètre, 32,500$000 ... ... 650,000$000

B. The construction of the platform for the location of the windlass, two walls of 5m. high, 5m. width, and 1m. 2 thickness, 1,200m. cubic at 100$000 for 20 platforms .. .. 12,000$000

C. The construction of jetty at the ends of the mortonas to the depth of two mètres below the level of the water, 7m. wide, 2m. high, and 20m. long, for 40 jetty, 11,200m. cubic metres of stone to quarry and transport @ 5$000 ... 56,000$000

D. Construction of 24 iron cars to receive the vessels, the weight of each one ten tons, and the price of each ton 400$000 ... ... 96,000$000

E. 24 windlasses of iron, of one ton each, at 400$000 ... ... 9,600$000

F Base or pedestal of windlass, constructed of first class wood, for each *mortona* four joists, with a transverse section of 0m. 16 and 6m. long, in all—24 × 4 × 6 = 576 mètres, at 2$000 1,152$000

G. Brass cables of a transverse section of 24 milli-
métres square, 200m. long (the slope being
1 : 12) gives for 2 4 *mortonas* 4,800m. of rope,
weighing 15m., 36 at 700$000 10,752,000

H. Planking of the platform of a superfice of 25m.
for each *mortona*, or in all 600m., with planks
of 0m. 1 thickness, at 1$500 per square métre 900$000

I. Direction and inspection of works, construction
of *ranchos*, utensils, &c. ... ... 50,000$000

Total ... ... 884,404$000

Or in round figures ... ... 900,000$000

II.—Approximate Calculation for the Construction of a
Navigable Canal on the right bank between Guajará and
San Antonio.

Such a canal having to admit small tow-boats designed for the
Upper Madeira and Mamoró, should be 20 métres wide at the
bottom and 1m. 5, to 1m. 7 of depth of water.

The difference of level between the terminal points being 83m.
and 3m. 5, the fall overcome by one of the projected locks, it follows
that the number of these should be 24.

The opening of the canal, the length of which will be 50 leagues,
and the construction of the locks will cost as follows :—

(*a*) Opening of the track and clearing of the
forest for a width of 50m., 1,545,700 square
fathoms, at 100 rs. ... ... 154,570$000

(*b*) Earthworks : excavations for a width of 20m.
deep, with *taludes* of 2 : 1, 1,483,200 cubic
fathoms at 7$000 ... ... 11,382,400$000

(*c*) Rock-cutting : approximate cubic fathoms,
7,416 at 20$000 ... ... 1,483,200$000

(*d*) Consolidation of the *taludes* by means of
pavement, staking, &c., approximating
154,500, at 10$000 the running métre ... 1,545,000$000

(*e*) Aqueducts (of 10m. width), approximately :

5 principal at 300,000$000 = 1,500,000$000
10 smaller at 100,000$000 = 1,000,000$000
300 drains at 170$000 = 50,000$000

2,550,000$000

(*f*) Twenty-four locks, in conformity with the detailed calculations which we have presented in former reports, average 100,000$000 ...   2,400,000$000

(*g*) Management of the work, trained staff :

(1) Survey of line, ten engineers, two years at 5,000$000 ... ...   100,000$000

The same engineers for the ten years of construction ... ...   500,000$000

(2) Labour for opening the track :

100 persons at 24$000 per month, 24 months   57,000$000

Foremen, 20, at 60$000 per month for ten years ... ...   144,000$000

801,000$000

Total ... ...   20,316,175$000
Or in round sum ... ...   21,000,000$000

III.—Approximate Calculation for the construction of a Railway on the right margin of the Madeira, along the line of the Rapids, in a total distance of fifty leagues.

A kilomètre of this road, adopting the Ahlburg system, will cost :—

(*a*.) Opening of track in virgin forests 30 m. wide, 1,000 m. long, 30,000 square metres at 10 rs. ... ...   300$000

(*b.*) In such lands it will be necessary to remove
for each corresponding metre, 5 cubic metres
of earth ; and in 1,000 m., 5,000 cubic
metres or 500 cubic fathoms at 5$000 ...   2,500$000

(*c.*) Average rock cutting per running metre,
9 m. 5 cubic, or 50 cubic fathoms in 1,000
running metres, at 25$000 ... ...   1,250$000

(*d.*) Drains and small bridges in each kilomètre,
average :—

1 wooden bridge	..	1,000$000
4 smaller	... ...	2,000$000
80 drains...	... ...	500$000

     3,500$000

(*e.*) Pavement between the rails and under the
same ; for each kilomètre, 36 cubic fathoms
at 30$000 ... ...   1,080$000

(*f.*) Rails :—

Each running metre weighs 35 kilograms
Or each kilomètre... ... 70   tons.
Traverses ... ... 6.6 ,,

Total ... 76·6

at 150$000 ... ...   11,499$000

(*g.*) Fastenings, 6,000 pieces, total weight 3 tons
at 300$000 ... ...   900$000

(*h.*) Laying and fastening per running metre
3$000, or for 1,000 ... ...   3,000$000

(*i.*) Management surveys, etc. ... ...   2,000$000

(*k.*) Tools, utensils, etc. ... ...   500$000

Total ... ...   27,449$000
Or for total length ... ...   8,481,741$000
In round figures ... ...   8,500,000$000

IV. --APPROXIMATE CALCULATION FOR THE CONSTRUCTION OF A MACADAMIZED ROAD OF A TOTAL WIDTH OF SIX MÈTRES, WITH FIVE MÈTRES OF PAVING FOR ONE KILOMÈTRE.

(*a.*) Opening of track 30 m. wide, in virgin forest, 1,000 m. long	300$000
(*b.*) Earth works, 700 cubic braças at 5$000 ...	3,500$000
(*c.*) Rock cutting, 50 cubic braças at 25$000 ...	1,250$000
(*d.*) Opening valletas 500 rs. per running mètre	500$000
(*e.*) Drains and bridges as above	3,500$000
(*f.*) Pavement 5 m. wide 0 m. 3 thick, in 1,000 m. long, 150 cubic braças at 45$000	6,750$000
(*g.*) Compressing the bed of the pavement	500$000
(*h.*) Tracing and measurement of waters	1,000$000
(*i.*) Management	500$000
(*k.*) Iron work, etc.	500$000
Total	18,300$000
Or for total distance...	5,654,700$000
Or in round figures	5,700,000$000

RECAPITULATION.

Approximate cost of the construction of a new route of communication between San Antonio and Guajará, on the River Madeira is:—

1. Inclined planes	900,000$000
2. Canal and locks	21,000,000$000
3. Railway, with cars drawn by animals ...	8,500,000$000
4. Macadamized road	5,700,000$000
And we still add that the cost of a road without being macadamized would be	3,090,000$000

F.—STATISTICAL IDEAS UPON THE COMMERCE AND THE PRODUCTIONS OF THE VALLEYS OF THE MADEIRA, MAMORÉ, GUAPORÉ, AND AFFLUENTS.

I.—LOWER MADEIRA.

In the lack of official documents relative to the population of the lower Madeira, we cannot determine the number of inhabitants of this vast region except approximately.

Let us suppose it, however, to be 5,000 to 6,000, including the Indians of the Mundurucú and Mura tribe, who, in a semi-civilized state are found in the old towns of Sapucaia-oroca, Manicoré, Canumá, Mané, Crato, etc.

The labour in which this population, so diminutive in proportion to the immense territory, occupies itself is only extractive; that is it takes advantage, without further work, of the riches which nature has strewed over those regions with prodigal hands.

In the first place it is to be noted that the *seringa*, caoutchouc, or borracha is an important article of exportation. It is a resin of the *ficus elastica*, which abounds as well on the river margins as among the adjacent lakes.*

The lands where it flourishes best belong to the formation of the *igapó*, which means the most recent alluvium, which in the season of the medium high water of the rivers is inundated. It is in such places, in locations frequently unhealthy, that in innumerable huts, is prepared annually the enormous quantity of 40,000 *arrobas* (32lbs. to the arroba in Brazil,) of Seringa (india-rubber) exported from the province of Amazonas, of which more than half comes from the Madeira.

Despite important discoveries of preservation by alum, and in a liquid state by Ammonia—even to-day the ancient routine is usually followed. The fruit of the Urucury and those of the Uauassú palms, whose smoke serves to solidify the milk, are, therefore, as indispensable for the rubber collector of the Madeira, for the making of rubber, as the milk itself.

---

* NOTE BY TRANSLATOR.—The present annual export of this article from the Amazon Valley is about 350,000 arrobas, or 5,000 gross tons.

4

We cannot fail to notice an interesting fact:—although magnificent rubber trees are found on dry lands in the vicinity of the rapids, no value can be realized from them as they give no milk, except in very small quantities. It appears, therefore, that the annual inundations of the *igapó* are necessary to produce the sap.

In the second place, as an article of export, is found the Cacáo, and the *Castanha do Pará*—Pará nut—(fruit of the *bertholetia excelsa*) which is found at all those points where *terra firma* reaches the margin of the river.

The cultivation of the Cacáo, despite its being wild in the table-lands, has not the importance that it might have in view of the facilities for planting it.

The cultivation of tobacco, formerly so important on the Madeira, has suffered a considerable diminution, the famed tobacco of Borba, being substituted for that of Maués.

The sad consequences of the pertinacity with which the population on the margins of the Madeira dedicate themselves to the gathering of the wild products, and principally rubber, is that true cultivation is found in a perfect state of abandonment.

The most important aliments in those regions, such as Mandioca flour, salt fish, (pirarucú) are imported from different parts of the Amazonas and even from Pará. As much in consequence of the enormous freights as by the gains of speculators, they frequently attain fabulous prices. This happens in face of the well known fact that said products can be produced in the place itself.

As a curiosity, we will state that the normal price of an *alqueire* (46 pounds) of mandioca flour from the town of Borba up, is $12,000 to $14,000 (six to seven dollars gold). Proportionate prices exist for iron goods, cloths, and the other products of European manufacturing industry. Thus the poor rubber collector, despite his working, so to speak, in a mine of gold, has gained nothing at the end of the year. He leaves immense gains in the hands of the monopolists who purchase the rubber for infinitely small prices, paying not in money but in merchandise.

The new Steam Navigation Company on the Madeira will soon commence its voyages as far as Crato and San-Antonio. The face of commercial matters will then entirely change in aspect, reducing itself so rapidly to free competition, that it will be to the interests of the Company to dispense with the monopolists and *regatões*.

II.—UPPER MADEIRA.

As we have stated above, the rubber tree in the region of the rapids beyond San-Antonio, does not pay the work of the rubber collector at any season of the year for lack of milk. In consequence of this the attempt of some Bolivians, who descended with Indians from the department of the Beni, to the extensive rubber districts, found near the mouth of the river Jata, was entirely frustrated. They were afterwards obliged to locate below San-Antonio, proving again that the region of the rapids cannot be molested by the Caripuna Indians.

The commerce of these latter with the navigators of the Madeira is limited to the selling them some roots of mandioca and a little corn in exchange for iron implements.

We do not doubt, however, that these inoffensive and good-natured Indians, if properly instructed and accustomed to work, could make great gains from the extensive cacáo woods which in that region are found on both margins. They would thus create a beginning for exportation in places which now are entirely lost to commerce.

III.—MAMORÉ.

In the neighbouring prairies of the old Jesuit missions, not only on both margins of the Mamoré, but also on the Itonama and Machupo, there exist at present some droves of horned cattle. They are, however, much reduced in number by speculators, who, authorised by the Bolivian Government, destroy on a grand scale, without any consideration whatever, a fountain of riches, which might be permanently established. With a little care on the part of Bolivia of these remains of old breeding, the immense prairies of the department of the Beni would have had a considerable increase. To-day there are drawn from them the loads of hides and tallow for the canoes which, from Trinidad and Exaltacion, descend the Madeira and Amazonas to Serpa.

To prove that the wealth in wild cattle in the prairies of the Beni has decreased at an accelerated pace, we alone state that ten years ago one might buy in Exaltacion, Santa-Anna, Trinidad, &c., a fat animal for 2$000, while to-day it costs from 12$000 to 14$000. (1$000 is 50 cents gold). Formerly nothing but the hide and tallow were made use of. The flesh was thrown to the birds of the fields. To-day the misery in some of the missions reaches such a point that the Indians eat a kind of great *minhoca* (*ver de terre*), which they find in damp ground.

Game, principally deer, was formerly very abundant, even in the vicinity of the missions. In consequence of a systematic persecution, without rest, on the part of the Indians, urged on by the buyers of skins, these animals are exterminated near the missions, and they are now found only in the most distant prairies.

While the riches afforded by nature are in this manner destroyed almost without work, on the other hand, no progress is made in agriculture. On the contrary, so soon as the patriarchal system, which existed in the time of the Jesuits and the first decades after their expulsion, was abandoned, the Indians scarcely planted enough to sustain themselves. To-day the only articles of exportation for Santa-Cruz are cacáo, gathered from the wild cacáo fields of the Mamoré, some tobacco and some fabrics of cotton (Macánas), which are manufactured yet on a small scale in some towns.

The active example of the fathers, the pomp of the religious ceremonies, an ever severe justice, and an unequal tact in the art of understanding and guiding the Indian spirit, were the means employed to give prosperity to the missions. When, however, the difficult task is confided, as it is to-day, to unskilful hands, and the infantile spirit of the Indian is placed in immediate contact with the corruption of speculators, the consequences cannot be other than the saddest.

The indigenous population of the Department of the Beni, distributed in fifteen villages or missions, to-day does not exceed 30,000 souls, and diminishes from day to day in consequence of the physical and moral misery in which it is living.

Perhaps the greatest impediment to progress in those regions is

the profound antipathy against the white race, which, in consequence of abuses and bad government, has unhappily been
engrafted upon the Indian mind. Were it not for this, cotton,
tobacco and sugar-cane would be cultivated. To-day nothing is
done except on a small scale, although the land and the climate are
the most appropriate possible. The exploration and cultivation of
the cacáo trees also offer lucrative branches of agriculture.

The introduction of spinning machines and improved appliances
in place of the primitive instruments with which the Indians actually work would be of immense advantage in view of the admirable
works which they execute without them. Perhaps in this method,
developing the natural gift possessed by this race for making all
kinds of fabrics, this branch of industry would take great proportions. Thus the missions, to-day in a state of decadence, would
acquire again that character for activity which they undoubtedly
had in the time of the missionaries.

Cascarilla, or quina, is an important article of exportation, which
does not, however, truly belong to the Department of the Beni, but
to the forests of the head waters of the Beni. In effect, the true
Calysaya, the bark of which contains the greatest proportion of
precious alkaloid, is found only upon the slopes of the high cordilleras at an elevation of from $1,000^m$ to $2,000^m$ above the level of
the sea. Up to the present the cascarilla has been generally
exported from La Paz through Tacna and Arica, on the Pacific.
The Bolivian Government, the better to fiscalize and monopolise
this important commerce, has established a species of bank or
deposit of quina at the capital.

Instead of carrying the bark on the backs of the bark-gatherers
over the intransitable roads of the Cordilleras, some attempts on a
large scale are being made to descend with it by the same water
near which it may be cut—that is, by the Beni and the Madeira to
Pará. The advantages of the second route over the first are clear :
the transportation by water, avoiding at the same time the ascent
of the mountains. However, the entirely unknown region of the
middle and lower part of the Beni, and the savage tribes said to
inhabit the banks of the rapids, cause the bark which up to the

present has been sent to Pará to take a more complicated road.
It comes in canoes by the strong currents and the rapids of the
Upper Beni to the mission of Reyes, and from there by land in ox-
carts to the mission of Santa-Ana on the Mamoré. Thence it is
again embarked in canoes by the Mamoré and Madeira to the
Amazons and Pará.

With all these embarrassments, difficulties, and landings, the
freight by this route is cheaper than that *vid* the Pacific, as is seen
from the following freights taken in Bolivia from trustworthy
persons :—

Freight of a load (250 lbs.) of quina from the
Beni to Tacna ... ... 50 pesos.

Freight of a load of quina from the Beni to
Santa Ana on the Mamoré ... ... 4 ,,
Beni to Santa Ana to Pará ... ... 6 ,,

Total ... ... 10 ,,

The above difference becomes more notable considering that, the
destination of the bark being a European port, the freight from a
port on the Pacific is greater than from one on the Atlantic.

BENI EXPORTS.

The amount of the actual exportation from the Department of the
Beni by the Mamoré and Madeira is sufficiently well calculated by
the number and tonnage of the canoes which descend, as follows :—

Fifty canoes of a tonnage of 300 to 400 arrobas average, loaded
with hides and tallow, give a total of 20,000 arrobas exports. The
imports by the same craft are, with reference to weight, equal to
the exports. In the value, however, they are superior, consisting
almost totally of the products of the manufacturing industry of
Europe. The imports are valued at 30:000 $ 000, while the
exports are 18:000 $ 000.

Nowadays there are few merchants who wish to risk their lives
and fortunes in the passage of the rapids with an Indian crew.
These at various times have given proofs of the hatred which they
possess against the whites by assassinating the masters. There is,

therefore, no competition, and the prices of goods which go by the Madeira retain the same prices as those viâ the Pacific. In Trinidad European fabrics *via* the Pacific route are found selling on a par with others by the Amazon and Madeira. This contradictory fact may be explained. The passage of the mountains is by bad roads, and the cost on the backs of animals is great in comparison to water communication; but this latter is happily embarrassed by rapids. We will state, aside from this, that the Bolivian Government does not collect import duties by the ports of the Pacific.* The difficulty also, which is general in every country, is found here, for innovations on the old routes create opposition, as it does in this case on the part of the muledrivers.

In view of all this, the actual commerce by the Madeira, which, as we have demonstrated, scarcely extends to Trinidad, and rarely to Santa-Cruz de la Sierra. So soon as a road past the rapids is made, it will assume proportions ten times greater, so that annually there would be exported 200,000 arrobas. It is indispensable however that the Bolivian Government should then undertake the construction of a cart road from Vinchuta to Cochabamba, and another from Trinidad or Cuatro Ojos to Santa-Cruz.

Again resuming the principal points which may have influence upon the prosperity of a new route of communication by the Madeira, and which depend upon the Government of the Republic of Bolivia, we may say that it will be necessary :—

1st. To give new impulse to cultivation and industry in the department of the Beni. Cotton should be planted on a greater scale. Machines to clean and weave it should be introduced. The new line of Brazilian steamers on the lower Madeira will contribute much to this purpose by facilitating the importation of machinery, etc.; also by diminishing the time over the canoe voyage, while the Indians employed as oarsmen to-day can devote themselves to agriculture.

2nd. New cattle estates should be created in the immense prairies, making use of the remainder of the wild cattle for this purpose.

---

* NOTE BY TRANSLATOR.—This is an error. An import duty of 20 to 25 per cent. on the value of goods is collected. The very small amount of goods taken up the rapids now are sold at the same price as those from the Pacific, giving about 100 per cent. profit to the very few daring traders who are willing to risk their lives in the passage of the falls of the Madeira.

These in the days of the missionaries were still found there, and the Brazilian citizen, Antonio de Barros Cardozo, in his estates near Exaltacion, has demonstrated that they can very quickly be domesticated.*

3rd. Cause the course of the Beni to be studied and explored, with the idea of exporting cascarilla by that route. Thus, will be avoided, not only the passage of the Andes, but also the delays and landings which take place when, as to-day, it passes from the Beni to the Mamoré.

4th. Make an easier route from Cuatro-Ojos to Santa-Cruz, and another from Vinchuta to Cochabamba, then establishing a steam line on the Mamoré.

IV.—THE GUAPORÉ.

Although neither our instructions nor the advanced season of the year permitted us to ascend the Guaporé from the mouth of the Mamoré, we may, however, from trustworthy information, give assurance that the margins of the Guaporé are almost depopulated. Even the ancient capital of Mato-Grosso, Villa Bella, is in a complete state of decay, which is also the condition of the fort of the Prince da Beira, an admirable work in view of the enormous difficulties of transportation with which its builders struggled at the date of its foundation. All these leave much to be desired.

The studies of the hydrography of that region, which might be extended to the Jaurú, might be the object of a new exploration. This, however, must be done at the lowest stage of water in the Guaporé; for the difficulties which it opposes to navigation probably result from a lack of water.

PRICES OF SOME GOODS APPROPRIATE FOR EXPORTATION FROM THE DEPARTMENT OF THE BENI.

1 arroba of cacáo (32 pounds)	...	...	2 pesos†	
1 ,, coffee	...		. 2 ,,	
1 ,, cotton in the seed	...	...	2 ,,	
1 ,, ,, clean	...	...	5 ,,	

* NOTE BY TRANSLATOR.—He had one herd of 3,500 when I was at Exaltacion in 1871.

† NOTE BY TRANSLATOR.—The Bolivian peso is worth about 3s. 4d., and contains 8 reals.

1 arroba of tobacco	...	...	...	3 to 4 pesos
1 ,, sugar	...	...	...	2 ,,
1 ,, mandioca flour	...	...		4 reals
1 ,, castanhas	...	...	...	4 ,,
1 ,, dried beef	...	...	...	1½ pesos
1 ,, tamarinds	...	...	...	6 reals
1 ox-hide	...	...	...	6 ,,
2 ounces of urucú in paste		...	...	1 ,,
1,000 good cigars	...	...	...	5 to 6 pesos
1 pot of chicha (beer)	...		...	1 ,,
1 garrafa (bottle) of chestnut oil	...		...	1 ,,
1 arroba of meal	...	...	...	3 to 4 reals
1 arroba of hogs' lard	...	...	...	12 ,,
1 pound of butter	...	...	...	3 ,,
1 arroba of cheese	...		...	1½ pesos
1 cow hide	...	...		...	4 reals
1 tiger skin	...	...		...	2 pesos
1 hog skin (4 pounds)	...	...		6 reals
1 arroba of tallow	...	...	...	2 pesos
1 ,, sheep's wool		...	...	½ ,,
1 yard of Macana	...	...	...	2 ,,

1 cut of calça (probably enough for trousers) of cotton

cloth	...	...	...	...	2 ,,
1 ordinary cotton poncho		...	...	2 ,,
1 fine poncho	...	...	...	6 to 10 ,,
1 towel, with 12 napkins		...	...	13 ,,
4 arrobas of wheat flour in Santa Cruz		...	2 to 3 ,,

(N.B. The same in Pará is worth from 25 to
35$000 ; from $12·50 to $17·50.)

1 fat ox	...	...	...	...	10 ,,
1 cow	...		...		8 ,,
1 horse	...	...	...	... 45 to 100 ,,
1 beast of burden	...	...	...	100 ,,
1 sheep in Santa Cruz	...	...	...	1½ ,,

N.B.—One peso is 8 reals = 1$600=80 cents , 3s. 4d.

PRICES OF SOME OF THE IMPORTED GOODS IN THE DEPARTMENT
OF THE BENI.

Worth 1$000 in Maués.	1 pound of guaraná	..	3 pesos.
	1 „ gunpowder	.. 1½ to 3	„
„ 22$000 at Pará.	1 quintal of steel	..	36 „
„ 2$000 „	1 „ shot	.. 24 to 36	„
„ 3$000 „	1 arroba of Swedish iron	6	„
„ 3$200 „	1 „ of English iron	3	„
„ $800 „	1 „ of salt	.. 2	„

G.—APPROXIMATE CALCULATION OF THE COST OF FREIGHTING BY THE DIFFERENT ROUTES OF COMMUNICATION PROJECTED.

It is of great interest for the *empresarios* of new lines of communication to know beforehand, even though approximately, the advantages which may result from such enterprises. They wish to make comparisons to see if the results are, or are not, in proportion to the capital employed. We undertake in the following to calculate the freights by the different routes. The indirect advantages, however, which result to the country from a new route of communication, by the development of commerce and agriculture, by colonization and in a stragetic sense, may, in many cases, be greater than those which result from a direct lowering of freights.

To calculate the freights by a new route, it is necessary to know the weight of the merchandise transported in a given time over the route, taking into account the future increase of traffic in consequence of the greater facility for transportation.

According to information which we were able to collect in Bolivia, in no case the weight of the imports *via* the valley of the Madeira, will exceed 10,000 tons yearly; even though all the indicated improvements were executed in the interior of the country.

such as the road to Cochabamba, etc.* In the following calculations, we will adopt the weight of 10,000 tons as a base. This will
represent the limit to which the traffic in this direction will reach in
the immediate future.

I.—CALCULATION OF THE FREIGHT FOR NAVIGATION IN CANOES OF 300
TO 350 ARROBAS BURDEN (6 TONS, MORE OR LESS,) AS ACTUALLY
CARRIED ON BETWEEN TRINIDAD AND SERPA. THIS, DESPITE
THE LANDINGS AND OTHER DIFFICULTIES, AND ON THE
HYPOTHESIS THAT THE TRAFFIC WILL ATTAIN 10,000 TONS PER
YEAR.

The time necessary for a round voyage from Bolivia to the
Amazons is calculated in the following manner :—

Average time ascending .. .. 110 days.
,, ,, descending .. .. 40 ,,

It being possible, therefore, to make two voyages by canoe in a
year, $2 \times 6 \times 2 = 24$ tons annually, and 417 canoes are necessary to
transport the total 10,000 tons per year. The number of oarsmen,
ten per canoe, will be 4,170 men ; and each one costing 4$000
salary per month, and 9$000 per month for food, it follows that the
salary and support of the total number of rowers per year will be :

1st. Salary of oarsmen per year ... ... 650:000$000
2nd. Interest of capital employed in canoes—417
  canoes at 300$000, 125$000 at 10 °/₀ 12:510 000
3rd. Repairs on the same per year 12:510 000
4th. Repairs and renewals of cables 8:000 000
  ─────────
  Total 683:020 000

And the transportation per ton costs $\frac{683,020}{10,000} = 68$302$

---

* NOTE BY TRANSLATOR.—It is to be regretted that the Messrs. Keller
never penetrated Bolivia beyond Trinidad, of the Department of the Beni, where
it is absolutely impossible to obtain a knowledge of the population, the
resources and the requirements of the numerous towns and cities which lie
within commercial reach of the Madeira river. The tonnage per annum which
he estimates, on the completion of a road from the Mamoré to Cochabamba, to
take advantage to the full extent of the Amazon route, will serve only as a
unit for the calculation of traffic. There are 60,000 tons of freight passing between Pará and Manáos yearly, at this date, 1873. It represents the traffic of
not more than 300,000 people. The trade of at least 2,000,000 of the inhabitants
of Bolivia must, perforce, pass the rapids of the Madeira, as the cheapest possible
outlet of the country.

It is to be observed that all the population of the province of Mojos would not be sufficient to furnish so large a contingent of oarsmen, and that, in consequence, commerce can never be carried to the point mentioned by the actual method of navigation.

II.—CALCULATION OF FREIGHTS BETWEEN TRINIDAD AND SERPA, ON THE SUPPOSITION THAT THE PASSAGE OF THE RAPIDS IS EFFECTED BY MEANS OF INCLINED PLANES OR MORTONAS, THE BOATS EMPLOYED HAVING A BURDEN OF THIRTY TONS EACH.

The round voyage in this case would last 130 days on an average; that is 20 days less than now required. Each craft being able to transport annually $2 \times 60 = 120$ tons, it becomes necessary for the transportation of 10,000 tons, $\frac{10,000}{120} = 90$ vessels, with a crew of $90 \times 25 = 2,250$ men.

The freight is calculated in the following method :—

1. Interest on the capital for the construction of
   Mortonas 900:000$000 at 10 °/₀ ... ... 90:000$000
2. Annual repairs of Mortonas ... ... 20:000$000
3. Guards for them—48 persons at 50$000 per month,
   makes per year ... ... 28:800$000
4. Interest on capital employed in vessels—90 iron
   craft of 30 tons at 3:000$000 = 270:000$000 at
   10 °/₀ ... ... 27:000$000
5. Repairs and renewals per year ... ... 27:000$000
6. Repairs of cables per year ... ... 10:000$000
7. Salary of crew—2,250 persons at 13$000 per month
   (for nine months) ... ... 263:250$000
8. Preservation of an *alagem* track in those points
   where in the time of floods it becomes necessary
   in a distance of 100 kilomètres ... ... 10:000$000

        Total... ... ... 476:050$000

The freight being for each ton $\frac{476,050}{10,000} = 47$600.

III.—CALCULATION OF FREIGHTS ON THE SUPPOSITION THAT A
NAVIGABLE CANAL BE CONSTRUCTED BETWEEN SAN ANTONIO AND
GUAJARÁ, FURNISHED WITH LOCKS TO OVERCOME THE DECLIVITIES,
FOR THE ENTIRE DISTANCE BETWEEN SERPA AND TRINADAD, WITH
STEAMERS AS MOTIVE POWER.

It is now known that the minimum of freight in fluvial navigation is obtained by tow-boats, with a velocity inferior to that of
steamers used for passengers.

In the present case the calculation shows that a steamer with
an engine of 50-horse power, drawing $0^{m.}$ 70, towing a vessel of
100 tons burdens at the rate of nine miles per hour ($4^{m.}$ 5 per
second) is the most appropriate for the Upper Madeira and
Mamoré, the water having a velocity of $1^{m.}$ per second.

The principal dimensions of such a steamer are as follow :—

Total length ...	...	...	...	$30^{m.}$
Maximum width amidships		...	...	$5^{m.}$
Depth	...	...	•••	...	$2^{m.}$
Draft of water	...	...	..	$0^{m.}$ 7

It should be constructed of iron, with a total weight, inclusive of
machinery, of 80 tons.

The weight of the launch, also of iron, for 100 tons of freight,
would be 30 tons.

Price of steamer	...	...	35:000$000
,, launch ...	...	...	10:000$000
Transportation of both to the place...			5:000$000
Total ...	...	50:000$000

The medium velocity up and down of these craft is nine miles per
hour. They can run 31.2 leagues in twelve hours; and the total
distance from Trinidad to Serpa in 12.5 days.

Counting, yet, a detention of fifteen minutes at each lock, and
twenty-four hours' additional detention at twenty intermediate stations, the necessary time for a round voyage would be $2 \times 14 =$
28 days.

It is only in medium waters that the steamer would be able to
run at night, on account of the danger for the boat, not only at low

water but at floods. During the latter, the river brings down a great number of trunks of trees, and therefore, we can only count on the days, adopting the round voyage of twenty-eight days or a month as a basis of calculation.

The weight that one of the above steamers can transport in a year is consequently $2 \times 12 \times 100 = 2,400$ tons; and four tug boats and four launches would be necessary for the transportation of 10,000 tons.

The expenses of this transportation are calculated as follows:—

(1.)	Interest on capital employed for opening the canal, 21,000,000$000 at 10 per cent. ...	2,100,000$000
(2.)	Preservation of canal, including salaries of forty-eight lock-guards	250,000$000
(3.)	Interest of capital employed in vessels, steamers and launches...	20,000$000
(4.)	Repairs of vessels	20,000$000
(5.)	Renewal of vessels in ten years, per year	20,000$000
(6.)	Salaries of six captains	12,000$000
(7.)	,, engineers, at 1,500$000 ...	9,000$000
(8.)	Salaries of eight firemen, 600$000	4,800$000
(9.)	Twenty sailors, 600$000...	12,000$000
(10.)	Oil, grease, &c.	5,000$000
(11.)	Management	10,000$000
(12.)	Combustibles—2,500 achas (bundles) of wood for twelve hours for each steamer, consequently, for four steamers, and 365 days, 3,650,000 achas at 20rs. each	73,000$000
	Total...	2,538,200$000

Each ton will consequently cost, from Serpa to Trinidad $\frac{2,538,200}{10,000} = 253$000$. This result, with that of the above calculations, is that the expenses of opening a canal, and the establishing of steam-navigation, retain no proportion with the weight of goods transported, which is insufficient, even adopting the greatest possible increase of traffic as a base.

In comparison, we will see what is the resulting freight for steam navigation between Serpa and San Antonio, and between Guajará and Trinidad. The distances and other circumstances between these points being more or less equal, it is sufficient to calculate it for the distance between Guajará and Trinidad.

This distance is approximately 142 leagues. The steamers can run it in 4-5 days, or nine days for the round voyage. The number of voyages per year would therefore be thirty-six, and a vessel could be able to transport the weight of $2 \times 100 \times 36 =$ 7,200 tons.

For the transportation of 10,000 tons two steamers will be required, and the freights are calculated as follows :—

(1.)	Interest on capital employed in two tug-boats and two launches, 10 per cent.		10,000$000
(2.)	Renewal of these in ten years, per year ...		10,000$000
(3.)	Repairs of same per year		10,000$000
(4.)	Salaries of three Commanders, at 2,000$000		6,000$000
(5.)	Salaries of three Engineers, at 1,500$000 ...		4,500$000
(6.)	Salaries of six Firemen at 600$000 ...		3,600$000
(7.)	Ten Sailors at 600$000		6,000$000
(8.)	Oil, grease, &c.		2,500$000
(9.)	Management		5,000$0(0
(10.)	Combustibles, as in the above calculation ...		36,500$000
	Total		94,100$000

Which gives per ton $\frac{94,100}{10,000} = 9$410$ for the freight between Guajará and Trinidad. Two steamers being also sufficient for the service between Serpa and San Antonio ; the freight between these last two points will be the same as that calculated above.

IV.—Freight resulting from the Navigation of Craft of Thirty Tons burden, overcoming the Rapids by means of Mortonas, between San Antonio and Guajará.

One launch under the above conditions, manned by 25 oarsmen, would pass the total distance from San Antonio to Guajará in five days, without counting the delay of the rapids and currents. The passage of each rapid by means of a *mortona*, occupying one hour, there will be required two days for this service.

In consequence, the round voyage would be 14 days, and two voyages per month might be made from San Antonio to Guajará and *vice-versâ*.

The weight annually transported by a launch would consequently be 1,440 tons, and to effect the transportation of 10,000 tons seven launches would be necessary. The total crew would be 175 men, and the freight is calculated as follows :—

(1.) Interest on capital employed in the construction
    of the inclined planes, 10 per cent.   ...   90,000$000

(2.) Preservation of the same per year...   ...   9,000$000

(3.) Renewal of rails for the mechanical planes each
    25 years, per year   ...   .   20,000$000

(4.) Interest on capital employed in vessels, 21,000$000
    —10 per cent.   ...   .   2,100$000

(5.) Repairs of vessels, per year   ...   2,100$000

(6.) Renewal of same, per year   ...   ...   2,100$000

(7.) Renewal of cables, per year   ...   ...   2,100$000

(8.) Payment of 175 oarsmen at 20$000 per month   .   42,000$000

(9.) Administration—

             (*a.*) Director   ...   2,000$000

             (*b.*) Three Assistants   3,000$000

             (*c.*) Six Clerks   ...   3,000$000

                                      8,000$000

                   Total ...   ...   177,400$000

The freight, in consequence, costing for each ton $\frac{177,400}{10,000}$—17$740 between San Antonio and Guajará.

Employing the tow-boats, as above calculated, for the lower and upper Madeira and in the Mamoré, the cost of freighting per ton becomes 18$820 + 17$740 = 36$560 from Serpa to Trinidad.

V.—CALCULATION OF FREIGHT BY A RAILWAY BETWEEN SAN ANTONIO
     AND GUAJARÁ.

The expenses of the construction of the track, are in conformity with the estimates above, of 8,500,000$000. The data for the calculation of freights are as follow :—

(1.) Interest on capital employed in the construction, 10 per cent. ... ... 850,000$000

(2.) Maintaining road, 100 men at 13$000 per month for the term of one year ... ... 15,600$000

(3.) Renewal of rails, the new system projected, requiring only 1/5th part of the total weight to be substituted in 25 years—per year ... 27,000$000

892,600$000

(4.) The calculation of the number of cars required is as follows :—Annual transportation, 10,000 tons, or 33·3 tons per day. One car loaded with six tons, drawn by two animals, with a medium velocity of three miles per hour, and working seven hours per day on the round voyage, will carry 12 tons ... ... 892,600$000

It will require 50 of these cars to carry annually 10,000 tons. Each car will cost 3,000$000, and the capital employed in the 50 cars, 150,000$000.

Interest on this capital at 10 per cent. ... 15,000$000

(5.) Renewal of cars for 20 years, per year ... 7,500$000

(6.) Repairs of same, per year ... 7,500$000

(7.) Grease, etc., per year... ... ... 500$000

(8.) 100 draught animals, at 150$000 each, 15,000$000 at 10 per cent. ... . 1,500$000

(9.) Renewal and preservation of animals ... 1,500$000

(10.) Care and forage, 15$000 per month ... 18,000$000

(11.) Salaries of 50 carmen, at 20$000 per month ... 12,000$000

(12.) Ten stations, or open *ranchos*, at 500$000 each, at 10 per cent. ... ... 500$000

(13.) Administration : 1 Director and 3 assistants ... 7,000$000

Total ... ... 963,600$000

And the freight per ton $\frac{963,600}{10,000}$ = 96$000.

5

The total freight from Trinidad to Serpa, per ton, is calculated therefore in the following manner :—

1. Navigation from Trinidad to Guajará...   ...   9$000
2. By rail to.San Antonio...   ...   .   96$000
3. Navigation from San Antonio to Serpa   ...   9$360

Total   ...   .   114$000

VI.—SUBSTITUTING A MACADAMISED ROAD FOR THE RAIL, THE CONSTRUCTION OF WHICH IS ESTIMATED AT 5,700$000, THE FREIGHT WOULD BE AS FOLLOWS :—

1. Interest on capital for construction, per year   ...   570.000$000
2. Maintaining road, at 3,000$000 per league, employing Bolivian Indians as labour   ...   ...   150,000$000
3. Making a round voyage in 18 days, transporting five tons, always in a cart drawn by four animals, there would be required, in all, 83 carts for the transportation of 10,000 tons annually. The 83 carts would cost, at the rate of 1,500$000, 124,500$000, and the interest on this capital at 10 per cent.   ...   .   12,450$000
4. Renewal of same, per year   ...   .   6,225$000
5. Repairs   ,,   ,, ...   6,225$000
6. Grease, etc.   ...   .   500$000
7. 415 animals at 150$000=62,250$000, at 10 per cent.   ...   .   6,225$000
8. Renewal of animals per year   ...   ...   6,225$000
9. Care of same at 15$000 per month   ...   ...   7,470$000
10. Wages of 100 carmen at 20$000 per month, per year   ...   .   24,000$000
11. Ten stations, or *ranchos*, at 500$000, at 10 per cent.   500$000
12. Administration—One director and three assistants   7,000$000

Total   ...   .* 864,000$000

Or the freight of one ton transported on the macadamised road

* NOTE BY TRANSLATOR.—Should be 796,820$000.

from Guajará to San Antonio is therefore 86$400,* and for the total distance between Trinidad and Serpa.

1. By the navigation of Trinidad and Guaporé ...   9$360
2. Macadamised road to San Antonio   ...   .   86$400
3. Navigation from San Antonio to Serpa   ...   ...   9$360

      Total   ...   .   105$120

VII.—To THE ABOVE-CALCULATED FREIGHTS, WE WILL YET ADD THOSE WHICH WOULD RESULT FROM MULE-TRANSPORTATION. THIS WOULD BE OVER A ROAD THAT MIGHT RUN MORE OR LESS IN A STRAIGHT LINE AMONG THE RAMIFICATIONS OF THE SERRA GERAL, PASSING THE SERRA OF PACCA-NOVA AT ITS SUMMIT.

This freight, as one may judge beforehand, is greater than that by the actual navigation.

The calculation is as follows :—

1. Opening the track in a straight line of, more or less, 45 leagues and a width of 30 mètres, at 50 *réis* the square fathom   ...   .   83,470$000
2. Earthworks where the level may be imperfect, say 1,000$000 the kilomètre ...   .   278,000$000
3 Construction of bridges 3 mètres wide over streams and ravines, with a minimum of 807$000 per kilomètre, for the entire distance   ...   ...   224,346$000

      Total   ...   .   585,819$000

For the calculation of the number of animals necessary, we have the following data :—

An animal will carry eight arrobas (32 lbs. each) four leagues per day. In consequence, the round voyage would be 22·2 days, in which there would be transported 0·26 tons. Each animal would carry in 300 available days 3·5 tons.

For the transportation of 10,000 tons, the base adopted, 2,857 animals would be required. In round numbers 3,000 pack-mules. Their total price, with harness, would be, at 150$000 each, 450,000$000.

<hr>

* NOTE BY TRANSLATOR.—Should be 70$082.

The following, then, is the cost :—

1. Interest per year on capital employed in opening road ... ... 58,500$000
2. Maintenance of road, per year, at 500$000 per league ... ... 22,500$000
3. Renewal of animals, 300 annually ... ... 45,000$000
4. Support of 3,000 animals, forage, etc., at 20$000 per month ... ... 720,000$000
5. 375 mule tenders, 30$000 per month ... ... 135,000$000
6. 75 mule drivers, one for each five tenders, at 45$000 per month ... ... 40,500$000
7. Forage for 75 saddle mules for the same ... 13,500$000
8. Adminstration :—
   (a) 2 Managers ... ... 6,000$000
   (b) 3 Book-keepers 4,500$000
   (c) 30 Stable boys 18,000$000
   (d) Lights... ... ... 3,000$000
   (e) 10 *Ranchos* ... ... 8,000$000
   _____ 39,500$000

Total ... ... 1,074,500$000

This freight by mules, in every way the most imperfect and dearest, is 107$000 per ton.

We will add that in the province of Minas-Geras, this kind of transportation per ton per league is 6$000 ; in the province of Paraná, 3$000 on an average—this latter being still a little dearer than the minimum as above calculated.

_____

H.—CONCLUSION.

The better to understand the results reached in the above calculations we give, in addition, the following table :—

The advantages offered by the different plans, in this manner, become more evident, and the selection from them becomes easier.

Table of Freights from Serpa to Trinidad of the Beni over the different Routes, projected in the Valley of the Madeira and Mamoré.

All Calculations based on an annual transportation of 10,000 tons.	Expense of Construction.	Freights per ton.
1. 10,000 tons transported as to-day in canoes of 6 tons, with river in its present state, would require 417 canoes and 4,170 oarsmen ...	...	68$354
2. Constructing mortonas at the rapids, procuring larger boats of 30 tons, and employing 90 boats and 2,250 oarsmen for annual traffic of 10,000 tons...	900,000$000	17$600
3. With steam navigation above and below, a canal at the rapids, with 4 tug-boats of 80 horse-power each, running 9 miles per hour, towing 100 tons	21,000,000$000	253$800
4. Navigation above and below, with mortona system at the rapids, 4 steamers, 7 launches of 30 tons each, with 175 oarsmen ...	900,000$000	36$580
5. Steam above and below rapids, railway round rapids, 50 cars of 6 tons each, and 100 animals	8,500,000$000	114$000
6. Steam above and below, a macadamized road at the rapids, with 83 carts and 415 animals	7,500,000$000	123$225
N.B.—The construction of an improved road would be of little use, for the transit would only be free in dry weather. The preservation of it would therefore be very costly.		
Opening a mule track for the transportation of 10,000 tons per year, which would require 450 mule drivers and 3,000 pack mules	585,000$000	107$000

For the above table we give the following elucidations:—

1st. The actual traffic to-day does not exceed 1,000 tons. It is difficult to obtain the comparative small number of rowers for 60 or 70 canoes. Therefore, it will certainly be impossible to find the necessary number of oarsmen, corresponding to the future development of the commerce. For this reason the improvement of the actual route becomes indispensable, and neither commerce or industry can be developed unless this necessity is attended to.

2nd. By the construction of inclined planes or *mortonas*, a craft of 30 tons burden becomes admissible, while the present one, which is with difficulty rolled overland at places on rollers, only carries at most eight tons. By this means a considerable diminution of freight is obtained: the construction of an *alagem* track, until it becomes necessary to employ still larger vessels, shows that 30 tons is the most appropriate.

3rd. The opening of a navigable canal on one of the margins, and the construction of locks at the most appropriate points, would offer, without doubt, the most perfect means of communication. The small actual commerce and the future of but little brilliancy, if the Bolivian Government does not at the same time complete a network of communications in the interior of the country, retain no proportion whatever to the enormous outlays of such a canal.

Actually, therefore, we cannot recommend the construction of so great a work.

1st. The project which offers the greatest advantages among all is that by inclined planes or mortonas and small craft at the rapids, and small steamers in the Mamoré and Madeira. The same craft that navigate among the rapids can be towed by tug-boats above and below them, as there are no bars.

The crew of the vessels being only 175 men. The indigenous population of the Beni, which is to-day physically and morally lost in a useless navigation, can return for the greater part to their accustomed occupations of agriculture and industry.

5th and 6th. Although not only the expenses of construction, but the freight by a railway are greater than by a macadamized road, we still prefer the first on account of its ease of repairs.

In any case, but principally if the new route is one of the last two, it becomes indispensable to colonize that region, hitherto uncultivated, for the purpose of producing the food necessary to sustain the operatives and animals.

From the above considerations, among the projects it is easily deduced that No. 4 recommends itself best for execution.

It appears to us, however, that until there are in that part of the province of Mato-Grosso and Bolivia, which have a more immediate interest in the projected line of communication, a denser and more industrious population, that neither the realization of steam navigation between Guajará and Trinidad, nor the construction of Mortonas between San Antonio and Guajará can result in great advantages.*

God guard your Excellency. Rio de Janeiro, May 20th, 1869. Very illustrious and very excellent Sir, Dr. Counsellor Joaquim Antão Fernandes Leão, most worthy Minister and Secretary of State for Affairs of Agriculture, Commerce and Public Works.

(Signed) JOSÉ & FRANCISCO KELLER.

---

* NOTE BY TRANSLATOR.—The Messrs. Keller overlooked the fact that, without an outlet, the rich regions they describe can scarcely offer inducements to the settler or to increased population. Within three years, 1,500 to 2,000 Bolivians have descended the rapids of the Madeira to find employment on the banks of the lower river, where their energies might be productive.

APPENDIX.

[TRANSLATION.]

Carlsruhe, 30th November, 1870.

Dear Friend and Colleague,

I wish, first, to try to answer the official part; that is to say, the technical questions, in your kind letter of the 21st instant, concerning the project of the Madeira Railway.

Before commencing, however, I shall beg you to call to the notice of our English colleagues that, for my part, I am far from treating our work as a complete work, but that I defy any one to do it more completely in the same space of time and under the circumstances in which we have done it.

It was necessary, then, to make that exploration in the manner in which we have made it, or else not to do it at all.

To your first question I must answer—Banks of large gravel are rather scarce, and it is only fine sand that is found in large quantities, and almost everywhere, on the whole line.

A cubic mètre might cost, with an average transport of 1,200 mètres, 3-4 francs.

To the 2nd—The quality of land is, as everywhere in Brazil, red clay, mixed, more or less, with sand, and covered with a bed of soil of a thickness of 30-40 centimes.

The price of 7 milreis, or 20 francs a cubic fathom (*braça cubica*), or 2 francs a mètre, is that of the " União e Industria" line.

To the 3rd—In consequence of the formation of the soil, which on the banks of the affluents always consists of alluvium, I have no doubt, although we made no special researches with this view, that iron screw-piles—more simple and economical, in this case, than in any other kind—can be employed.

These affluents have not a strong current, and do not drift many trees. They come principally by the Beni, which, happily, we have not got to pass.

The width of these affluents varies from a few mètres to 150 mètres. Of this latter and considerable size there is, however, only one.

To the 4th—The rocks of the banks of the Madeira in the cataracts belong to metamorphic formations, and resemble gneiss, mica-schist (Glimmerschiefes) and granite.

They are very hard, and the extraction of these stones would be very costly, but their quantity is luckily very small, the track of the railway coming close to the foot of the hills in a few places only.

It will be more profitable in all cases to construct all works of any importance of iron, and to avoid as much as possible stone-constructions.

Close to the mouth of the Mamoré only, stone more easily worked is found—a kind of ferruginous sandstone (pedra canga), of which the walls of the Fort do Principe da Beira, in Matogrosso, are built.

To the 5th—The forests of Madeira are—like all the forests of tropical America—very rich in building-woods, which greatly surpass in quality those of temperate climates.

If there were no iron, covered wooden bridges might be constructed that would prove more durable than even the *covered* oak bridges among us, which last almost for centuries.

But the building of the bridges of the "Companhia União e Industria," where at first we tried to use wood (in trellis, etc.), showed us that the cost of the transport of selected woods, the opening of roads across the forest, which is full of thorny brakes, and the high wages of European carpenters, made a wooden bridge come almost as dear as an iron one.

As to sleepers it is another thing, for neither long tracks for their carriage are required, nor good carpenters to prepare them.

To the 6th—In this alluvial soil it is not difficult to find clay fit to make bricks.

To the 7th—Although the right bank of the Madeira, at San Antonio as well as Guajará, may be below the level of extraordinary floods, the difference is not great, and it seems to me more profitable to make the few embankments necessary for the station on the very edge of the river, than to place the station five hundred or more mètres from the bank.

To the 8th—At San Antonio, at least, it will be easy to place the station so that there shall not be too much current along the jetty or quay during high water; and even at Guajará there is nothing to fear on that account, since a more favourable location can always be found higher up by prolonging the railway.

About the topographical plan, it should be remembered that if it be wished to speak of the average height of the two banks, the left, or western bank, is rather more hilly and elevated than the opposite one; but, even if it were the contrary, the *local* formation of the bank, the *width, depth,* and *fall of the river,* are the data for determining whether, during the floods, the current at a given point will be strong or not.

Generally speaking, absolutely nothing can be said on this point, unless it be that the current on one side is, on the whole, quite as strong as on the other.

I hope that these replies are fairly satisfactory; and I have only to speak about one other matter. You ask me if, on receiving a telegram from you, I can come immediately to London.

Although I am in a position to respond without delay to your summons, I should not, however, be willing to make the journey to London to no purpose, and I should like to know before coming there the offers the new Company intend to make me.

I thank you for the review that you have kindly sent me; and as to your article, so ably expressed, I have nothing to remark, except that neither my father nor I are Prussians, but thorough Germans.

* * * * * * * * * * *

Thanking you again for your photograph, which has pleased me greatly,

I am, Sir,

With the most perfect consideration,

Your Friend and Colleague,

(Signed) F. KELLER.

COL. GEORGE EARL CHURCH,

19, Great Winchester Street,

LONDON (E.C.)

EXPLORATION

OF THE

RIVERS AND LAKES

OF THE

DEPARTMENT OF THE BENI, BOLIVIA,

By JOSÉ AGUSTIN PALACIOS,

FROM

1844 TO 1847;

ALSO, HIS NOTES RELATIVE

TO THE DEPARTMENT OF MÓJOS.

TRANSLATED FROM SPANISH BY

JAS. WM. BARRY, SECRETARY,

MADEIRA AND MAMORE RAILWAY COMPANY,
LIMITED.

1874.

PROVINCE OF MÓJOS.

(From *El Constitucional.* 14th January, 1869).

NOTES RELATIVE TO THE PROVINCE OF MÓJOS, IN THE DEPART-
MENT OF THE BENI, TAKEN BY DON JOSÉ AGUSTIN
PALACIOS, IN THE YEARS 1844 TO '47, DURING WHICH HE
WAS EMPLOYED THERE AS ADMINISTRATOR GENERAL OF
TAXES OF THE DEPARTMENT.

POSITION.

This province called Musu (nowadays Mójos, which was
conquered by the first Inca, Yupanqui, is situated between
10 deg. and 16 deg., south latitude, and 64 deg. and 70 deg.
longitude, west of the meridian of Paris, representing an oblong
superficies containing 13,750 square leagues of 25 to the degree.

BOUNDARIES.—On the north, by Perú; east, by Brazil; south,
by Chiquitos, with the departments of Sucre and Cochabamba;
and, on the west, by the province of Caupolican, in the department of La Paz.

TERRITORY.

The land lies very low, and the greater part is generally
inundated for a certain time every year.

It is very probable, judging from the appearance of the
land-surface, that, not many centuries ago, this territory was
covered by the sea. Of this there are many proofs—for instance,
there are no other heights therein than the banks of the rivers
and lagoons, except the elevations which will be hereafter

indicated; and that the rivers—the Mamoré particularly—form, every year, deposits of mud. These deposits are known among geologists by the name of *"alluvial earth,"* which is very fertile, and, according to travellers, resembles that which the Nile deposits, in Egypt. Finally, that, in the towns of the interior, near San Joaquin, there are *pampas* recently covered with trees, of which the regularity in the order wherein they have been placed is surprising, and appears, at first sight, as if the hand of man had intervened; so that nature advances continually, little by little, according to the conditions of each year.

The inundations, of which mention has been made, follow a direction from north to east as far as the fort of *Príncipe Imperial*, in Brazil, where the hill-ranges commence.

MOUNTAINS.

Only four hills, or insignificant heights, are to be seen, namely :—

1. That of El Cármen, at fifteen leagues to the south-east, between the rivers Blanco and San Miguel.

2. El Colorado, on the right bank of the river Machupo, near San Ramon.

3. At five leagues from Exaltacion, to the right of the Mamoré, on the brink of the Iruyane ; and

4. The range of San Simon (sketches of which I preserve) is discernible to the east of Magdalena and Báures, and is very rich in gold-mines.

RIVERS.

The Barbádos, which rises in the province of Chiquitos and gives birth to the Guaporé or Iténes, is navigable as far as Casalbasco, and could be united with the Paraguay by opening a canal of 4,800 yards in a flat soft soil. By this means, the Plata and Amazon would be connected, giving a navigable extent of 1,200 leagues.

The Verde has its origin in San Ignacio de Chiquitos, and,

after coursing north, west, and east, joins the Barbádos in longitude 64 deg. and latitude 14 deg.

The Serre rises to the north of Concepcion de Chiquítos, is incorporated with the Guaporé twenty-five leagues lower down.

The Blanco, or Báures, also rises in Concepcion de Chiquítos, and like the two foregoing it courses north, west, and east, passing close by Concepcion de Báures, discharging itself into the Iténes, near the fort of Beira.

The Machupo enriches its waters with the streams of San Juan, is navigable from San Pedro. The Moocho, the Mohino, the Machupo, and the Chananona, all pass together in front of San Ramon and San Joaquin, and combine with the Itonona, and, jointly with the latter, flow into the Guaporé, also near the fort Beira.

The Guaporé runs to the west, east, north, west, east, and then incorporates itself with the Mamoré, at 12 deg. south latitude and 68 deg. longitude, west of the meridian of Paris.

The Mamoré receives all the waters of the eastern slope of the cordilleras. Its tributaries, beginning on the east, are as follows:—The Ibare flows west and east from the country of the Guarayos, and, coursing north, west, east, receives, on the left, the waters of the Tico and San Antonio, and joins the Mamoré a little above Trinidad.

The Rio Grande or Zara is situated in the north of the Province.

The Piray rises in Samaipata; passes close to Santa Cruz and unites with the Rio Grande at 15 deg. south latitude.

The Ibabo flows, at first, under the name of Yapacani; passes close to San Cárlos and enters the Mamoré, with a south-easterly direction, near Zara. Between these two rivers are found the well-known Sirionós indians, the terror of the Mojeños, whose insecure boats continually expose them to become the victims of the ferocity of those barbarians, of a cruel and obstinate character. They are true children of nature, since both sexes go completely naked. Most of them are rather fair, and have beards and Roman noses. The frontiers of this side of the capital are always exposed to the incursions of those wild indians, one of which happens regularly every year,

The Mamoré rises to the east of the Ibabo upon the eastern slope of the *Cordillera de los Yuracarés*. Its bulk is increased by the Chimoré. It flows northwards, and, for some degrees, deflects west and east. This river preserves its name as far as 10 deg. south latitude, where, combined with the Beni, it takes the name of the Madeira.

The Chaparé, formed by the rivers Coni, San Mateo, Paracti, and many others, rises in Yuracarés to the west of the Mamoré, flowing into it rectangularly, on the south, in 15 deg. south latitude.

The Sécure, formed by the rivers Chipiriri, Samusebete, Isiboro, Yaniyuta, Sécure, and Sinuta, receives the torrents from the eastern slope of this watershed, between 68 deg. and 70 deg. west longitude. It unites with the Mamoré in front of Trinidad, towards the north, in 150 latitude.

The Tijamuche rises westward of the Sécure. It receives the waters of the Taricuri, traverses the north-easterly part as far as the Mamoré, into which it flows in latitude 14 deg., a little above San Pedro.

The Apere rises westward of the former, receives the tribute of the San José, courses north-east, flowing into the Mamoré half a degree from the aforesaid, at a distance of ten leagues.

The Yacuma rises to the westward of the Apere, near Réyes, and, augmented by the river Rapulo, passes by the town of Santa Ana, joining the Mamoré in latitude 14 deg.

The Iriyane rises in the pampas of Réyes, and, swelled by the river Boroca, flows into the Mamoré, in 13 deg. latitude, towards the north.

The Mamoré, after receiving these eleven affluents, unites with the Iténes, or Guaporé, in latitude 12 deg., and continues northwards until its confluence with the Beni, both forming the Madeira. Between the Mamoré and the last chain of the Andes there is a large portion of unexplored land, which commences near the towns of Buena Vista and San Cárlos, whence the mountain-range takes another direction towards the north-north-east, until opposite Exaltacion, a town situated more to the north of Mójos. The distance between the range and Exaltacion exceeds eighty leagues. A large portion of this land is the best in the whole of Mójos; is more elevated, and only in some parts subject to inundations. Here reside wild indians, many of whom are already known; amongst them the Toromónas, who extend as far as the river Purús, or Cuchibare.

TRIBUTARIES OF THE BENI.

The river Beni empties itself in the plain at the point of Rurenabaque, in 14 deg. latitude, receives the rivers Yungas, Ayopaya, Inquisivi, Larecaja, and Muñecas (as shown in the plan which I have made), together with the Tuiche of Caupolican; and, moreover, the rivers Undumu, Madidi, and others issuing from the east and from Carabaya. It continues its course towards the north as far as 11 deg., where it changes its direction, swerving to the north-east, and blending at length with the Mamoré in 10 deg. latitude.

They calculate 18 degrees, or 10,000 square leagues, for twenty-four rivers, all navigable for steamboats.

LAKES.

There exists in Mójos a lake called Rogo-Aguado, or Domú, the length of which is seventeen leagues; depth, two fathoms and a half. For further information with regard to this lake the reader may consult the diary of my voyage in the year 1845.

The lake of Ibachuna, or Lago del Viento (Windy Lake), which will have the length of four leagues of latitude and eight of longitude from north to south, drains into the Rogo-Aguado.

The little lake of Yapacha towards the north-east.

On the east is found another little lagoon called Puaja, the waters of which, with those of Rojo-Aguado and Yapacha, form the river Yata-chico, a tributary of the Mamoré. I take the Yata-grande to be a branch of the Beni, from the clearness of its waters and the stratification of the soil, since, in the plains, there is no appearance of its source.

The lagoon of Chitiope, which is situated further up than Cármen, at the very head of the Rio Blanco.

The Itonama is found placed on the river of the same name, and is five leagues in length, and two in breadth.

Near San Ramon, two lakes are to be seen, the one at a distance of half a league, and the other at two leagues. Both are of an oblong shape. Near San Joaquin, also, there is another lake.

CLIMATE.

Only two seasons are known in the province—summer and winter; the former is the wet season, the latter the dry. The average temperature throughout the province is 70 deg. Fahrenheit, and does not descend lower than 35 deg., for which reason it is tolerably healthy.

PREVAILING WINDS.—This province being situated at the tropic of Capricorn, the prevailing wind through the year should be the south-east; but it is noticeable that the north-east is the only wind in summer: and in winter, from May to September, the north and north-east alternate, with intervals of the south, which only lasts at the most three days. Of these, the first, which comes from some snowy mountain-ranges, is very dry, cold and piercing, with heavy rains and very violent hurricanes, which cause ravages, and even the death of old men, children, and cattle. The second, traversing an immense tract of forest, is humid, very healthy, and at times strong; and without this it would not be possible to inhabit the place, by reason of the mosquitos and other insects which infest the place.

POPULATION.

According to the statistical returns sent in by the curates to the bishopric, the population exceeds thirty thousand inhabitants. A great difference is observable in the deaths of males and females. The number of widows in all the towns exceeds that of the widowers. The reason is that the men, in consequence of their profession, perish insensibly in navigation, through the dangers incident to their canoes on their voyages, and by many other accidents to which the opposite sex is not subject. Thus it is that the population has not increased.

It is undeniable that the Mojeños possess great natural talent. A paternal solicitude on behalf of the Government of Bolivia to make them acquire a competent knowledge of arts and the necessary branches of education would be one of the first elements of action which would contribute to render Mójos what it now is not, and which it only rests with the future to develop.

TRIBES WHICH INHABIT THE SOIL.

The Mójos have originally occupied from 13 deg. to 16 deg. south latitude, and 64 deg. to 69 deg. west longitude.

The Itonámas occupy from 13 deg. to 14 deg. south latitude, and 65 deg. to 67 deg. longitude west of Paris.

The Canichánas are comprised within 13 deg. and 14 deg. south latitude, and 67 deg. and 68 deg. west longitude.

The Mobínes dwell to the west of the Mamoré, on the banks of the Yacuma, in latitude 14 deg., 68 deg. to 69 deg. west longitude.

The Iténes are at 12 deg. south latitude, between the rivers Mamoré and Iténes, and are known by the name of Guaráyos. They are bearded.

The Pacagudras are at 10 deg. latitude, and 67 deg. to 68 deg. west longitude.

The Chapacúras are 15 deg. south latitude, and 64 deg. to 65 deg. west longitude.

The Marópas are on the river Beni, and are called Reyesános.

The Sirionós are found on the Rio Grande and the Piray, between Santa Cruz and Mójos, from 17 deg. to 18 deg. south latitude, and 68 deg. longitude west of Paris.

LANGUAGE.—The principal dialect is the Mójo; nevertheless, there exist many other distinct ones of the different tribes which populate Mójos, such as the Canichana, Mobima, Itonama, Guarayo, Cayubaba, and others,

CANTONS.—The province is divided into fourteen cantons, which are as follows:—*Trinidad*, at three leagues from the Mamoré, and two from the Ibare. *Loreto*, at twelve leagues from the aforesaid, towards the south-east. *San Javier*, at six leagues from Trinidad, between the mouths of the Tijamuche and Apere. *San Pedro* is found at six leagues to the north of San Javier. *San Ramon*, to the north-north-east of San Pedro, distant from it thirty leagues. *San Joaquin*, situated at eight leagues to the north of San Ramon. *Magdalena*, at twenty-five leagues to the east of San Ramon. *Guacaraje*, at a distance of nine leagues from *Concepcion;* and the latter at twenty leagues to the south of Magdalena. *Cármen*, lying along the right bank of the Rio Blanco. *Exaltacion*, to the north of Santa Ana, distant some fifteen leagues. *Santa Ana*, at a quarter of a league from the river Yacuma. *Reyes*, to the westward of the aforesaid, at a distance of seventy to eighty leagues. *San Ignacio*, to the westward of Trinidad, at fifteen leagues' distance. There also exist many *estancias*, situated half-way between one town and the another.

RAPIDS.—*(Besides Rivers)*.

It has been believed that one of the insuperable obstacles to the navigation of the Maderia are the rapids (cachuelas). This inconvenience is one that might easily be surmounted, so as to throw open the navigation at the same time. There are three classes of rapids, divided in the following manner:—

1st Class.—Ribeirão, Girão, and Theotonio are dangerous.

2nd Class.—Guajará-Guassú, Bananeiras, Pão Grande, Madeira, Aráras, Pederneira, Paredão, Caldeirão, Morrinhos, and Santo Antonio. At four leagues distance from the last named lies the island of Tamanduá, where the Brazilians assemble every year to make turtle-butter, and carry away immense loads of these animals. This being the only part in which the turtles abound in the river Madeira, as, also, in the Maranhão, where a similar one is found, so great is the abundance that it is impossible for a man to take away the eggs, each of which contains from 90 to 110 of spawn. The egg is round, rough, and elastic, might do for a "*pelota*," agreeable to the palate. These little oviparians are, at the time of their being hatched, the victims of the crocodiles ; but, notwithstanding this, they increase rapidly.

3rd Class.—Guajará-Merim, Lages, Misericordia, Periquitos Tres Irmãos, and Macacos.

PRODUCTIONS.

It appears that Providence has wished to make a perfect paradise of the eastern part of Bolivia. Nature has taken pains in adorning this place, giving it all the attractions of its benefits. In a little sketch like the present, it is difficult to recount all that Mójos possesses in the three kingdoms. It only leaves its inhabitants to wish, because they find whatever is produced in the three zones.

Animals of infinite variety inhabit these fertile regions, and among them horses and cattle abound. From the fierce tiger or jaguar to the meek lamb, from the stately American eagle to the imperceptibly-small *organillo*, and from the fair sportive butterfly to the minutest denizen of the microscopic world—all are to be seen there. Delightful abode, where man enjoys and admires so many marvellous animals created for him !

It would take a large volume and not the pages of a newspaper to give a catalogue of the vegetation. Immense woods, fertile valleys intersected by important streams, are everywhere to be met with. The encircling air is perfumed by various flowers. How close must be the vegetation, where one sees everything, from the bulky cedar and the lofty palm to the lowly moss ! where many precious and varied cabinet-woods are found, such as mahogany, *tirbeti*, lignum vitæ, jacarandá, the strong *chonta*, the *bibosi*, the famous rose-wood so esteemed in Europe, various resin-yielding trees. The india-rubber tree grows abundantly, and proves of great value, being employed for manufacturing purposes, and even for ships.

Vegetable products are found which are useful in dyeing,

such as Brazil-wood, *tartaguillo*, (spurge), *añil* or indigo, *achiote* (heart-leaved bixwort), *nopal* or prickly pear-tree, and others. For medicinal purposes there is a multitude of plants, from which chemical manipulation extracts balsams and therapeutic drugs; among these are the valuable Peruvian bark—*quina* or *cascarilla*; the *chepereque*, good for curing any sort of wound; many others of great value in commerce, such as vanilla, coffee, cocoa, coca, and all that man needs for his consumption. Still further, minerals of every kind in abundance, and principally *gold*, constitute the riches of this place.

It would be diffuse to advert to numerous other indigenous productions of this central portion of South America, where the Supreme Creator has been pleased to encasket so many treasures for the enjoyment of Bolivia and her sons—power, wealth, and civilisation ; and, by opening her gates to foreign commerce, she would become the foster-mother to thousands of exiles, whom starvation banishes from Europe.

CONCLUSION.

In closing this sketch, so imperfectly executed, it is necessary to point out that the immense territory which we occupy is best adapted for foreign emigrants, on account of its rich and innumerable productions of every description, unsurpassed in any other part of South America, and by reason of the multitude of its navigable rivers, the fertility of its soil, which yields a bountiful return to the sower; and, lastly, because of its peculiar situation, it comprises advantages vastly superior to the remaining districts of Bolivia.

The project of emigration, as well as that of navigation, unfortunately, cannot be carried out until the cessation of the dissensions and internal feuds which distract the attention of the governing authorities from an object of such great importance, which is to place the whole of Oriental Bolivia in contact with the great markets of the United States of North America, whence their commodities would be imported with considerable utility. Under this influence the sciences, commerce, arts, and all the useful manufactures would advance.

I pray to Heaven that this wished-for day may arrive, when, beneath the shadow of peace, Bolivia might become the precious gem of the Continent, and add new glory to the immortal genius who established it.

(Signed)   JOSÉ AGUSTIN PALACIOS.

DEPARTMENT OF THE BENI.

NAVIGATION OF THE RIVER BENI, ROGO-AGUADO, MADEIRA, ETC., IN THE DEPARTMENT OF THE BENI, BY JOSÉ AGUSTIN PALACIOS, IN THE YEAR 1844.

The vast and rich territory which Bolivia possesses towards her orient attracts, nowadays, the attention of energetic Europeans; and the navigation of the mighty rivers which water its whole area is the thought which predominates with several joint-stock Companies capable of realizing so great a project.

In view of this, it seems to me opportune to re-publish my "Explorations of the Madeira," in order that I may contribute something, little though it may be, to the noble work now in contemplation for the welfare of Bolivia and the glory of our patriotic Government, which so zealously studies the interests of the nation.

In the year 1846, I had the fortune of paying a visit to all the affluents that swell the majestic Madeira, a tributary of the Amazon. I am, therefore, convinced that no other traveller can furnish more abundant and reliable data than those recorded in my work alluded to, and others yet in MS. I determined to make another voyage for commercial purposes, and carefully observed everything I met with.

In the brief account I give of my daily adventures, as a guide to those who are attracted to follow in my steps, my countrymen will not, I hope, overlook the difficulties which I encountered.

The want of a lithographic press renders this publication incomplete ; for I should wish it to be accompanied by a map of those parts, illustrating the courses of all the rivers. I procured a few copies to send to persons to whom they will prove useful, and am prepared to lend the original to all who may desire to copy it.

RIVER BENI.

When the Supreme Government of the Republic was pleased to appoint me Administrator-General of Revenue in the Department of the Beni, and afterwards to commission me to navigate the great river of that name, and the others with which I occupy myself further on, I felt an earnest desire to be useful to my country. I knew the risks and dangers I would have to encounter in my arduous task, but the conviction that the successful issue of my commission would furnish Bolivia with the means of prosperity and aggrandisement made me cheerfully encounter all.

On one occasion, when instructions were given to me relative to the communications which should be established between the towns of the Beni and those of the province of Yungas, I had the pleasure of informing the Government as to that which would be useful in connection with the explorations I made from Reyes to the canton of Libertad, capital of the said province of Yungas. This comprehends a distance of one hundred and thirteen leagues, but, by land, the distance would be considerably less. The country is very fertile, covered with high mountains, on which are to be found various kinds of Peruvian bark.

There are, moreover, level and spacious slopes, which offer every convenience for a road, with the exception of twenty leagues, from Tamampaya to Ebenay, in which are to be found seven rugged spurs of hills of a good size. Two bridges are needed in the rivers Tamampaya and Totorani, although there exists a road opened up by Señor Revuelta. These would cost but little, because the rocks in that vicinity are of the kind used for grindstones.

From the capital of this department to that of Yungas there is a distance of thirty leagues, by different roads, all accessible to traffic of every description. From the capital of the Beni, or Trinidad de Mójos, to the town of Reyes, there are one hundred leagues of good land and water communication.

There is yet another route from the town of St. Borja, in the province of Mójos, by way of the mission of Chimenes, to Santa Ana de Mosetenes, situated on the bank of the river Beni, and sixty-two leagues from Reyes and fifty-one leagues from the capital of Yungas. I inspected this route with the Superintendent of San Borja, in company with twelve Chimenes, who descended the river Beni as far as Reyes. Here no impediment occurs to the opening of a good road whereby the merchants of both departments might convey their goods with convenience.

The navigation of the river Beni presents this difficulty, that it only admits the passage of light and narrow wood-rafts, in consequence of the sandbanks and strait channels. Moreover, the bends of the river form rapids or strong currents, which, on the slightest want of care in manipulating the raft, would tear them asunder. Nevertheless, with flat-bottomed iron boats, the navigation would be greatly facilitated. The most dangerous rapids are those of Charia, Guachivó, Sipna, Wayaniboco, Sitipti, Chanami, Napañati y Poraqui, as far as Magdalena, a mission-town at $14\frac{1}{2}$ degrees beyond Iripachiqui, Bopinay, Mitti, Puñuya, Bohoy, Piñechi, Toracaya, and Sira, which are below and do not offer much risk.

Further down stream is to be found the rapid of Beú and the currents of Sibava, Quendique, Sambé, Torre and Chaguacala. There is a good depth in some parts, while in others it is but small. The rivers of the province of Yungas flow into the Beni, under the names of the Bopi, Sorata, and Muñecas. Other tributaries are—the rivers Mapira, or Caca, those of Caupolican, the Tuichi, and the Cochabamba, with the Beni, and many other important ones, flowing by the mission of Cabinas, situated behind

B

Carabaya, extending towards the wild tract of the Toromonas, as far as the bank of the river Magno, Purús, or Cuchivare. No one had explored the remainder of the river Beni until the year 1846, when Burza, a Prussian, was sent, but without much success. I pushed up-stream from the confluence of the Madeira with the Mamoré, and explored that rapid, believed to be an immense cataract, of which I shall speak in due course. In some parts bordering the river are found veins of silver and gold, and beds of salt, coal, lime, etc.; while diamonds exist in the Tequejé. There are, moreover, many valuable fossilized remains, and a variety of rare and wonderful objects, animal, vegetable, and mineral, wherewith to enrich a museum of Natural History.

LAKE ROJO-AGUADO.

A.D. 1845.

The Supreme Government being desirous of knowing whether the extensive lake Rojo-Aguado had any communication with the Beni, or if it flowed out of it, in order to facilitate its navigation by the Mamoré, ordered me to make explorations for this purpose. I accordingly started on this expedition. I set out from the town of Exaltacion, which is the nearest in the vicinity, and shaped my course W.N.W. five leagues, as far as the station of La Cruz. Half-a-league before arriving there we crossed the river Iruyané, which runs in a north-easterly direction. It contains an abundance of water, and is capable of being navigated. Its exact source is not known, but it is supposed to flow out of the Beni, or from some marshes or pools in the plains of Reyes. At the above-named station there is a flat-topped hill, about three hundred yards high, with square base. It is composed of white " *soroche*" stone, highly auriferous, and is everywhere covered with coarse grass and forest, amongst which is found the rubber-tree.

Thence I continued my march westward as far as the station of San Cárlos, a distance of eight leagues. The site is enclosed

by marshes and hill-slopes, which afford good pasturage to large
herds of cattle. Continuing to the N.E. for three leagues, I
came across the lagoon of Ibachuua, or Del Viento. This piece
of water measures four leagues in width, and eight in length,
from north to south. The channel is surrounded by marshes as
far as lake Rojo-Aguado. I then proceeded N.E. ¼ N. for two
leagues, changing to the east three leagues, north-east two
leagues, to the east two more, through lower tracts as far as lake
Rojo-Aguado, known also by the name of Domú. On its banks
there still exist traces of the ancient dwellings of the Cayubabas,
who now-a-days constitute the population of Exaltacion. This
town is surrounded by a ditch, or moat, probably as a protection
against the incursions of the Chacobos, Caripunas, or Pacaguaras. From this place, finding that the boat I was relying
on was not completed, I made an excursion in a small canoe to
the two islands midstream, a league distant. These are covered
with impenetrable thickets. The surface is slightly higher than
the lake—in that spot not more than a fathom deep. On the
following day I launched the boat, measuring twelve yards in
length, one and a quarter wide, and one deep; but as it would
sway about considerably I had two little canoes tied on, which
served as ballast. I sailed from the port in the direction
N.E. ¼ N. At five leagues' distance, I found a brook which
serves as an outlet, and is connected with another small lake
towards the north-east, called Yapacha.

I now changed my course, coasting E.N.E. for three leagues,
continuing for three leagues more towards S.E. ¼ S. Thence I
steered due south for eight leagues, one-and-a-quarter to the
south-east, and four-and-a-half to the S. ¼ E. I sailed with
the wind on the quarter, at the rate of six miles an hour, over
a mean depth of two-and-a-half fathoms. I landed on some of
the promontories, and observed that the forest extended only to
a short distance; but the prairies are so vast that they extend
to the horizon. I set them alight. We descried, to the north-
east, some smoke from the camp-fire of the wild Chacobo
indians. We afterwards saw them. They numbered more
than three hundred, including some fair and ruddy faces.

I continued to the E. ¼ N., and, having sailed four leagues, the wind, blowing from the N., became so impetuous that large waves were raised, and the boat was swamped, so that, more than once, I was nearly wrecked. I, therefore, landed and remained on shore for twenty-four hours, until the violence of the wind abated. This not occurring as soon as desirable, I, wishing to employ the time profitably, explored the mouth of the stream Ibachuna with great success.

I started the following day, rowing against the wind, the water being so boisterous that it threatened every moment to submerge the boat. I steered N.N.E. for six leagues, until I reached the port I started from. The lagoon is of clean and good water. The bed is composed of oxide of iron, with a depth of two-and-a-half fathoms. There is a large quantity of ray and other fish, alligators, and *bufeos*. There also exists a large number of wild fowl, including the bird called *toro*, the size of a partridge, with black plumage, in the form of a parasol, on the head. It has, moreover, a pouch on the chest, with black feathers of the same general character as those on the body, but much finer. Its note is like the lowing of oxen.

In the forest are almonds of various kinds. To the east there is another small lagoon called Puaja, of which the waters, joined to those of Rojo-Aguado and Yapacha, form the Yata Chico, or "black river," which flows into the Mamoré. I think that the Yata Grande is merely a branch of the Beni, on account of the clearness of its waters, the sloping away of the land towards the Mamoré, and because in the prairies its source is imperceptible. The only one traceable is the Rio Negro of the lagoon Rogagua, in Reyes, which, also, is a confluent of the Beni.

The navigation of the Yata Grande is interesting, and I should have undertaken it when I went down the Madeira if I had had a body of armed men. This is absolutely necessary, on account of the large number of savages swarming its banks. Nevertheless, I ascended as far as its rapid, where a large quantity of tar is to be found. The Iruyané is worth exploring for the same reasons as the Yata.

RIVERS MAMORÉ AND MADEIRA.

A.D. 1846.

The high importance to Bolivia of my subject obliges me to call the attention of my fellow-townsmen, of the Government, and of all men interested in the progress of mankind, to a matter which has hitherto been looked upon with indifference. Our State Administrators have busied themselves solely with the acquisition of the port of Arica, deeming this to be the only channel for ameliorating the condition of Bolivia, and even for establishing her political existence on a solid and firm basis. But the navigation of the Madeira offers an opening the beneficial results of which are incalculable. This is the first point our Government should consider.

Once embarked on this expedition nothing could deter me—not even the vast and dangerous rapids of this famous river—the numerous wild indians that inhabit its banks—or the tragic end of the unfortunate Señor Tadeo Gorriti, who fell a victim to the fury of these barbarians. I did not shrink from the rickety little canoes, which, measuring only about 12 yards in length, $1\frac{1}{4}$ in width, and $\frac{3}{4}$ of a yard deep, are, therefore, risky vehicles to travel in. Add to this the disinclination of the boatmen. In the expedition sent out by the Prefect Don José Barja, the crew fled at the sight of the first rapid they encountered. Such, then, were the difficulties I had to contend against, which obliged me to act with firmness towards the crew, composed, as it was, of men who took but little interest in the enterprise, which they regarded as impracticable or highly dangerous.

Moreover, the very authorities of the Beni Department, instead of working to aid our interests in that quarter, and observing a pacific and conciliatory bearing, would seem to have principally consulted their own convenience, and, for this purpose, to have resorted to the most despotic means. This course would tend to act as a deterrent to immigration, notwithstanding that those fertile regions invitingly offer their riches to all who determine to work to obtain them.

At length, after great efforts, I was able, on the 7th of October, 1846, to start. I took with me my son Gregory and six armed men. The Vicar, Dr. Eustaquio Duran, also accompanied me, with the intention of endeavouring to convert the barbarians, and infuse into their minds the sublime tenets of our holy religion. His retinue consisted of twenty boatmen in a *garitea*, and fifteen in a canoe. The crew consisted entirely of Cayubaba Indians from Exaltacion. I had a canoe with fifteen Canichanas, from San Pedro, another with fifteen Trinitarios, and a small pilot-boat, manned by six Cayubabas. The only things I stood in need of were practical technical knowledge and instruments. However, all was compensated by my ardent desire to be the first to achieve for my country a discovery which, I doubt not, will, one day, be the germ of her prosperity and happiness.

SEVENTH DAY.

Set sail from the port of San Martin, distant a quarter of a league from Exaltacion. We steered north, the wind being contrary. It blows from that quarter uniformly throughout the river. The depth of the Mamoré is from five to six fathoms, the width 300 yards, and current half a league an hour. The heat, by Fahrenheit's thermometer, was 84 deg. We shifted east or west, according to the windings of the river, making four leagues north, as far as the confluence of the River Iruyana, which lies on the left. For a league we met with rocks of oxide of iron. At 4 P.M. the thermometer rose to 90 deg., and we had a heavy fall of rain for half an hour. Continuing on our way, we arrived at a place called Navidad, having sailed seven leagues. There one sees the last of the wretched huts erected by the natives of Exaltacion. The banks of the Mamoré lie low, and are heavily timbered, the water's edge being lined with willows and long grass, or reeds.

EIGHTH DAY.

The atmosphere became cloudy. We continued our journey, the current running one mile an hour; depth, six fathoms; dead calm; thermometer, 81 deg. At mid-day, the wind blew roughly from the north. We saw to the north-east the fires from the dwellings of the Chacobo Indians. Depth of river, ten fathoms. Upon the banks are several abandoned cocoa plantations belonging to the State. At 3 P.M. the thermometer rose to 87 deg., and it commenced to rain. Half a league from the river Matucaré, towards the right, are found strata of oxide of iron. Continuing for another half a league, we stopped at an island, having sailed ten leagues with the same course as the day previous. The Mamoré is two hundred yards wide.

Ninth Day.

Atmosphere still cloudy. We followed the windings of the river, finding ten fathoms of water. Run thus for a league. Thermometer, 84 deg.; same bearings. The stream Achichuru (the name signifies *agitation*) flows in on the right; another, called Boroboro, on the left; a third, the Tanarupi, also to the right, equal in volume and width to the Matucaré. There are here rocks of the same oxide, and a deserted cocoa-plantation belonging to the State. On the left enters the Mayosa stream. In this vicinity we met Maba, chief of the wild Chacobo Indians, Bora, and two companions. The Vicar urged the first-named individual to persuade his people of the advantages they would acquire by forming a mission. We delivered to the chief the message sent for him by the President of the Republic. Having sailed ten leagues, we passed the night in a place we called Posancos.

Tenth Day.

Cloudy : thermometer, 77. Mamoré three hundred yards wide; soundings, good five fathoms. Three streams—Yona, Pejo, and Toro, enter on the left. At 10 o'clock the thermometer stood at 89 deg. In all these parts there are high slopes with palm trees. At noon the thermometer marked 100 deg. We caught sight of the high ground on which stands the fort of Principe de Beira, to the north-east. At 4 P.M. the thermometer fell to 80 deg., and we had a heavy rain which continued till we reached the confluence of the Iténes, or Guaporé, the width of which is 500 yards, and depth 3 yards at mid-channel. It has another affluent 100 yards wide and 3 feet deep. Both flow from south to north. The width of the Mamoré, at this part, is 300 yards, with 7 fathoms depth. The direction is towards the east, but that of the Iténes points north. Both are 800 yards wide, and 6 fathoms deep; current, half a league. From this part onwards there are convenient sites for building towns, particularly along the left bank. Here the Brazilians once had a port which, after having half-built, they abandoned. We travelled eleven leagues that day.

ELEVENTH DAY.

Being Sunday, the Vicar performed Mass for us. Meanwhile, the baggage and provisions were dried. We then continued to the N., leaving on our left three low hill-ranges clothed with rich foliage. After journeying two leagues, we encountered the oxide strata, and a pleasant spot which was well adapted for a township, with a convenient anchorage and a stream which has its source among the haunts of the wild Sinabo Indians. Thermometer, 90 deg. At four leagues, the opening of the Yata Chico is encountered. This stream is also called the Rio Prieto, on account of the blackness of its waters, which flow out of the lagoons already mentioned. The natives of the country call this stream Jibo. The width is 12 yards, and the depth 2 fathoms. Superior almonds of various kinds grow in the neighbourhood. At 5 P.M. we had some rain, with a southerly wind which blew so violently that we had to take shelter on a sand-bank, which we called South Bank, since we were sailing southwards. Distance traversed, 7 leagues. Now that I have spoken of the south-wind, I will here give some account of this phenomenon, which causes no small damage to men and cattle. It arises from the clouds descending considerably, and forming a spacious horizon from the cordillera which lies to the south. This wind being continuous, and the only passage being through this mountain-chain, it must either be driven back or become impregnated with nitrous particles, which render it extremely cold; and, encountering opposition from the north, it forms a terrible hurricane, accompanied by rain or parching blight, which, though it lasts but three days at most, causes sudden death and destruction. The north-wind is mild and fresh, although it proceeds from the arid coasts of Africa, for it traverses the Atlantic and the forests which extend from the mouth of the Amazon. It is thus cooled in such a manner that in Mójos the temperature is moderated by this cause, the thermometer never marking more than 70 deg., and descending to one-half that figure on those days when the atmosphere is charged—that is to say, during the prevalence of the south-wind.

TWELFTH DAY.

We started during a heavy rain, shaping our course towards the S. Thermometer, 70 deg.; current, 1 league; soundings, 7 fathoms; width of the Mamoré, 1,000 yards. On the left, the river Soterio discharges itself, 25 yards in width. The slopes are steeper, and covered with dense forest-growth and palm trees. The lands are fertile—could not be better—and do nôt suffer from inundations like the Mójos district. We proceeded for 7 leagues, and stopped at a landing which we called Elvira, on account of the abundance of trees of that name, from the bark of which are made tow-ropes for the boats.

THIRTEENTH DAY.

We set out with a south wind, steering E.; depth, 16 fathoms; current, a league. The slopes at the Bolivian frontier are higher than those of Brazil. They are profusely covered with almond-trees, which are the loftiest trees of the forest, which, itself, is of colossal proportions. At a short distance towards the left, a brook empties itself. To this, too, we gave the name of Elvira. At midday, the wind shifted into the N. Thermometer, 85 deg.; soundings, 8 fathoms. The land rises to more than 50 yards in height, clothed with almond and ipecacuanha-trees. Having proceeded 8 leagues, we stopped to have a rest; but, in the middle of the night, we were disturbed by a tiger, which, fortunately, did us no harm.

FOURTEENTH DAY.

Off again, with a steady wind from the N. We discovered a low range of hills ahead. Three leagues further on, towards the right, we found three small islands and a large one mid-stream. We stopped there just long enough to build a hut, in which we deposited our provisions for the return trip, giving to the island the name of La Provision. We continued our journey, with the thermometer at 90 deg. Passed the rivers Cacanova and Soterio, leaving both on the right. At 5 P.M. passed the rapid of Guajará-Merim, marked with 1 deg. of latitude, on my map. The rocks which form this rapid project two yards above the surface of the water. The declivity is one yard. At high water this rapid no longer exists. We made a halt here; having travelled six leagues.

To form a correct idea of these *rapids*, it would be necessary to have in view the details that are in my possession. But as this is merely a brief sketch of my journey, I shall content myself with explaining what the rapids really are, in order that they may not be confounded with the *cataracts*. The rapids are produced by a barrier of rocks running east and west, across the river, thus forming a link between two low hill-ranges flanking both banks. The impetuosity of the current has opened up several tortuous channels, with sharp points of rock sticking up here and there. The navigation thus becomes dangerous. The risk is lessened by choosing the deeper channel, in which, at high water, the rocks are completely covered and the rapids disappear. The currents in both the rapids and cataracts are very violent and jerky. For the ascent of the river it is necessary to tow the boats or to unload them, except in the case of flat-bottomed steamers, drawing little water, and capable of carrying a cargo of fifteen to twenty tons. There is abundant wood for fuelling. Steamers might pass through any of the channels, but not the canoes, which are lightly and badly built, without rudders and scarcely a ton burthen, liable at any moment to turn over or be swamped. The beds of the rapids might be considerably improved by clearing away the fragments of rock and *débris*, deposited by the river—which become hardened through amalgamation by means of the oxide of iron —leaving open only three channels with locks, without which they would be impassible, as I will show hereafter.

FIFTEENTH DAY.

Started from the island of Guajará-Merim (signifies *little*), which ought to be called Ipecacuanha Island, from the abundance of that plant. A quarter of a league further on, near a cape formed by that island, we encountered the rapid of *Guajará-Guassú*, which signifies *great*, and is represented in the first plate of the map I have commenced. This rapid is formed by banks rising four yards above the level of the river. The banks are flat and thickly wooded. When we passed, the channel to the extreme left was dry. We, therefore, chose the second, which had but little water in it. Our boats were much tossed, but we had no further mishap than breaking the rudder of the boat in which the Vicar travelled. At noon the thermometer marked 90 deg. To the south-east we saw the hill-range which helps to form the two above-mentioned rapids. The mosquitoes exist here in large numbers, and occasioned us considerable annoyance. At three leagues' distance we found the rapid of *Bananeira*, which forms the second plate of my map. It takes its name from the plantain-tree, which they call "banana." The boats had to be unladen for the passage. We made the descent between two parallel rock-cliffs six yards high, through the fifth channel, which contains less water than the others. The decline is four yards in fifty. Many veins of silver exist. To the left there is a good channel, almost dry, which, if cleaned out, would be the best of the lot. In the centre stands an island abounding in *butua*, a medicinal plant.

SIXTEENTH DAY.

We proceeded for a quarter of a league between rocks, the current running rapidly. This part is called the "tail-end of the rapid." At a distance of two leagues we met with a low range of densely-wooded hills, at the foot of which we found the rapid of *Pão Grande*, or the "huge stick," shown on the third plate of the map. The Yata Grande here empties itself, to the left of another spur of hills, adjoining the former, with a direction from east to west. At the mouth of this river are rocks fifteen yards high, covered with beautiful foliage. The width equals 105 yards, and the depth 3½ fathoms. The water is clear, and the banks attain a good elevation. For a league we were urged along by strong currents, and passed the rapid, which is formed by rocks twelve yards high. The decline is 3 yards in 150. The central channel is the best, but has a bank in the middle which makes it dangerous. To the north there is a thickly-wooded hill, and to the north-west a higher one is visible, situated on the river Beni. Various hieroglyphics and a cross between two pillars have been inscribed upon the rocks. Thermometer rose to 90 degrees; depth, 7 fathoms; current, 2 leagues an hour. It would be very easy to open a channel along the left bank. At a league's distance is the rapid of Lages, named after the *laja* stone. This forms the fourth plate of the map. This rapid is formed by a bank running east to west. The bed drops 1 in 50. Descended the principal opening without any occurrence of importance. At two leagues the Beni flows in. We stopped at the island in the north of this river. Latitude, 10 deg. 25 sec. south.

RIVER MADEIRA.

SEVENTEENTH DAY.

After measuring the width of the Mamoré—there 1,000 yards, with 7 to 8 fathoms depth—I proceeded to do the same with the Beni, divided into two arms, and found that one of these measured 800 yards in width, and 7 to 15 in depth; and the other 450 yards across, and 7 fathoms deep. Both these branches have a northerly direction, the Mamoré losing its course to the N.N.W. We started thence, and at a short distance to the left we met with a stream. After proceeding for about 600 yards we perceived a party of Caripuna Indians, consisting of twenty men, eight women, two girls, and nine children. The men lived in a hut called a *malocca*, oval in shape, about 25 yards long, 15 wide, and tolerably lofty. There are little compartments all round, each of which is tenanted by a man, whose furniture consists of a hammock, a bench, a stick, and a small basket to hold the feathers of the birds he kills in the chase. All held darts in their hands. They responded most stiffly to our salutation; but their hostile tone changed when we made them presents of cutlasses (*machete*), knives, fish-hooks, glass beads, liquor, and other articles. They laid aside their weapons, and treated us with genial familiarity.

The women live in another *malocca*, apart from the men. The former take charge of all the domestic utensils. Nothing of this kind is to be found in the dwellings of the men. The greatest cleanliness prevails. The chief, who was named Pachú, had but little real power, as unbounded licence existed throughout the community. All dispensed with the use of clothes, and wore eye-teeth of the wild boar as ear-rings. The women have their lower lips pierced, through which they pass a topaz-coloured languet eight inches long, made of a resinous substance. They also pierce the cartilage of the nose, inserting into the aperture two small feathers, arranged in the form of a moustache, and paint the forehead scarlet, the lips and eyebrows black, and stain the rest of the face with the juice of the aromatic myrrh-tree. Bracelets are worn on the upper part of the arm, the wrists, and the calf of the leg; and, out of decency, they cover the private parts with a leaflet suspended by a string. Moreover, the men bind with a cord the part which constitutes their sex, the same operation being performed with their dogs. Dead bodies of men are buried in the houses; those of women are interred in the fields.

In the delta formed by both rivers there are several islands and loose rocks forming the *Madeira* rapid, named from the vast quantities of wood (*madeira*) deposited on the banks of the river when swollen. This is done to such an extent that the wood-heaps appear like hills. The river doubtless takes its name from this circumstance. The last-named rapid is represented in Plate 5 of the Map. The headland between the two rivers could not be better adapted as a site for a large town. Almond-trees of a superior quality abound; also two kinds of excellent cocoa, vanilla, and other equally-valuable productions The scenery, too, is appropriately beautiful. At the close of the day we placed on a suitable spot a cross, which the Vicar blessed, and we directed our prayers to the Supreme Being, to the accompaniment of Nature's solemn music in those silent regions. The ceremony produced feelings of lively pleasure in the minds of the indians.

EIGHTEENTH DAY.

At daybreak, to the same weird music, the Vicar celebrated the Holy Sacrament of Mass. The wild indians were present, and looked on with cold indifference at some of the august rites of our religion. After this ceremony I left the rest of the party, and, with four men in a canoe, I made for the immense cataract of the river Beni. We paddled four leagues up-stream, closely observing on our way some spacious and fertile localities, beautifully situated, and free from inundation of the river. The latter is here five hundred yards wide, and nine to twelve fathoms deep; current a league and a half; thermometer, 95 deg. As it rained in torrents, we were obliged to halt at an island opposite the hills we perceived two days ago.

NINETEENTH DAY.

The day broke with a serene and clear atmosphere. At two leagues we encountered the rapid which corresponds with that of *Bananeira*, in the Mamoré. The rocks are of the same height, and bar the river from east to west. The western channel which appears to be the best, was, at that time, without water, and we noticed that it might easily be cleared by means of mechanical appliances. The declivity does not exceed four yards in fifty. In the centre the current cannot be stemmed, but this is easily feasible near the banks, by towing the boats; or else a circuit by land might be made. Mosquitoes swarm here. Thermometer, 95 deg.; current, 2 leagues; depth, 10 fathoms. After sketching the rapid, we returned to our bivouac of the night previous. Numerous rocks and strong currents exist in this, as in the rapid of Pão Grande. One league lower there are other smaller rocks and currents, similar to the current of Lages. Thus it arises that the prolongation of the hill-ranges cause the rocks and little hills to coincide in both rivers. We landed in order to fraternize with some wild Indians we met. The band, under the leadership of a chief named Sonó, consisted of eighteen men, eight women, and six young persons of both sexes. I gave orders that my canoe should accompany the rest of the party as far as the *malocca* of Pachú, while I made my way thither on foot with the Indians. I met the Vicar there, who was delighted because the natives had expressed to him their wish to become Christians, and build a town; and had offered to collect a large population for the purpose. We made them a present of a dozen *tipoyes*, or women's dresses, and some implements, which made them pleased with our visit. In return, they regaled us with a repast, the details of which I must omit for want of space in this work.

At nightfall I received information that the Trinitarios, indians of my crew, had exchanged their provisions for birds and other articles, intending, no doubt, to desert when the night was well advanced. I avoided such a misadventure by sleeping on the canoe, without pretending to know anything of their intentions.

TWENTIETH DAY.

We set sail at 6 A.M., and, after going a quarter of a league, encountered the *Madeira* rapid (Plate 5), thus named for the reason I have already stated. This rapid drops three yards in about three hundred. The rocks are lower than in the preceding ones. There are several channels, easily accessible for every kind of craft. The river is half a league wide. Thermometer, 90 deg.; current, a league; depth, 12 fathoms.

At two leagues is the rapid of *Misericordia*, which is shown in Plate 6 of the Map. There is an island in the middle, and a good channel. It has taken its name from the danger arising from the impetuosity of the current. The least want of care might cause the boat to be whirled helplessly along, and engulfed in the rapid of *Ribeirão*, a quarter of a league further down. This rapid, represented in Plate 7 of the Map, drops 4 yards in 150. As the channels are very dangerous, it is necessary to unload the boats and drag them overland about

450 yards, as far as the river Ribeirão, in which, by means of a strong sluice-gate, a good lock might be constructed. The advantages of this arrangement need not be demonstrated here. The granite rocks are also low, with a direction from east to west.

Twenty-first Day.

We went in search of the wild Indians who live there, and at half a league's distance we met an apostate woman called "Mariana," who has acquired considerable celebrity and influence among these her untutored subjects. She was with her father, her husband, and sister, and was also surrounded by eight men, three women, and four young persons of both sexes, among whom was a blue-eyed, rosy-cheeked girl. I presented Mariana with an under-garment, dress, blanket, necklace, and *aretes;* to the others I gave knives, fish-hooks, cutlasses, necklaces, medals, etc. We treated them to some liquor, which they accepted with the greatest complacency, manifesting much gratitude and friendliness. I met there a Brazilian, who had been wrecked a few months previously, with two companions. They were going down-stream in a canoe. I engaged him as my guide for the rest of the journey.

Twenty-Second Day.

We had mass, at which Mariana and her followers were present. I proposed that they should form a town at the con-fluence of the Beni with the Mamoré, whither they were about to make a move. They readily acquiesced to the suggestion, and offered also to collect many more nomad indians who frequented the neighbourhood. We remained among them the whole day, during which time our men were employed in twisting hawsers and caulking the boats.

Twenty-Third Day.

The Vicar, who had prolonged the journey more than he anticipated, parted from us with much mutual regret. The boatmen bid each other farewell, as if they never expected to see each other again alive. The crew of Trinitarios wanted to abandon me, fearing imminent perils. They urged, by way of excuse, that they had no provisions; but, by persuasion and energetic measures combined, showing them the commercial benefits accruing to them, I succeeded in getting them to go on; and to make them more contented I gave them some of my provisions. After the much-regretted separation, we set sail at 7 A.M. At a quarter of a league we had to unload the canoes, in order to tow them through a bad channel, by reason of the impetuosity of the waves and the strong and boisterous current. With larger vessels this inconvenience would no longer occur. Mariana journeyed down in a kind of canoe, made of bark, and offered to wait for us at some convenient halting-place on the river. We continued, for a league, to encounter violent currents. This makes the navigation very dangerous, when performed in canoes such as ours. The descent is barely eight yards in more than a league, including the channels and gradients which are there to be found. The forest, which is dense and lofty, affords an abundance of cacao, almonds, ginger, and other productions. There also exists rock-crystal, veins of silver, and some stones which, by their brilliancy and other qualities, indicate an approximation to the diamond. Moreover, the eye everywhere meets with birds so beautiful and of such rich and varied plumage that, added to all the other objects which constitute the enchantingly-magnificent scene, these make one forget the inconvenience and danger which beset the path of the navigator. Thermometer rose to 105 deg.; the sounding-line marked fifteen fathoms, and the current was about two leagues. We made three leagues during the whole of that day, and passed the rapid of *Periquitos*, or paroquets, which offers no danger. It forms Plate 8 of the Map. Halted at a malocca of wild Indians, near a stream called Pocel, and met Mariana with eight men, eight women, and four boys. Still on Bolivian territory.

TWENTY-FOURTH DAY.

Another half-league down-stream brought us to *Araras* (Plate 9 of the Map), named from a species of parrot.*

* Called by the Brazilians *Parabas*, and *Caques* by the Bolivians.

C

The descent is two yards in about three hundred. The channels are good. On the coast of a large island hereabouts dwell other indians, also under the rule of Mariana. The current runs at the rate of two leagues. The lead-line marked fully twelve fathoms, and the thermometer stood at 95 deg. The wind here blows steadily from the north. For six leagues we followed a very safe passage, as far as the confluence of the river Abuna, which flows in on the left; latitude, 9° 40′; width, 140 yards; depth, 3 fathoms. Above the point of confluence there reside some indians, whom I proposed to visit on my return; also two numerous and brave tribes on the Abuna. There are roads from them to the Beni and the last-named rapids.

TWENTY-FIFTH DAY.

We left Abuna with the south wind, and at four leagues we encountered the rapid of Pederneira (Plate 10), so called from the abundance of flint. The bed drops four yards in about three hundred. A good passage already exists, which might easily be improved by cleaning. There is a small hill close by, and at a short distance are others of greater elevation. The rocks are also low, the direction being chiefly north-east, afterwards changing to east.

Three leagues further lies the rapid of Paredão (Plate 11). Before passing it one leaves on the left a spur of little hills, tolerably high, and others, lower and more isolated, can be discerned. This rapid has but one large channel, though there are several narrow passages. Its current is very swift, running between rocks of granite and flint, which rise perpendicularly to a height of thirty feet. The descent is three yards in about one hundred and fifty. There dwelt there a batch of indians, composed of five men, six women, and five boys. One of these was of a reddish copper-colour, and was called "Bermejo" (Rufus). We made them some presents, and they did not leave us all night. I recognized the fruit of the guaraná, which abounds here, and is largely used by the Brazilians.

TWENTY-SIXTH DAY.

Off again with a stiff breeze from the south. After proceeding a league we were obliged to stop, as the fog was so dense that we could not see the rocks and other obstacles in the way. But it soon cleared up, and, continuing our journey, we came to the confluence of a small arm of the river. This stream branches off just above the rapid, having its source amongst the little hills which we had previously seen, and which run parallel with the river. We shaped our course eastward towards the Bolivian frontier. From the top of one of these hills we descried immense prairies, covered with patches of long grass. At the foot of this height the site is well-adapted for building a town.

At a distance of six leagues we found the rapid of Tres Irmãos, so called from three detached hills of the same size which rise in the vicinity. The current is rapid for about a league, but not dangerous; rate, about six miles an hour; depth, 15 fathoms; thermometer, 95 deg. Here the river Mutumparaná empties itself, flowing out of Brazilian territory. Having sailed up it for six leagues, I came across an indian named Masini, with five men, nine women, and six children. Below the rapid there is a large quantity of rock of the kind used for grindstones. After going three leagues we stopped for the night at a pretty sand-bank. The rapid of which I have just spoken is represented on Plate 12 of the Map.

TWENTY-SEVENTH DAY.

In the space of six leagues the channel of the river, which has a depth of three feet, is very good. This distance takes in the rapid of Girão (Plate 13). Here the boats have to be drawn on shore, and dragged overland for some six hundred yards. The river flows due south, dropping five yards in about five hundred, and the bed can easily be cleaned out, so as to render it convenient. The cliffs are thirty-six feet high, and, along the Bolivian frontier, there are several detached hills. I left the canoe, which was manned by the Trinitarios indians, and proceeded with only that of the Canichanas, so as to be in light marching order.

TWENTY-EIGHTH DAY.

From Girão to the rapid of the Caldeirão do Inferno (Plate 14) are two leagues. We passed it, rowing through the middle channel, but with great danger, in consequence of the frail nature of our boat, and the exposure to the violence of the currents, while the waters in large volumes broke across the canoe, and caused it to rock violently. One of these concussions nearly engulfed us in the waves. There are channels running all ways; and thus, when the waters meet, the foam is lashed up by the winds. Near the coast of Bolivia, there is a good channel, which can be vastly improved. For eight leagues we continued, without any fresh occurrence of importance, as far as an island which we called Encallados (stick fast), as the river here widens considerably, and contains many sand-banks. In some parts the sounding-line shows four fathoms.

TWENTY-NINTH DAY.

At six leagues from Encallados island is the rapid of Mor-rinhos (Plate 15), skirted by a low hill-range. The rocks attain an elevation of eighteen feet. The declivity is three yards in about nine hundred. The channel might be improved. The thermometer rose to 95 deg.; current, 2 leagues; depth, 10 fathoms. After navigating for five leagues we came upon the rapid of Theotonio, which is the largest of all, and is shown on Plate 16.

THIRTIETH DAY.

The rapid of Theotonio is produced by three parallel lines of granite and flint rocks, with a height of forty-five feet. They cross the river with a west-south-westerly direction, and termi-nate at the foot of a hill situated towards the Bolivian frontier. In the first line the bed drops five yards, three in the second, and two in the third. The boats have to be dragged overland for about four hundred and fifty yards. Many veins of white and blue *soroche* are found here. At a league we passed the rapid of Macacos (Plate 17), which signifies *monkies*. It does not present any difficulty. The rocks are low, and the current equal to that of Guajará-Guassú.

Four leagues further on is the rapid of Santo-Antonio (Plate 18), which has a good channel on the Bolivian side. The declivity is the same as in the preceding rapid, and the rocks are lower. Thermometer, 100 deg.; current, 2 leagues; depth, 10 to 15 fathoms. From there the river for a league has a good channel, and would admit vessels of a large draught; but the only advisable craft in which to traverse the rapids would be flat-bottomed boats of 15 to 20 tons, as I have already intimated.

Note.—The Beni rapid is shown on Plate 19; that of Abuna on 20, and Yata Grande on 21.

RETURN JOURNEY.

We re-crossed the rapids of Santo Antonio and Macacos without any special incident, although the water had risen considerably. Slept at Theotonio.

SECOND DAY.

By 10 A.M. we had passed the canoe and baggage over the Theotonio rapid. Meanwhile, we stopped to shed a tear on the graves of the unfortunate Don Tadeo Gorriti and his companions in misfortune, who were murdered by the indians. We placed a cross upon the spot, with an appropriate inscription. After bidding a last long adieu, we continued our journey, and slept on an island two leagues before Morrinhos. Thermometer, 94 deg.

THIRD DAY.

Re-passed the rapid of Morrinhos, and stopped for the night at the island of Encallados, near the mouth of the river Yaci-Paraná, on the coast of Brazil. This part is inhabited by twenty indians, twelve women, and eight children. Amongst these I recognized the assassins of Gorriti and his companions,

FOURTH DAY.

After sailing four leagues we stopped and entered an Indian *malocca*. The inmates were four men, an equal number of women, and three children. We were informed that at no small distance from where we were then conversing there existed a numerous horde of indians, on Bolivian territory. The Maparaná flows in here. An abundance of sarsaparilla grows in this vicinity; also ipecacuanha. We pushed on, and slept at the entrance to Caldeirão.

FIFTH DAY.

We passed the rapid of that name in two hours, through the channel on the Bolivian side. Spent the rest of the day in Girão, after having dragged the canoe, and rejoined our crew of Trinitarios indians.

SIXTH DAY.

Traversed the rapid of Tres Irmãos, and halted at the mouth of the Mutumparaná.

SEVENTH DAY.

Ascended the Mutumparaná, in search of Masini, to obtain from him some pitch, of which there is an abundance. We asked him to conduct us to the people of Abuna, as he had relations there, having married amongst that tribe.

EIGHTH DAY.

Returned with as much pitch as we required, and fell in once more with our party. It rained hard, and we were obliged to stop for the night.

NINTH DAY.

Reached Paredão. Crossed that rapid by towing the canoes. The river Ferreiros flows in at this point, off the mouth of which we passed the night. We were visited by indians.

TENTH DAY.

Passed Pederneira, hauling the canoes on this occasion also. Slept at the mouth of the Abuna. On the way we passed three springs of water, bubbling up from an iron-lode, which extends for fifty leagues,

ELEVENTH DAY.

I proceeded with a single canoe to Abuna, where several hieroglyphics are cut on the rocks. At four leagues distance stands an abandoned *malocca*, and, after having passed it, we encountered, at three leagues, a very bad rapid, which has a declivity of six yards in about one hundred and fifty. I learnt that the tribes I meant to visit were three days' journey off and, as provisions were becoming scarce, I gave orders that we should return the following day.

TWELFTH DAY.

We continued our journey as far as the *malocca* of Tupí and Guaycurú, where there are fifteen men, twelve women, and an equal number of children. We slept there, after paying them many compliments. They proposed to accompany us at daylight, and would have sacrificed us in cold blood had we not been forewarned.

THIRTEENTH DAY.

Passed the rapid of Araras, and slept at the place where Mariana was.

FOURTEENTH DAY.

Passed Periquitos; heat 105 deg. Slept at length at Ribeirão.

FIFTEENTH DAY.

Recrossed this rapid, and slept at the spot where we said "Good bye" to the Vicar.

SIXTEENTH DAY.

Hauled up the boats, and occupied ourselves the rest of the day in caulking them and twisting ropes.

SEVENTEENTH DAY.

Repassed the Madeira Rapid on the Bolivian side, the boats being laden as usual, but they had to be towed, as the current was strong, and no progress could be made with the oar. By mid-day we reached the headland which marks the confluence of the rivers Beni and Madeira, and amused ourselves till night-fall with reconnoitering the country, which could not be better adapted for the site of a large city. It offers all the advantages which render life agreeable; is free from inundations, and has a splendid anchorage. Cacao of the best quality grows in profusion, as well as almonds of a superior kind, vanilla, nutmeg, and other valuable productions of the vegetable kingdom.

EIGHTEENTH DAY.

On this glorious day we commemorated the anniversary of Ingavi with repeated volleys of musketry, and unfurled our national banner. My son, who had been among the indians, now rejoined us. He brought with him two of them with their sons, who were desirous of paying a visit to the people of the Mójos district. With great pleasure, I permitted them to join our party, and supplied them with some articles of clothing.

Recrossed the rapid of Lages by rowing, and that of Pão Grande by towing. By noon we reached Yata Grande. I devoted the remainder of the day to exploring the river as far as the rapid in it, which corresponds with those of Bananeiras and the Beni. An easy ascent might be made by keeping near the right bank. We retraced our steps, and slept at the confluence of these rivers.

NINETEENTH DAY.

Retraversed Bananeiras, the canoes being towed. On the island, near the falls, I collected a little *butua*, which is held in high estimation by the Brazilians as an excellent dissolvent for bruised blood and internal hæmorrhage, caused by a blow. We pushed on, and slept above the falls.

TWENTIETH DAY.

Repassed the Rapids of Guajará-Guassú and Guajará-Merim. In the former we found that the channel we had selected on the down-trip was, on this occasion, dry. We accordingly chose the third, after having picked a large quantity of ipecacuanha.

TWENTY-FIRST DAY.

Journeyed eight leagues. Removed the provisions from Store Island, and slept at the place where the tiger assaulted us.

TWENTY-SECOND DAY.

Arrived at Elvira, where one of the Canichanas died of a disease called *bicho*, or *mal de valle*.

TWENTY-THIRD DAY.

Reached the *Playa del Sur* (south bank) without any occurrence of importance.

TWENTY-FOURTH DAY.

Proceeded as far as the junction of the Iténes, and halted two leagues higher up.

TWENTY-FIFTH DAY.

Slept at Posancos.

TWENTY-SIXTH DAY.

Reached the place in which we encountered the four Chacobos, and in the middle of the night there arrived a canoe laden with provisions sent by the Vicar, which we would not accept, as we already had enough.

TWENTY-SEVENTH DAY.

In the Mayosa stream, which rises in the territory of the Chacobos, we saw the canoe *Illampu*, from Exaltacion, run aground. In the stream Tanrupi is the landing-stage of the

Guarayos indians, who are wild and have strange customs. Slept at the mouth of the Matucaré.

TWENTY-EIGHTH DAY.

Reached Navidad, where the Guarayos had killed, with arrows, two Brazilian carpenters who were going down stream for wood in the canoe *Illampu*.

TWENTY-NINTH DAY.

Arrived at Exaltacion, all well and in good spirits. Nothing further of any importance occurred than the events above recorded.

OBSERVATIONS.

[CONCLUSION.]

The rapids may be divided into three classes:—1st, the dangerous ones; 2nd, the moderate; and, 3rd, those in which, at high flood, the violence of the current disappears. They may be arranged in the following order:—

I.—HAULING OVERLAND is necessary at Ribeirão, Girão, and Theotonio.

II.—TOWING is requisite at Guajará-Guassú, Bananeiras, Pão-Grande, Madeira, Araras, Pederneira, Paredão, Caldeirão, Morrinhos, Santo Antonio.

III.—Of the third class are Guajará-Merim, Lages, Misericordia, Periquitos, Tres Irmãos, Macacos.

Note.—The channels vary according to the volume of the waters, and it is necessary to use some caution in selecting the best channel for the ascent and descent.

It results, then, from the above, that the navigation of the Madeira is feasible, if the Government adopts decided measures for its accomplishment. The obstruction offered by the rapids is not an insuperable obstacle, since it may be remedied by dredging and excavating the channels, or, at worst, by opening a roadway, where the foregoing expedients cannot be adopted. The use of steamers is absolutely necessary, so as to render the transit of the rapids both safe and expeditious.

The importance of this is incontestable, since it would throw open the portals of Bolivia to foreign commerce and the Atlantic, thereby introducing all the elements of civilisation, power, riches, and aggrandisement. It seems as if Nature strove to hide her richest treasures in those unfortunately pent-up regions. Here are combined an amazing fertility of soil, lofty timber-clad mountains, rich in vegetable productions and different species of beautiful and valuable wood, wealth in minerals, fishing, and the chase—in short, not only the conveniences and comforts of life, but also the means of gratifying the most capricious taste. Our statesmen, instead of thinking about Arica or Cobija, should direct all their energies and attention to the navigation of the Madeira.

This done, Bolivia would need nothing else to make her successful, and secure an existence, which, unpleasant as it may be to confess it, has been, and, without such measures, will be, precarious.

The roads of the departments of Santa Cruz and Cochabamba are short, for the capital of the former is distant only two days' journey from the port of Cuatro Ojos. The same may be said of the roads to Guarayos and Chiquitos. From the capital of the latter department to Chaparé, or Chimoré, is only a distance of fifty leagues. There is another road from Loreto as far as Chaparé or Todos Santos. All these roads can be improved at trifling expense.

A good road might, also, be opened up from this department to Reyes, by way of Mosetenes, branching off to the province of Yungas. The work would be easy, divided into three sections, by which means operations could be simultaneously carried on. The first section would comprise the distance from Chulumani to Ebenay; the second from that point as far as Mosetenes; and the third from this, or Magdalena, to Reyes. The workmen of the two last-named sections could be paid a real a day, and beef-rations, for which great facilities exist in the country by reason of an abundance of cattle.

But I must conclude. I must omit many remarks which would seem out of place in a diary. Nor have I referred, in all cases, to the course steered, since it is sufficiently indicated on the Map that I have prepared for the purpose, a copy of which I have presented to the Minister of War for preservation in the

Topographical Department; another copy to the Prefect of the Beni; and a third to Señor Ildefondo Villamil; reserving one for myself. I shall be glad if, one day, the Government, swayed by patriotic administrators, resolves to extend to this fluvial question the requisite protection, which will confer on the Republic a service of the highest practical importance.

JOSÉ AGUSTIN PALACIOS.

La Paz, 12th April, 1849.

APPENDIX.

ON THE

NAVIGABLE RIVERS WHICH FLOW INTO THE MARANON, RISING IN THE CORDILLERAS OF PERU AND BOLIVIA.

BY THADDEUS HAENKE, MEMBER OF THE ACADEMY OF SCIENCES OF VIENNA AND PRAGUE, ETC., ETC.

THE provinces of Perú, conquered and occupied to the present by the Crown of Spain, form but a small part of the South American continent. They stretch along the Pacific, but with a breadth which is small when compared with that of the continent. For the most part, they extend no further inwards than the Cordilleras or Andes. The abrupt slopes of the snow-peaks on the eastern frontier; the wild and rugged nature of the roads, without exception; the labyrinth of dense forest, spreading over immense tracts, as yet but little known;—these, then, are the causes and obstacles which have hitherto impeded the natives from penetrating and exploring the interior of these extensive regions. If to this be added the risk among the wild and uncultivated tribes who inhabit those tropical climes, the insupportable hot season, the inconvenience of innumerable insects and other poisonous animals, and the multitude of the mighty and impassable rivers, one cannot wonder that, for the most part, in Perú its conquerors restrained the march of progress at the Cordillera. Certain it is that, in consequence of the very serious impediments already referred to, and the greed of conquest so dominant in the past, but now reduced, if not entirely extinguished, there have remained whole tracts unknown amongst the Spanish and Portuguese possessions. Of this class are—the Gran-Chaco, the district between Paraguay and Chiquítos, and that which stretches from Mójos and Apolobamba as far as the banks of the river Amazon and Ucayale; and many others between the rivers Purús and Huallaga, and as many more on the north bank of the river Amazon, between the river Orinoco and the mountain-range of Quito and Santa Fé de Bogotá.

The rivers, numerous and great, which descend from the vast barrier of mountains throughout their whole extent, have, when explored, been employed as the only means of intercourse, being the outlet opened up by nature—a wilderness of woods and inaccessible mountains. Assuredly the names of Chiquítos, Mójos, and Apolobamba would have been buried in forgetfulness if the rivers Paraguay, Grande, and the Beni had not opened up a path to the early Conquistadores, bearing them on their waters to remote regions, isolated and most difficult of access by the other route. Without the least doubt, among the districts of Perú, those of Chiquítos, Mójos, and Santa Cruz are those where the Spanish rule has most advanced eastward ; but these conquests did not follow the line of the Cordillera from west to east, but from south to north, by means of the long and toilsome ascent of the Conquistadores, *viâ* the river Paraguay ; and, many years after the pioneer establishments were formed, attention was first turned to the means of communication with the towns of Alto-Perú, by way of the rivers Beni, Mamoré, and very many others which ramify through an extensive intervening area. Here the astute zeal of the Portuguese, aided by the transit of the different rivers and the intermediate tracts, which are less rugged than those of the Cordilleras, advanced their positions along different roads. This they did, not to populate and cultivate the lands lying between the western frontier and the coast of Brazil, but merely in order to check the progress of the Spanish arms in this quarter, and prevent them from extending their conquests further into the interior of the continent. The above-named provinces, as well as very many more, situated to the east of the cordillera of the Andes, have, in fact, a common misfortune, however fertile the soil or valuable the productions. This misfortune, this great drawback to the happiness of the numerous tribes who inhabit those districts, is the celebrated mountain chain of the Andes, a range unique of its kind, not only as regards the elevation of its peaks, but also the enormous extent they traverse with the main body and the manifold lateral offshoots. It would seem that nature had reared this barrier to divide the inhabitants of the eastern and western slopes, giving to each a varied proportion of its bounties. It might be said of this immense mountain-region what Horace says of the ocean :

Nequidquam Deus obscidit
Prudens oceano dissociabili
Terras.

Thus it happens that, what with the infinite dangers accompanying the transit, the absolute impossibility of exporting the produce of those eastern tracts, the difficulties, which, if overcome, would augment the cost of carrying to such an extent that the conveyance even to the towns of Alto-Perú would equal the intrinsic value of the commodities,—such being the case for the most adjacent towns, it would be quite impracticable to effect the exportation to Spain, by reason of the great distance between the localities where these articles are produced and the seaports where it is intended they shall be shipped ; also, on account of the excessive cost. From the provinces of Mójos and Chiquitos, goods destined for Lima will have to traverse a double spine of the Andes, two hundred leagues by land, and six hundred more by sea. If for Buenos Ayres, in addition to the Cordilleras— so wide-spread in the neighbourhood of Jujui—exports will have to travel overland for at least six hundred leagues.

Except for precious metals and minerals, it would not pay to undertake mule-back traffic at exorbitant rates. These unavoidable obstacles to the western outlet for the districts already alluded to, and others lying east of the Cordilleras, naturally exert a restraining influence on the inhabitants. They regard the cultivation of their most valuable crops with feelings akin to apathy ; for, in view of the difficulties of egress, they content themselves with raising merely a small quantity, necessary for their domestic use; whereas, were substantial inducements afforded them, they might supply the wants of other lands. But these difficulties of exportation are, in reality, only apparent, depending only on the mode of exit adopted ; for, by changing the route, the inconveniences spontaneously disappear; the natives, no longer deterred, will receive a new incentive to cultivate their fertile lands; Church and State will win new conquests, and firmly cement previous acquisitions ; while commerce will be invigorated with produce gleaned from an immense area.

Nature appears to have intended that South America should be surpassingly great in every respect. Here, only, towers up the giant mountain-range of the Andes ; here sweeps down the unique Amazon and the river Plate; here extend forest and prairie almost illimitably, and without a parallel in other countries. Amidst a seeming chaos of riches, Nature herself appears to indicate most markedly the best-adapted and shortest lines of communication between the vast provinces thrown together in this monstrous fragment of the world, whereby the varied and abundant productions may find an outlet. The numerous rivers, all of them rich and navigable, which descend from the Cordillera, are the fluvial pathways which Nature herself has opened up, defying mountain-chains and penetrating forest-depths; thus providing in the wilderness a highway for the benefit of mankind.

The Amazon, or Marañon, the chief river of the globe, is the principal channel, and is, without exaggeration, a fresh-water sea, stretching a distance of about a thousand leagues, almost across the continent, and communicating with all the provinces of Perú—which extend from the equator to the eighteenth degree of south latitude—by means of a network of navigable streams, that, at length, flow into the main body of waters.

And here it is desirable that I should give a brief account of the principal navigable rivers which, from the highlands of Perú, flow southwards, along the eastern slope, to the Amazon.

Going from east to west, from the celebrated Angostura del Pongo de Manserriche, the first river is the Huallaga, the uppermost cascades of which are in the vicinity of Lima, not far from

D

those of the Marañon itself—lat. 11 deg. south. One of its
principal branches descends from the mineral springs of Pasco
to the east of Lima, through a wide and craggy fissure, as far as
the town of Guánuco; then passes under two peaks of the
Andes—Chinchao and Cochero. Here, in June, 1790, when I
made my first entry into these parts, I discovered the subterra-
nean entrance at its confluence with the river Chinchao. The
Huallaga now turns northward among the various spurs of the
Andes, through the tract called Lamas, enriched with the waters
which descend from the mountains of Humalies, Moyobamba,
and Chachapoyas. All of these abound in excellent specimens
of Peruvian-bark. In lat. 7 deg. south, the river contracts,
forming an *angostura* or *pongo*, similar to that of Manserriche,
but much shorter. From there it serpentines among the mountains, through a plateau, as far as its confluence with the
Marañon, near the missions of Laguna, in latitude 5 deg. south,
and about 77 deg. longitude west of Paris. This river was
descended, in the year 1560, by Pedro de Ursoa, who was
despatched by the viceroy of Perú, Don Antonio Hurtado de
Mendoza, Marquis of Cañete, in search of the famous gold-lake
of Parrima and the town of Manoa del Dorado. The celebrated
missionary, Samuel Fritz, often ascended this river on his way
to Lima.

The second—taking them in the said order—is the river
Ucayale. Its size and volume of water rival those of the
Marañon, at their point of confluence. For this reason, it has
been confounded by many writers with the Marañon itself. It
issues from the lagoon of Chinchaicocha, in the pampas of
Pombon, thirty leagues eastward of Lima, in latitude 11° 30'.
This important river waters an extensive tract, and is one of the
largest in the whole continent. I have traced it to its source,
and have made many explorations on this river during my
journey from Lima to Cuzco; and, also, in the year 1794, when
travelling from the rivers Yauli, Jauja, Mayoc, Mantaro, Canaire,
Tambo, Pachachaca, Apurimac, Paucartambo, Vilcanota, as far
as the Partido de Cailloma, belonging to the municipality of
Arequipa, and in the east as far as the confines of the Partido
Carabaya. Flowing from a narrow portion of the Cordilleras,

this river is swelled by the Perrene, and, in latitude 8 deg., by the Pachitea. It meanders through the wide pampa of Sacramento, and on amidst a labyrinth of thickets and countless streams, which all pay their watery tribute. The banks are peopled by numerous tribes, whose names alone would compose quite a vocabulary, and who are eager for religious instruction. After traversing an immense distance, it joins the Marañon close by the missions of San Joaquin de Onaguas, in south latitude 4° 30′, and longitude 73° west of Paris.

Descending from this mission, at distant intervals along the same bank, the rivers Yavari, Yutay, Yuruta, Fefe, and Coari empty themselves. They are of secondary importance, it is true, but light craft can, at all seasons, ascend them a long way—as far as the confines of Alto-Perú.

In longitude 63 deg. west, and latitude 4 deg. south, we find the river Purús, there known as the Cuchivara. It is a river of the first order, and, according to the reports of the indians, is equal in size to the Marañon. Its true source has not, as yet, been ascertained. However, I have sufficient data whereby to fix this spot, with tolerable accuracy, somewhere between the Cordillera de Vilcanota and the eastern portion of the Carabaya mountains, from which issue many important rivers, most productive in gold. The wild tribes of indians, such as the Chuntachitos, Machuvis, and Pacaguaras, dwelling to the west of the mission of Apolobamba, told me, in October, 1794, that about ten days' march westward from the Beni, flowed a vast river, through a prairie clothed with trees of a colossal size. They explained most intelligibly that, on the banks of that river, lived some of their own tribe, and many others; also, that this river—named in their language *Mano*—was longer and wider than the Beni, with which it was united. Since, in the space between the rivers Ucayale and Madeira, no river of this magnitude discharges itself, I am strongly inclined to believe that the Purús and Mano are one and the same, and that the discrepancy in name is due to the different tribes inhabiting its banks within the wide area as far as its confluence with the Marañon, each of which tribes gives it another name.

Fifty leagues eastward from the aforesaid river the well-known Madeira empties itself, in longitude 60° 30′ west, and latitude 3° 30′ south. It takes the name of Madeira from the logs of wood and trees which it washes down during the inundations from November to April. It rises in a wide plateau of the Andes, from the heights of Pelechuco, Sorata, La Paz, as far as the inmost of the Spanish dominions, and the cordillera of the Chiriguanaes. I will describe this river at some length, by reason of the great space over which its ramifications extend, the safety while navigating on its principal tributaries, its proximity to the Mar del Norte, and the fact of its affording the shortest line of water-communication with the Amazon and the Portuguese settlements, as well as the Spanish colonies.

The internal cordillera, or that of the Andes, which, from Quito, with but slight exception, has a direction from north-west to south-east, before arriving at the boundaries of the province of La Paz, in 16 deg. south latitude, first makes a considerable bend, or detour, and, thenceforward, varying its original direction, inclines more to the east, thus receding from the coast-line, and penetrating deeper into the continent. This variation has the effect of producing, in a short space, the spot, or marked line, which determines the direction and course of the waters on both sides—I mean to say, from north to south—and the two common watersheds of the whole continent, namely, those of the rivers Amazon and La Plata. This important line falls somewhat further than the eighteenth degree of south latitude, and separates the waters of the two sides according to the declivity and the fall which the hills present on the north or the south. And the river Amazon receives now, by the inclination of the cordillera to the eastward, not only the waters of the west, but also of the south, and, moreover, a large portion of the eastern rivers. The principal branches which form the river Madeira are—the Beni, the Mamoré, and the Itènes, most of them navigable from but a short distance downwards from their sources.

Of the three, the Beni is the most easterly, and is formed by numerous very large rivers, which unite with it within a short distance from each other. They together form a good body of waters. All flow down from the heights of the cordillera, and their basin extends from Pelechuco, Suches, Sorata, Challana, Songo, La Paz, and Suri, as far as even the province of Cochabamba. The easternmost is the Tuche; then comes the Aten, the Mapiri or Sorata, the auriferous Tipuani, Challana, and Coroico, which flow together. With another, of the name of Chulumani, are united the Tamampaya, the Solacama, the La Paz, Suri, Cañamiña, and the river Cotocajes, the most easterly of all. I had the good fortune to explore the sources of all these, in my frequent voyages; and, on the 22nd September, 1794, I embarked on the river Tipuani, descending by it to the Beni, being piloted by the Indians as far as the missions of Apolobamba and Mójos to the town of Reyes, near Isiamas and Tumupasa. The journey did not take more than four days, on account of the rapidity of the current. This river takes its course along the same fissures of the cordillera, which for a considerable distance, is depressed in height. There are many bad places, but the dexterity of the indians in the management of the craft removes all apprehensions of danger on the part of the navigator. Below the town of Reyes, on the west, it also receives other rivers, such as the Tequeje, the Masisi or Cavinas, etc. From its junction with the Mamoré, in about 10 deg. south latitude, they both lose their names, and their combined waters are known as the river Madeira. In the prairies this river flows quietly and majestically along, presenting no danger whatever. It contains islands of considerable size, and its breadth in several parts exceeds a quarter of a league. It abounds with an amazing quantity of fish of all kinds, and various amphibious animals, but particularly the crocodile, or *caiman.* Both banks are thickly wooded with lofty trees, and are peopled by a large number of uncivilized indians, such as the Cavinas, Pacaguaras, Bubues, Torromanas, Nahas, and Tobatinaguas, on the west; and, on the east, by the Bulepas and many others. These are now beginning to be visited by the missionaries of Apolobamba. Communication might easily be established between the Beni and the Mamoré, by way of the river Yacuma, which rises in the vicinity of Reyes, and, from that town, crosses the wide intervening prairie, flowing from west to east, and uniting with the Mamoré near the town of Santa Ana. The descent is so gradual, and the elevation above the sea-level so slight, that in sixty leagues the ground drops scarcely twenty feet.

The second, or intermediate branch, is the Mamoré—a river in no respect inferior to the Beni—which divides the extensive tract adjoining the missions of Mójos into two considerable portions, coursing between them with a direction from south to north. The river Chaparé—including the rivers Paracti, San Mateo, Coni, Chimoré, Sacta, and Matani—descends from the cordillera and mountains inhabited by the tribe of Yuracarés in the vicinity of the town of Cochabamba. The Rio Grande, which divides the province of Cochabamba from that of the Charcas, lying along another branch, into which are discharged the rivers of the mountain-range adjoining the town of Santa Cruz; and, from the confluence of both, in latitude 16 deg. south, the common stream receives the name of the Mamoré. The Mójos indians ply on this river, against the current, with fruit and other products of the country, more than a hundred leagues, from the town of Exaltacion to the outskirts of Santa Cruz. The same year, 1794, in October and November, I pursued my investigations from the river Beni to the Yacuma, and then continued my course on the Mamoré and Rio Grande as far as the port of Forés, near Santa Cruz.

The third, or most eastern branch, is the river Iténes, rising in the low hill-range, in the centre of Brazil. These hills have been but little noticed by the Portuguese, in whose possession they are. The river Iténes flows from east to west. Its waters are more transparent than those of the Beni or Mamoré. At a short distance up this river stones are to be found which are as precious as diamonds in the lowlands of the Beni and Mamoré. The volume of water is less than in the two rivers mentioned. It passes by the fort of the Prince of Beira, one of the most advanced outposts of the Portuguese, situated in latitude 12 deg. south, and longitude 66° 30′ west of Paris. The river unites with the Mamoré almost in the same latitude, but half a degree westward of the said fort.

These are the three principal branches of the celebrated river

Madeira, the most available of all the channels I have referred to for communication with Spain by way of the Atlantic, and as an outlet for the productions of all the countries east of the Andes. It is to be lamented that the inhabitants of the richest and most fertile of the Spanish possessions in this continent—which are placed as above indicated—should be forced to adopt, with stupendous efforts, a roundabout way, *viá* the settlements on the Pacific coast, for the export of its wealth, battling with all the elements in a most toilsome ascent against the current of rivers, which, in the neighbourhood of the Cordilleras, descend with greater impetus and violence at every moment of the navigator's course up-stream. Then there is the transit of the Cordillera itself, so fatal to the wretched indians, who, accustomed to the mild temperature of their native tracts, and with no stouter covering than a thin blouse, suffer in that frozen atmospheric region all the privations and rigour of a Siberia or Kamschatka. On the other hand, however, by following an easterly course, and letting their barks float down with the favourable current of the rivers, without anything else to do than to steer them, the journey would be expedited to a considerable extent. Condamine says, in his travels, that the Cordilleras may be regarded as a hindrance equal to a thousand leagues by sea.

With the exception of the lands of Guayaquil, situated on the western declivity of the Cordillera, the eastern slopes and plateaux of the Andes are the most productive portions of South America. All the gold—and that, too, of the best quality known,—is the exclusive product of these regions, and I venture to say with confidence that there is not a river or any broken ground in that locality which is not provided with this metal, although the labour of its extraction may, in some parts, be greater than in others, success, of course, depending on the depth, in an inverse ratio.

The cocoa of Apolobamba, Mójos, Yuracarés, and all the forest tract stretching from there to the banks of the Marañon, far surpasses that of Guayaquil in quality. The superior kinds of *quina*, or *cascarilla* grow exclusively on this side of the Andes. What shall I say of the cotton, the extensive fields of indigo, the balsam of copaiba, sarsaparilla, China-root, gum-elastic, and most fragrant vanilla, all of which are prodigally produced by Nature in these countries? The dense, lofty forests on the banks of all the rivers contain woods of remarkable strength, beauty, and divers colours, not only useful for the construction of houses, but also of vessels of considerable burthen. Many of the trees distil odorous resins and medicinal gums. There is also gathered in those parts a species of bark called *corteza de clavo* (clove-bark). The outer portion of it resembles the cinnamon, but it is much thicker and darker, on account of the age which the trees attain here, than that of the East Indies. The odour and taste are those of the clove.

The communication with Perú by this portion of the river Amazon and the Atlantic would be a powerful agent for the advancement in civilisation of those countries. This result would be brought about by the growth of traffic with other nationalities, which they now lack. The missions would acquire new vigour, and win fresh triumphs, while, at the same time, they would open up tracts as yet unknown. If the products of Perú be conveyed along this route, and Spain would have the acumen to form some establishment or station in one of the mouths of the river Amazon, what advantages would accrue to navigation, with an immense saving of distance; what difference in a voyage from Spain to the mouth of this river—which takes about a month—when compared with the other route round Cape Horn to Lima, or even Guayaquil! At least three thousand miles would be saved on the round voyage out and back. The indians are excellent boatmen for river-navigation. A few men will briskly and dexterously handle a launch of fifty or sixty feet in length ; nay, even vessels of greater capacity ; or even a ship. They are indefatigable as sailors, although called upon to work for months at a time. They hardly need to carry provisions, for everywhere there is an abundance of fish, tapirs, deer, monkeys, and other animals, which may be killed with bow and arrow. Such game will provide all that is necessary for the sustenance of the party. Moreover, there is a vast quantity of wild fruits and roots, which can from time to time be gathered.

The sole difficulty in realizing this project consists in the tenacious opposition of the Portuguese, who most jealously guard what they consider to be their rights; but, no sooner would a permanent peace be established than these obstacles could be removed. Further, the powerful influence of France might be brought to bear in making the rivers Amazon and Madeira a common fluvial highway for the two nations who are mutually interested in the adjoining districts, and between whom this immense continent is divided. I have no other object in proposing this arrangement than the earnest desire I entertain of contributing, to the best of my ability, to the happiness and welfare of the Spanish people, whose generosity has furnished me the means of visiting those remote countries, and to utilize for their benefit the experience which I have gained during long and arduous journeys in those parts. Nobody can assert that the proposed project is a chimera, the hallucination of a theorist, or an idea impossible of execution. I, indeed, confess there are difficulties in the way, but this is only due to the opposition of the Portuguese. Let but the Government bestow on this subject the attention it merits, and, I doubt not, means will be found to induce the Portuguese nation to relax somewhat the rigour of its pretensions to the absolute monopoly of the river Amazon, and many others which have their origin within the Spanish dominions.

France, whose enthusiasm in protecting the various privileges of mankind—that mighty power, the ally and friend of Spain, is now insisting that the Cape of Good Hope shall be thrown open as a common harbour and halting-station for all nations trading with India. In like manner she might, by her influence, moderate the assumptions of the Portuguese, and cause the banner of Spain to be unfurled upon the waters of the Amazon and Madeira, by virtue of international right. I offer myself as the first to attempt this new route, to make this journey to Spain by way of the above-named rivers, if the Government will provide me with the necessary passports, letters of recommendation, and the astronomical instruments specified in the annexed list, so that I may push on without delay or annoyance at any of the fortified positions which the Portuguese have taken up along both of these rivers. This preliminary journey would serve the purpose of exploring and methodically detailing the whole course of the river Madeira, its depth, the dangerous parts, the tributary streams, and the precautions to be observed while navigating upon it; also to obtain some idea as to the general features of the country through which these rivers flow the characteristics of the indians, and the nature of the products. The east winds—which, according to the statement of Condamine, in his " Travels," prevail from October to May— favour the up-voyage against the current in both rivers, although in the interior of the continent the south and north are the dominant winds, which, in the rainy season, alternate uniformly the one with the other.

The two accompanying plates illustrate the interesting geographical features. No. 1, particularly, which shows the new settlement of Santa Cruz, projected by his Excellency, will serve to explain the rivers formed by the Madeira; while No. 2 will show the continuation of its course as far as the confluence with the Amazon, and the latter to its outlet into the sea.

In view of the close relation subsisting between the mission and the subject of which I have just been treating, I take the opportunity of making some remarks as to the actual condition in which they now are. Since the conquest of both Americas the religious feeling of the kings of Spain has ever regarded the conversion of the numerous tribes of docile inhabitants as a subject of the highest importance. They have generously expended large sums of money in these spiritual victories, without any pecuniary recompence, but with a success and progress which varied at different periods of their history. Now-a-days— since the enthusiasm which in former times incited masses of mankind to aid in the triumph of religion has evaporated—one cannot regard these missionaries as fulfilling the nature of their office in the true sense of the word. They sway with temporal as well as spiritual power, and are the foremost agents in the acquisition or loss of political alliances with the uncivilized tribes, and, thereby, of the countries and provinces which these tribes inhabit. From a populous and well-directed mission increased by the neophytes, grows up a town, and an assemblage of towns constitutes a province. It is an erroneous principle

and one which has caused much mischief, that any friar whatso-
ever is fit to engage in the work of converting infidels and
preaching the Gospel. The scrupulous and successful discharge
of the duties of the ministry unquestionably requires men of
superior talent and education, great firmness of character, and
extraordinary prudence. They should be men unmistakably
destined by Providence for this work—men blest with a constitution of uniform strength, so as to withstand the heat of the
torrid zone, the stings of insects, and the immoderate rains of
the season ; and a good memory, in order to acquire with ease
the multifarious dialects of the indians. The chief subject
engrossing the consideration of a missionary should be the
study of man—that being who exhibits himself under more
varied aspects than the chameleon—and here, more especially, of
man in an uncivilized state, as he was made by the hand of
Nature, unrestrained, without other law than that of brute
force, agitated by violent passions, the only outlets for his
actions,—in a word, one of the brute creation, with only the
outer resemblance of humanity.

None of the qualities referred to shine forth in the most earnest
of the religious brethren of San Francisco, who now assist in this
object at an immense expense to the State. They persuade
themselves that they have fulfilled all their obligations in hur-
riedly reading the customary prayers every day. The love of
riches makes them forget all the striking rules as to poverty
which their order prescribes. They derive incredible benefits
from the simplicity and hard work of the converts, whom
they fetter with tasks which they could not perform if they
were beasts of burden. In their temporal government they
rule despotically, being ignorant of all knowledge which apper-
tains to economy and industry; and fortunate it is if they
only stop at this, and do not commit faults which moderation
must hush up, out of respect for their office, and because there
is no doubt that a religious body is worthy of the first considerations when it observes the rules of its institution, and when its
members do not abuse their privileges. On the contrary, the
indian, guided by these masters, although for thirty and more

years, has learnt nothing else than to recite like a parrot a few
prayers which he cannot understand. He has not acquired the
slightest sound idea of the Supreme Being, that should be the
principal aim and end of their actions. Their industrial arts have
remained the same as they were previous to the arrival of their
converter; and, after so many years, the indian remains as much
a Gentile as before, and, at length, casting off the chains of an
imprudent servitude, he returns to the wilderness. This is the
deplorable state of the missions in charge of these evangelisers.
This perverse conduct is the chief cause that, since the expulsion
of the Jesuists, not only has nothing advanced, but also very many
of the missions have ceased to exist. Instead of advancing they
have fallen off; and the Portuguese proceed, step by step, occupying more territory, and approaching nearer every day to the
Spanish dominions.

The happiest era of the Spanish missions, situated on both
banks of the river Amazon, was towards the close of the last
century. The celebrated missionary, P. Samuel Fritz, a German
Jesuit, endowed by Providence with all those gifts which adorn
this ministry, penetrated, in the year 1686, to the towns of the
uncivilised tribes of this river. He reduced, in a short time, the
numerous nations of the Omaguas and Cocamas. By his
example, the neighbouring tribes—the Yurimaguas, Aisuares,
Banomas, and others—voluntarily assisted, drawn merely by the
kind treatment whereby he taught them to live with just laws
and a policy till then unknown to them. By this method he
triumphed, in a few years, in all the countries which run from
the river Napo as far as the mouth of the river Madeira, without
any other arms than those of his gentle manner and singular
prudence. With the wide-spread triumph over so many tribes
he secured a large tract of country, under the proper dominion
of the sovereign of Spain, on both banks of the river Amazon.
But it is a cause for regret to see the actual condition in which
they are. From the mouth of the river Madeira, situated,
approximately, in longitude 61 deg. west of Paris, they have
been retiring and abandoning these missions, as far as that of
Pebas, which is, in reality, the last of the Spanish possessions,

situated in longitude 71 deg. This gives a loss of territory of 10 deg. in longitude, which, reckoning even in a straight line, would be equivalent to two hundred leagues, and the Portuguese have advanced theirs as far as San Pablo, near Pebas, including the conquest of all that territory and the rivers which communicate with Perú. I am convinced that the Portuguese had greater luck in the selection of the religious emissaries whom they destined for these conquests—such as Carmelites, men of different education and behaviour to the inhabitants of Perú, and who regard with patriotism the interests of their country.

The learned Jesuit, Samuel Fritz, not only had talent, prudence, and aptitude for such conquests, but also, at the same time, was well informed in mathematics and astronomy. He was the first who executed a map of the whole vast course of the river Amazon; and Condamine, the Parisian academician, did not hesitate to insert it, by way of comparison, in the same map which accompanies his work. Some superficial acquaintance with geography and the use of the needle should be the concomitants of the evangelic office, so as to enumerate to the governor of the district, together with a relation of his adventures, the hill-ranges, rivers, lagoons, and other characteristic features of those localities in which he exercises his apostolic functions. But these most useful branches of knowledge are ignored by almost all our missionaries, and it is difficult to find one, here and there, with enough education to keep a rough diary of his travels. The organisation and regulation of the missions on the banks of the rivers Amazon, Napo, Ucayale, Purús, Madeira, Beni, and in the part further north than the Mamoré, is a subject which, by all means, merits the attention of the Government, by reason of the proximity of the Portuguese, who take advantage of the least want of vigilance, aggrandizing themselves with new tracts, and rapidly approaching the Spanish dominions. The measures which the Government might deem opportune refer particulary to the disseminating colleges (colejios de propaganda) of Quito, of Ocopa, and that which has recently been founded in the town of Tarata, in the province of Cochabamba.

Such is the information I am able to give your Excellency on this weighty and important question, in virtue of the office which was kindly conferred upon me on the 11th of March last.

May God protect your Excellency many years,

THADDEUS HAENKE.

Cochabamba, April 20th, 1799.

To DON FRANCISCO DE VIEDMA,

Governor in command of this Province.

FINIS.

THE

RAPIDS OF THE RIVER MADEIRA.

EXTRACT FROM THE

"EXPLORATION

OF THE

VALLEY OF THE AMAZON,

MADE UNDER DIRECTION OF

THE NAVY DEPARTMENT,

BY

WILLIAM LEWIS HERNDON AND LARDNER GIBBON,

LIEUTENANTS UNITED STATES' NAVY.

PART II.

By LIEUTENANT LARDNER GIBBON.

WASHINGTON, 1854."

DUNLOP AND CO., PRINTERS, NEW STREET, CLOTH FAIR, E·C.

1874.

THE RAPIDS OF THE RIVER MADEIRA.

SEPT. 19.—A turn in the river brought us in sight of high land to the north. The negroes blew two cow's horns, and shouted at the sight of it. Laying down their horns, they paddled with a will to their own musical songs, by which they kept time. We met a north wind, which created a short wave as it met the current of the stream, increasing its speed. The land has become low on both sides, and is swampy, with signs of being all flooded in the rainy season.

At 9 a.m., thermometer, 82°; water, 81°. At 3 p.m., thermometer 87°; water, 80°. We passed an island rocky and wooded. Flowers bloom and decorate the richly green foliage on the banks. The current is quite rapid, and we dash along at a rate we have not been able to do before on the Mamoré, passing the mouth of a small river—Pacanoba—which flows from the Brazils, and through several islands. We came alongside of one of them for the night. Within the death-like, mournful sound of the "Guajará-merim" falls, our raw hides were spread, hair side up, as table and chairs. While the men made a fire I was listening to the roaring waters, and thinking what sensible fellows those Cuyavabos Indians were to run from it The night was starlight, but the mist arising from the foaming waters below us was driven over the island by the north wind, which prevented my getting the latitude. Small hills stood a very short way back from the islands in Brazil. The land appears to be above the floods on both sides. As we are free from mosquitoes at night, and the savages do not inhabit our little island, we sleep soundly.

SEPT. 20.—By daylight we were up and off, pulling across to

E

the Bolivian shore to the head of the falls. We were in doubts how our boat would behave in the rapids. After taking out part of the baggage, which was passed over a rocky shore below, the boat was pulled through without any difficulty. The channel was about fifty yards wide, with very little fall; the whole bed of the river was divided by wooded islands and black rocks, with large and small channels of water rushing through at a terrible rate. A steamboat could, however, pass up and down over this fall without much trouble. We embarked, and found our little boat, which had been named "Nannie," gliding beautifully over the short waves formed by the rapid motion of the water. The rocks are worn away in long strips, and cut up into confused bits by the action of the river constantly washing over them. On the islands quantities of drift-wood and prairie-grasses are heaped on the upper side. One of these islands occupied the middle of the bed for three-quarters of a mile in length. We follow the channel down on the Bolivia side to its low end at a rapid rate; when we came to the foot of the first fall we looked back *up hill*, to see the number of streams rushing down, each one contributing its mite to the roaring noise that was constantly kept up. We saw no fish, but last night met large flocks of cormorants flying in a line stretching across the river, close to the surface of the water; this morning they came down again. These birds spend the night over the warm bed of the Iténes, and return here in the day to feed. No sooner had we cleared these falls than we found ourselves at the head of another rapid, more steep, called " Guajará-guassu." Pedro took us to the upper end of a path in the woods, on the Brazil shore, where Don Antonio had transported his cargo overland, three hundred and fifty paces, to the foot of the falls. His large boats were hauled through the water by means of strong ropes rove through large blocks.

Our cargo was landed, and while Richards, with one man, was engaged carrying the baggage down, I took the boat over on the Bolivian side, and we hauled her three hundred yards over the rocks and through the small channels, down an inclined shelf of about twelve feet fall. The main channel is in the middle of the river, with waves rolled up five feet high by the swiftness of the current, through which a steamboat could pass neither up nor down.

The river cuts its way through an immense mass of rock, stretching across the country east and west like a great bar of iron. The navigation of the river Mamoré is completely obstructed here; the river's gate is closed, and we see no way to transport the productions of Bolivia towards the Amazon except by a road through the Brazilian territory. On the east side of the river, hills are in sight, and among them a road may be found where a cargo might pass free from inundations. The navigable distance, by the rivers Chaparé and Mamoré, from near the base of the Andes at Vinchuta, to Guajará-merim falls, is about five-hundred miles. We anxiously pulled across towards the baggage, as the division of a party in this wild region is attended with great risk. This day's work gave us some little experience in the new mode of navigation. The sun is powerfully hot, but the negroes strip themselves and ease the little boat gently down in the torrent between rough rocks. Don Antonio's advice was of the greatest importance to us in the choice of a boat and men. The long canoes of Bolivia would have been broken to pieces in this first day's travel among the rapids. There are no paths through the wilderness by which we could travel in case of an accident, and rafts we had seen enough of at the head of the Madre-de-Dios. Embarking our baggage, we continued under a heavy thunderstorm, which came up from the north-east and whirled over our heads, sending down heavy drops of rain. The banks of the river are twenty feet high. The country on the Bolivian side is level, and there the lands are overflowed half the year; but the Brazilian side is hilly; the ridges appear to run at right angles with the river, which passes over the toes of the foot of them. The whole country is thickly wooded with moderate-sized forest trees. The river below these falls is occasionally three-quarters of a mile wide, with a depth of from twelve to thirty-six feet. The current is rapid as we leave the foot of the falls, gradually decreasing in speed until the boat enters the backed water which is dammed up by the next ridge of rocks which thwart the free passage of the river.

SEPT. 21.—At 3 p.m., thermometer, 83°; water, 81°. The south wind blew all last night, accompanied with rain. Early this morning we arrived at the head of " Bananeira " falls, distance eight miles from the upper shelf. I find Pedro useful in pointing out the ends of the paths over the land cut by Don Antonio; his services as pilot, however, are not to be depended upon. Titto seems to be perfectly at home in the management of a boat among rocks, and assists me most of the two. The cargo was landed on an island near the Bolivian shore. The path led through bushes and trees down hill, near four hundred yards. The work of transporting the boxes amidst the annoyance of swarms of sand flies was harassing, and with difficulty Richards could make the ill-natured member of the crew carry as many boxes as he did himself. The river flows windingly; the baggage could be sent straight across, but the boat had to be dragged, towed, lifted, and pushed through the rough rocks and rushing waters for over a mile; this was trying work; the heat of the sun was very great; the negroes slipped, and it was with great difficulty at times they could hold the boat from being carried from them by the strength of the waters as they heavily pass through the choaked passages. The men stand easing down the boat up to their necks in water. The rocks are only a few feet above the water level; they are smoothed by the wearing of the water and drift wood. It is not easy for the men to keep their feet under water. These negroes are good men for such service—they crawl among the rocks like black snakes. Bananeira falls take their name from quantities of wild banana trees formerly discovered here, but we saw no traces of them. The fall is about twenty feet; the islands, are generally very low, a few feet above the present surface of the river; all the rocks, and a great part of the islands are overflowed in the rainy season; large heaps of drift-wood lodge against the trees. On the highest rocks we found pot-holes, worn down to the depth of eight and ten feet by the action of small pebbles, put in motion by the current as it passes over and whirls down, boring into the solid mass of coarse granite. These pot-holes are generally half-full of stones, the large stones on top; gradually descending towards the bottom, they were smaller, until at the very last they were composed of bright little transparent, angular-shaped stones, less in size than a pin's head; among these the diamond-hunter looks sharp. Some of these pot-holes are three feet wide at the mouth, decreasing in edge size uniformly towards the bottom. When we gained the foot of these falls, over which it is utterly impossible for a steamboat to pass at any season of the year, we had to ascend a channel on the Bolivia shore for the baggage. Mamoré lay by a part of it as watch, while the rest of the party were on the other side of the island; we were nearly exhausted; the men had nothing to eat half a day, and the dog looked thin and sick. There were no fish, birds, monkeys, or Indians to be seen, nor were the men successful in finding castanhas, Brazil nuts, which they very much needed, as they had nothing to eat but their allowance of farinha. The negroes were very tired, but I observed the life improved them; they looked stronger and were getting fat. This was a great relief, for we were the worse for wear. I was kept in constant excitement lest some accident should happen to our boat, or that an attack would be made upon our baggage party by the savages. At 3 p.m., thermometer, 85°; water, 81°, and less muddy; dashing over the rocks appears to filter it.

The boat was carried along at a rapid rate by the current, which boiled up and formed great globular-shaped swells, over which the little boat gaily danced on her homeward way. The satisfaction we felt, after having safely passed these terrible cataracts, cheers us on. We were nearly the whole day getting two miles. We were prevented from the danger in our path to proceed at night. The boat was fastened to the Brazil bank, and, after supping on a wild goose Titto was fortunate enough to shoot, we slept soundly until midnight, when we were suddenly aroused by the report of a gun. The men were lying by a fire on the bank, near a thick tall growth of grass which skirted the large forest trees; Richards was close by me; I heard Titto's voice immediately following the report, saying " The devil;" we were all up in arms. Titto said he had shot at a tiger which was approaching the men as they slept. Mamoré had been faithfully prowling in the woods, keeping close watch over us while we all slept; because he gave the men some trouble in the boat they laid this plan to put our trusty friend to death. Richards found the dog shot in the heart, close by the heads of the men, four of whom were in the secret, while Pedro and the Indian were sleeping. We placed great confidence in the watchfulness of Mamoré ; from him we expected a quick report of savages or wild animals. With him on watch we slept without fear, as the Indians were more afraid of the bark of a large dog than of the Brazilian soldiers.

From what we had seen of the men we were convinced they were a rough, savage set, who would put us to death quite as unceremoniously as the dog. They expressed an impudent dissatisfaction when I ordered Titto to put a man on watch, and keep sentinel all night. We lay till daylight with our pistols prepared for an attack from any quarter. The negro murderers on the highways of Peru are more desperate and unmerciful than either Spaniard or Mestizo ; so it is with a half-civilised African negro. At daylight I was particular to let every man of them see my revolver. We kept a close watch upon them both by day and by night. They had, for some reason or other, unknown to us, taken a dislike to Richards, who never gave them an order except when he was left on shore to attend the portage of the baggage. They were under an impression we were ignorant of what they said when speaking their own language, as Titto and Pedro spoke to me in Spanish. On one occasion, after the loss of Mamoré, I overheard the ill-natured one, after Richards spoke to him about tossing water into the boat with his paddle, say to the rest of the crew, " I don't know whether I won't put a ball through that fellow yet, by *accident !*" After which I had no confidence in any of them, and told Richards our safety remained in constant watchfulness, and the good condition of our firearms.

SEPT. 22.—The river below Bananeira falls is seventy-eight feet deep and half a mile wide, passing through rocks and islands, where we found the wild Muscovy ducks. With a rapid current we soon reached the mouth of the Yata river, a small stream flowing from the territory of Bolivia, not navigable for vessels larger than a ship's boat. At " Pão-Grande " rapids, the country is hilly on both sides, and wooded with large trees, from which fact the rapids derive their name. These rapids are about five miles from those above, with a fall of fifteen feet in one hundred yards. The boat was carefully passed through narrow channels among rocks fourteen feet high. Don Antonio came up over these falls, when the river was flooded, by keeping close along shore ; he fastened the upper block of his tackle to large trees, or heavy rocks, and by hard pulling, inch by inch, dragged his boat along. No steamer could pass up or down "Pão-Grande." At 9 a.m., light northerly breezes ; thermometer, 81° ; water, 81°. Two miles below brought us to Lages rapids. The boat was kept in mid-channel, and paddled with all the might of the men ; we passed through the rocks at such a swift rate hats had to be held on. This was a glorious passage ; the little boat seemed to fly through a channel that might be passed by a steamboat.

A bark canoe lay by the Bolivian shore. Our negroes blew their horns, which brought four savages and a black dog to the bank. Two of them wore bark frocks, and two were naked— real *red* men. As we floated along by the current, the following conversation took place between the savages and the negroes :— Savage : " Oh ! " Negro in the bows : " Oh ! " Savage : " Venha ca "—come here—very clearly pronounced. We told them to come to us and they ran away, while we paddled slowly on. These Indians are of the " Jacares " tribe ; they were soon paddling after us fast. We waited but a short time ; their swift canoe was constructed of one piece of bark, twenty-feet long, and four feet beam. The bark was simply rolled up at each end and tied with a vine from the woods ; between the sides several stretchers, four feet long, were fastened to the edge of the bark by small creepers, and a grating, made of round sticks fastened together with creepers, served as a flooring, which kept the bottom of the canoe in shape when the Indian stepped into her. Two young men, dressed in bark dresses, sat in the stern, or one end, with well made paddles. On the other end sat two naked women, each with a paddle lying across her lap. As they came alongside, amidships sat an old chief with a basket of yuca, a bunch of plantains, a large lump of *pitch*, and several small pieces of superior quality, called by the Brazilians "breu." The

Indians use it for securing arrow-heads, we find it serviceable in sealing our bottles of fish, or fixing the screw to our ramrod. Besides which the old man brought one richly green parrot for sale. We bought him out with knives and fish hooks. One of the women was good looking, the figure of the other was somewhat out of the usual shape. On being presented with a shaving glass they expressed great pleasure, and one after the other looked as far down their throats as they could possibly see by stretching their mouths wide open. Their greatest curiosity seemed to be to explore the channel down which so much of the results of their labour had passed. When they saw their dirty half-worn teeth, the holes in their ears, noses, and underlips, one of them poked her finger into her mouth through the lower hole, and brutally laughed. They wore long hair behind, and clipped it off square over the forehead, which gave them a wild appearance. The women were very small ; their figures, feet, and hands resembled those of young girls. Their faces proved them to be rather old women. They appear cheerful, laughing and making their remarks to each other about us, while the men wore a surly, wicked expression of face. One of the young men became very much out of temper with Pedro, because he would not give *all* the fish-hooks he had for some arrows. The old man seemed very much excited when he came alongside, as though he half expected a fight. He was a middle-sized person and chief of all the Indians in his tribe who inhabit the Bolivian territory. He represents his tribe as few in numbers and scattered over the country. Like the women, the men have great holes in their noses and under-lips, but nothing stuck in them. We supposed they were in *undress* on the present occasion. The chief enquired the names of the different persons, and wanted to know which was the " captain " of the party. The women begged for beads and assumed the most winning smiles when they saw anything they wanted. We invited the chief to accompany us to the next falls and assist us over. He shook his head, pointed to his stomach, and made signs with distressed expression of face that he would be sick. He was then told we had more fish-hooks and knives ; if he brought yuca and plantains, we would trade at the falls. To this he consented, but said his people and the Indians below were not friendly, and that the enemy generally whipped his people. Three miles below Lages we came to the mouth of the Beni river. This stream resembles the Mamoré in colour and width; but while the latter has a depth of one hundred and two feet, the former has only fifty-four feet of water. Temperature of Mamoré water, 81° ; of Beni, 82°. Near the mouth of the Beni there are islands. The whole width of the river is about six hundred yards ; the junction of these two streams forms the head of the great Madeira, which is one mile wide.

In the month of October, 1846, Señor José Augustin Palacios, then governor of the province of Mójos, explored the falls in the Mamoré and Madeira by order of the Government of Bolivia. We find the map of Señor Palacios a remarkably correct one. He ascended the Beni for a short distance, finding a depth of seventy feet of water to the foot of the falls, beyond which he did not go, but returned and continued his course down the Madeira to the foot of its falls, when he retraced his steps to Mójos by the way he came. We have accounts of many falls on the Beni river from the province of Yungas down to the town of Reyes, between which falls the river is navigated by the Indians in wooden balsas. The Beni has never been explored throughout its length, but with the falls above Reyes, and those seen by Señor Palacios near its mouth, which appear to have prevented him from ascending this stream on his return, we have reason for saying the Beni is not navigable for steamboats. The outlet for the productions of the rich province of Yungas is to be sought through the country from the gold-washings of Tipuani to the most convenient point in the Mamoré between Trinidad and Exaltacion. The distance from the latter place to Reyes, on the Beni, is not very great. From the general conformation of the bottom of the Madeira Plate, we are of the impression that the road would have to be cut high up towards the base of the Andes, so as to clear the annual floods. The Mamoré, therefore, is the only outlet for the eastern part of the department of La Paz, as well as a great part of the department of Santa Cruz. The ridge of hills and mountains at the base of which the Beni flows, stretching from the falls of the Madeira to the source of the River Madre-de-Dios, or Purus, separates the Madeira Plate from the Amazon Basin, and divides the department of the Beni from the Gran Paititi district in Brazil, which extends north to the Amazon river. Paititi, it may be remembered, was the name given by Padre Revello to our favourite dog, lost on the road from Cuzco to Lake Titicaca.

We are about to pass out of the Madeira Plate, having arrived at the north-east corner of the territory of Bolivia. The lands about the mouths of the Beni and Mamoré are now inhabited by wild Indians ; some parts of them are free from inundation. Cacao grows wild in the forests. The head of the Madeira contains a number of islands. Here we find the outlet of streams flowing from the Andes and from the Brazils collected together in one large river. Water from hot springs and cold springs, silvered and golden streams joining with the clear diamond books, mingled at the temperature of 82° Fahrenheit.

The Madeira river flows through the Empire of Brazil, and keeps the northerly course pointed out for it by the Mamoré. The first falls we met were close to the junction of the Mamoré and Beni, called " Madeira," three-quarters of a mile long. It is difficult to judge the difference of level between the upper and lower surfaces of the river. As the falls are shelving, and extend a great distance in length, the distance we run during the day is not easily estimated. At one time we go at the rate of fifteen miles an hour, and then not more than one mile in half a day. This fall is not less than fifteen feet. Large square blocks of stone stand one upon another in unusual confusion. The boat was paddled through for a quarter of a mile, and by passing half the baggage out over the rocks she was slided and floated through narrow channels close along the eastern bank. The whole bed of the river, as we stand at the foot of the fall and look up, is a mixture of rough rocks lying in all positions on the solid foundation of granite, surrounded by foaming streams of muddy water. While we loaded our boat again at the foot of the falls, Titto discovered some Indians approaching us from the woods. They came upon us suddenly, from behind a mass of rocks, with bows and arrows in their hands. Don Antonio had warned me before I left him to be on my guard when the savages came up in this way. He said when they send women and children to the boat in advance, then there is little chance of a difficulty with the men; but when the women and children are kept in the rear, and the men come with bows and arrows in hand, the signs are warlike. We were, therefore, prepared. We, however, recognised our friends, the Jacares. An old chief brought a woman along loaded with roast pig and yuca. She carried a deep, square willow basket on her back, suspended by a strap of bark cloth round her breast. The chief and his two men were dressed in bark cloth frocks and straw hats, while the only thing on the woman's back was her basket. One hand bore an earthern pot, which she also offered for sale. Titto traded with the party, and they gradually became much more easy in their manners towards us. For the want of an interpreter I could not make out what customs were observed among them. These Indians bear the name among Brazilians of great thieves. They, however, appeared to be perfectly satisfied when we left them with the reasonable exchange. The passion expressed by one at Pedro for not giving him all his fish-hooks for a few arrows rather leads us to believe that, if they had outnumbered us, they would have been troublesome. We gave them no opportunity to treat us unkindly, for we were exceedingly polite, and so well armed with all, that they very justly acted their part in a spirit of reciprocity. There is great difficulty in knowing how to meet the savage. Treat him as a civilised man and his better feelings are touched. It won't do to approach him indirectly, letting him see that, while willing to trade, there is a prudent readiness for a fight. They took a polite leave of us by shaking hands all round. We introduced the custom, which they seemed to like, though the stiffness of their elbow-joints proved they did not understand the matter. They sauntered up the rocky bank on the sand to where they had left their bark canoe at the head of the falls, and we went dashing on through the rocks in the rushing current.

Sept. 23.—The river was seven hundred yards wide, and one hundred and five feet deep.

We passed " Misericordia " rapids, or swift current, but not a ripple was to be seen. The channel was clear of rocks, and we soon came to the Riberião falls, which are two miles long. The baggage was carried five hundred yards over a path on the east bank. Don Antonio transported his vessels on wooden rollers here. I think he said he was nearly one month getting up these two miles. The men were anxious to see whether they could not pass this fall with the boat in the water. They launched her down one shoot of twenty feet nearly perpendicular by the rope painters in the bow and stern.

Our boat was beginning to give way to the rough service, and as she leaked, it became necessary to lighten her load; then, too, the men began to fag. After they succeeded in getting the boot safely over a dangerous place, the boxes had to be carried one by one. The heaviest box was that in which were planted three specimens of Mójos sugar-cane. I had just cut my first crop, and found the plants were doing well, when it became necessary to relieve our little boat, and we were unwillingly obliged to leave behind what might have proved of importance to a Mississippi sugar-planter. Our baggage was taken out and restowed a number of times. Once the boat was on the top of a rock, at another half under foam. The sun was scorching hot and we had the full benefit of it. When the water is thrown on the bare rocks it hisses as if poured upon hot iron. The sides of the pot-holes are ridged like the inside of a female screw; some of them are nine feet deep. The water in them is quite hot; one of the negroes seemed to be fond of lowering himself into the pots of hot water; his face had rather a distressed expression, and while standing with his head above the edge of the pot, he looks as though undergoing a hot water cure. The river appears to have worn away the rocks less than above. It flows over a solid mass, in which there are many gutters cut, from four to six feet deep, of the same width. Our canoe safely passed through one of these by the ropes, as the crew walked along the level rock. There were numbers of these gutters cut parallel to each other. The rock was worn as smooth as glass. After descending some distance in the middle, we found the channels so large and dangerous, that we must gain the east side of the river; the only escape for us, besides retracing our steps, was to cross a wide channel with a furious cataract above, and another close below. We hugged the foot of the upper as close as possible, and the men pulled with such force that one of the paddles broke when we reached half the way. With the remaining three we made a hair-breadth escape; the boat could not have lived an [instant had we been carried over the lower fall. The rollers formed by the swiftness of the current are five feet high; large logs are carried down so fast they plough straight through the waves, and are out of sight in an instant. The men came near upsetting the boat in a dangerous pass. They seemed to be giving out through pure exhaustion. They have very little to eat; far-inha adds not much to their strength, and jerked beef spoils. No fish are to be found, nor birds; a monkey would be a treat. Night overtook us half way down the falls, and we came to, on a barren rock, where there were two small sticks of wood, of which we made a fire, boiled water, and gave the men coffee. I observed a southern star, and turning for another in the north, was glad to find it had passed the meridian, as sleep was much more necessary than latitude. On the west side of the falls stood three small hills; on the east side a large white-trunked forest tree. This was the largest tree we had yet seen, though not quite equal to a North American huge oak.

SEPT. 24, 1852.—At daylight we *crawled* on; it would be a mistake to grace it with the name of travelling. The country is thickly wooded with Brazil nuts and cacao trees interspersed. Four miles further down we came to " Periquitos " rapid, which takes its name from the number of parrots inhabiting the woods. These parrots are green, scarlet, and yellow, with long tails; they fly slowly overhead in pairs, crying an alarm as we are seen approaching. We paddle through these few rocks without the least difficulty. Banks of the river thirty feet high; soundings, fifty-four feet. At mid-day a thunder gust with rain came from the north. As we are passing out of the Madeira Plate, we find the climate changing; northerly winds bring rain here, while southerly winds bring them farther south. At 3 p.m., thermo-meter, 86°.; water, 83°.

" Araras " rapids were passed with much toil, easing the boat down by ropes made of bark, which are best for such work as this ; the water has little effect upon them. The fall is small and the channel clear. While the men gather Brazil nuts from the woods, we bottled a young turtle, taken from among eggs found in the sand. Amphibia are poorly represented ; we see no alligators, snakes, or frogs. The water has become much more clear ; it has a milky appearance. The banks slope down regularly ; being covered with a light green coat of grass, they have the appearance of cultivation.

SEPT. 25.—At 9 a.m., thermometer, 84°. ; water, 82°. ; light north wind. At 2 p.m., thunder to the north-east. On the east bank were cliffs of red clay fifty feet high, breaking down perpendicularly. We passed the mouth of the Abuna river, which is fifty yards wide, and flows in from the south-west. At 3.30 p.m., thermometer, 86°. ; water, 82°. In the evening lightning to the south-west. We came to a number of rocky islands in the river, and took up our quarters on one of them for the night. We slept under blankets ; there is a heavy dew, and the nights are quite cool. Richards was aroused by a severe pain in his ear ; he was suffering all night long. The men told me it was common among the soldiers at the fort, caused by exposing the ear to night air and dew. The only remedy reported was " woman's milk," which was not at hand.

SEPT. 26.—For the eighteen miles between the " Araras " rapids and Pederneira falls, we found a current of only one and a half mile per hour, with a depth of sixty feet of water. We have observed, between all the falls passed, that the current becomes slow, and as there is very little damming up of the water by the falls, that the general inclination could not be great. We also found the land gradually getting higher, as though the river was flowing through a country which sloped against the current. We find at the Pederneira falls the strata perpendicular ; the river does not flow over a flat mass of rock as before, but cuts its way through a vertically grained rock ; so fair and square has the river worn its passage, that the gap resembles a breach in a stone dam. The river turns from its northern course at a right angle, and flows east, inclining a little south, as though it wanted to turn back and flow into the Madeira Plate again. We suppose this fall to be situated on the top of that ridge of hills and mountains extending across South America from the Andes to Brazil. We are now on the chain which fastens Brazil to the base of the great mountains, and the river is sawing across and cutting it gradually asunder. Part of our baggage was carried over, and our boat towed along the east bank with less difficulty than we expected ; we found a rapid current below.

On the south bank of the river we saw two bark canoes ; the negroes gave us music on their cow's horns, and two red women appeared on the bank at a path in the thicket; they belonged to the " Caripuna " tribe. We pointed down the river, and called for " Capitan Tupe ; " they ran away, and we continued to the Paredão falls. A whale boat might pass through the main channel with ease, but our boat was too small to attempt it. The baggage was landed on a sand beach near the rocks, which were elevated forty feet above the water level. In the rainy season the floods cover them all except ten feet. I climbed up to the top for a view of the country, and to seek a passage for the boat. The men had a short distance to paddle, and then tow her through a narrow channel by the ropes. The landing place was in the rapid current; they missed it, and the boat ran away with them through the rocks—they were carried at a frightful rate ; Titto shouting to the negroes at the top of his voice to pull for their lives, so that he might steer them safely, which he fortunately did. They were all so much frightened that it brought them to their working powers. The sight was an interesting one for me, as the smallest rock in their way would have dashed the boat to pieces. As I turned to go down I found myself surrounded by a party of savage women and children who had come up behind me. There were eight women, ten children, and two unarmed men, all, from external appearances, savages of the purest water. On taking out my handkerchief, the women and children all laughed. One of the men stepped before me, and, putting his hand into my pocket, took all the fish-hooks out, and appropriated them to his own use, by handing them to a homely woman who bore a sucking baby, and then cooly inquired whether I had a knife to give him. He was a short, thick-framed man, quite fat and hearty; the women were all ugly; the boys were the most cheerful, manly-looking Indians we ever met with. At my suggestion they walked to the boat with me. Their chief " Capitan Tupe," as they call him, was absent on a hunting excursion. Their huts were some distance from the falls, so that we missed seeing their houses. They were quite friendly with us. Some of the men who came afterwards, left their bows and arrows behind the rocks, and walked up unarmed. The women carried their babies under the arm, seated in bark cloth straps, slung over the opposite shoulder. The infants appeared terribly frightened at the sight of a white man; one of them screamed out when Pedro milked the mother into a tin pot, for the benefit of Richards' ear, which still troubled him. The woman evidently understood what was wanted with it, and stood still for Pedro to milk her as much as he chose. The boys are remarkable for large bellies. The older ones express a willingness to go away from their mothers; one was asked, by signs, if he would go with me; he shook his head, no; when he was made to understand that he could get a pair of trousers and something to eat, he then nodded his head, yes. Pedro tells me they swell themselves up by eating earth, which Indian children all do. One of the Caripunas got into the boat and examined the baggage; he soon found a knife, which he took, and came out with it in his hand, before everybody. It belonged to one of the negroes, who took it from the Indian. The savage appeared disappointed; he was then told if he would bring yuca or other provisions for the men, he should have a knife. They all declared they had nothing to eat in their houses. We made them a little present, and bought a bow with arrows from one of the boys. They were particularly desirous of getting fish-hooks and knives. The full dress of the men consists of beads of hard wood round their necks, with bands bound tight round the arms above the elbow, and round the ankles. The foreskin is tied up to a band of cotton twine, which is wound tight round the hips and under part of the belly. All

wear their hair long, and cut square off in front. In large holes
in their ears they carry pieces of bone or a stick of wood.
Through the hole in the nose a quill is pushed; the cavity being
filled up with different coloured feathers, gives them a moustached
appearance. These people are nearly all of the same height
and figure, but differ very much in the features of the face.
Some have thick lips, flat noses, and round faces; others are
just the reverse. The former very ugly, and a few of the latter
tolerably good-looking. The women are larger than those we
saw near the mouth of the Beni. There are not many of them;
they live about in small bands, and said they found few fish
in the river. They promised to plant yuca and corn, so that
the crew might have something to eat on their return to the
fort. As we embarked, they said "*shuma*," which Pedro informed us meant " good man ;" but probably referred to more
presents.

The lands on the south side of the river are inhabited by
the Caripunas. It is flat and a beautiful spot for cultivation.
Small mountains and hills are in sight, on the north side, as
we descend by a rapid current. The river seems to be creeping
along on a ridge, seeking an outlet to the north. At 3 p.m.,
thermometer, 90°.; water, 83°.; light northerly airs; thunder
to the north, and a rainbow to the north-east.

Sept. 27.—At " Tres Irmãos " rapids we found no difficulty.
A large island in the middle of the river chokes it, and the water
rapidly flows through two channels. As we dashed by, the men
blew their horns for " Capitan Macini," another Caripuna chief,
who lives on the south side of the river, with a small band of
his tribe. Pedro speaks of " Capitan " in complimentary terms.
He is represented as being exceedingly obliging ; we wanted his
services as pilot, but missed him. After passing " Tres Irmãos "
rapids the river turns north. A rapid current carries us through
a chain of hills on each side, tending east and west. The
foliage is unusually green and thick ; forest trees have been
broken by the action of violent winds. We scarcely are fairly
launched out of the Madeira Plate into the Amazon basin, before
we meet, at mid-day, a storm of wind and rain from the north-
east, accompanied with thunder. We find the sea-way in mid-

F

channel much too high for our little boat, and bring to, while the storm passes; the wind carries a cloud of dry sand before it. At 3 p.m., thermometer, 85°.; water, 84°. We are now being avalanched down an inclined plane. Arriving at the head of Girão falls, we find the true falls of the Madeira. They are short, but the rush of water through a confined space, between immense masses of rock, baffles large sized vessels, and prevents their passing either up or down the river. Don Antonio transported his boats over the land here.

Richards was suffering very much from his ear ; his under eyelid hung down, the corner of his mouth became drawn up on one side, while he seemed to lose control of the muscles of his face ; the pain was beyond endurance. All the men began to feel the effects of the change of climate; the nights cold, and mid-day sun very hot. They complained of headaches and pains in their backs ; the strongest of them were jaded. Before they went to sleep, I dosed the party with raw brandy all round, which cheered them up. They have been much more respectful lately, and work with a will.

SEPT. 28.—The men are all in better health this morning. They carried the baggage through the woods on the east side of the river, and with the greatest difficulty got the canoe through the rocks. The river has been turned to the eastward by hills on the north side. The fall cannot be estimated with any degree of certainty ; the descent is more precipitous, and the roaring of the foaming waters much greater than any we before met. We were from daylight until 3 p.m., making the passage from the upper to the lower side, before we got breakfast, which we took under the shade of trees, where the thermometer stood at 99° ; wind north-east.

Pedro shot a few fish with his arrows, and a negro caught one with a line. As the vegetable kingdom appears fresh and vigorous, under the strong breezes filled with moisture from the North Atlantic, so again do we find animal life in abundance. The trade-winds from the ocean cross the land from Cayenne, in French Guiana, and strike this side of the Amazon basin. The clouds roll up, and the waters are wrung out in drops of rain.

The Paititi district of country which we have on our west, and the Tapajos district on the east, are watered by the north-east trade-winds. They get their moisture from the North Atlantic, and here we find on the side of these hills the boisterous region again, and ʻthe trees are torn up by the roots. These acts of the north-east trade-winds are written upon this slope of the Amazon basin exactly as we met the south-east trade-winds as they struck the Andes on their way from Rio Janeiro. The Caripuna Indians we have just left told us they came *down* the Madeira for fish. They find little game and no fish, even in these mighty waters, above the boisterous region. The two Yuracares Indians we met on the side of the Andes, said they would catch us fish when we got further down the rapid Paraciti. Fish are just as particular in their choice of waters and *climate* as those animals which inhabit the dry land.

The foam that is produced by the water dashing over the rocks, floats aloft in the shape of mist; and in the calm, clear, starlight nights, the gentle north-east breezes cast a thin gauze-like veil around us and affect the glasses of our instruments. All observation of the stars seems to be forbidden. Early in the morning, as the sun's rays strike upon the river, they gradually absorb the mist, and first that portion which has been scattered by the night-winds, and looking just then up or down the river from an eminence, the traveller may see the position of each cataract, like the smoke of a line of steamers. The powerful sun soon evaporates this mist, which speedily disappears as it rises. One of the crew caught a small electric eel, which opened its galvanic battery and shocked the whole party. A rapid current and no bottom at twenty-five fathoms water.

SEPT. 29.—We got our baggage stowed and all on board ready for a long pull, but soon fetched up among the rocks again. "Caldeirão do Inferno" rapids are caused by three rocky and somewhat wooded islands in the river. We pulled part of the way through on the west side without discharging baggage; the boat was gently eased down by the ropes. At the foot of these falls, which could not be passed by a steam-boat, we discovered a bark canoe, manned with savages, paddling

with all their might away from us ; they seemed to be very much
alarmed, and were soon out of sight. As we came to a place
rather too rapid for safety among rocks, the men got out and
towed us along the north bank ; while doing so, three savage
men, three women, three children, and five most miserably thin
skeleton dogs, came to see us. The men laid their bows and
arrows behind the rocks, and approached us without fear, but
the slim dogs were disposed to show fight. They were weak
and slab-sided animals ; quite unsuccesful in their endeavours
to raise a bark at us, but coughed out a sickly sort of noise as
they hung around their masters' legs. One had his ears boxed
by a tiger, which gave him a perpetual stiff neck. They all
looked as though they had been vainly struggling with the beasts
of the forest. An unsightly old woman brought us a fried fresh
fish from the river. One of the men had bilious fever, but was
attended by a pretty girl, who took her paddle in one of the
canoes which kept company with us. The parrots swarm along
the banks of the river, but there are few other birds. The current runs at the rate of six miles per hour. River, three quarters
of a mile wide, with sand banks and islands in the stream. We
landed on the north bank with the Caripuna savages ; men,
women, and children, all seated themselves in a friendly way
round our cow-hide which was spread on the ground for
breakfast.

Richards was left in charge of the boat, while I, with
one of the negroes, armed with a musket, followed a
path through the woods single file for a quarter of a mile
from the river. As we came in sight of huts the men and
boys gathered under an open house at the end of the path ;
the women all seized their babies and ran into two enclosed
buildings in the rear. The savages did not take up their bows
and arrows, which, however, lay at hand, but several of them
held knives, and others picked theirs up. Thomas, the tall
negro soldier, came to a stand just outside of the shed, while
I walked under and took a seat in one of the grass hammocks
slung between the posts on which the roof was supported. The
boys all laughed and gathered round me. One man came up
and leaned against a post close by me with his arm elevated.

He held a knife in his hand; my hand was concealed under my jacket, where Colt's revolver rested in a belt. The Indian wanted to test me, as is their custom. A fine rooster passed by. Savage was asked to sell it by signs of hunger. He at once took down his hand, and called out to the houses, when the women came out with their babies. One of them, a good-looking squaw, came to him, and they had a consultation about the chicken. She nodded her head, and the boys gave chase to catch it for me. There were thirty savages living in this wild out-of-the-way place. One of the men was chipping off the outside of a hollow piece of log with his knife for a drum, two of which already hung up under the shed. They expressed no pleasure at seeing us. They looked as though they preferred we would go away. The roof of the wooden house, under which the men were collected, was beautifully thatched with a species of wild palm-leaf. The frame work was made of poles stripped of their bark, fastened together with vines or creepers. The whole rested on forked posts, set in the ground, between which there were slung a number of grass hammocks. Bows and arrows were their only home-made arms. The knives were imported. After making friends with them, they all came up, shook hands, and took a good look at me. The floor of the guard or men's houses was swept clean. It seemed to be kept in military order, clear of all household or kitchen furniture. One of the men and several women went with me to examine the dwelling-houses of the women. The roof extended within two feet of the ground. The sides and gable ends were also thatched in, with a doorway at each corner, and one in the centre next the guard-house; five entrances in all. The inside presented a confused appearance. Piles of ashes were scattered about the ground-floor as though each woman had her separate fireplace. The inside measured about forty feet by fifteen. Earthen pots and plates were lying about in confusion; dirty, greasy hammocks hung up; tamed parrots were helping themselves to plantains. An ugly monkey looked dissatisfied at being fastened by the hinder part of the body to a post. The unpleasant variety of odours drove us out. In the third house there were but two doors. Here the miserable dog kept up a terrible noise. The women took me to the hammock of an old sick Indian, who they made signs was dying, by laying their heads in the palms of their hands and shutting their eyes. He was covered with a dark cloth blanket, which was cast off by him that I might see his thin legs and body. He was very much reduced. By the whiteness of his hair, I judged he was dying of old age, or suffocated inside the damp, filthy house ; where he seemed to have been turned to the dogs. There was one house in which the women slept. The open house was the sleeping apartment of the men and boys. There was great order among the men ; the grounds about were swept. Where the women were seemed all confusion and want of cleanliness. Their faces were covered with dirt. As to their clothing, we could better describe what they did not wear.

We saw no signs of a place of worship, nor of what was worshipped, though the Brazilians say they have seen among them " wooden images," figures of head and shoulders in shape like a man. A Catholic priest once visted these people, but found no encouragement. They looked on indifferently, taking more interest in the music of a violin and the singing than anything else. The lofty forest trees shade the little huts ; a path leads farther inland, where they cultivate patches of yuca and corn ; though they have little to eat from the land at present, and take to the river for food. The children of these Indians strike us as being remarkably intelligent, compared with those on the tops of the Andes. All Indian children seem to be in much brighter spirits than the older ones. They have yet to be taught the art of using chicha, which the women are said to give their husbands here in the woods. We gave the multitude an invitation to join us at breakfast. A little boy walked by me with the rooster under his arm, and all followed, single file, with the music of crying babies, to the bank of the river, where they seated themselves round. Some presents were made to them in exchange for the offer of several chickens and a large partridge. To the little girls we gave earrings, to supply the place of fish or beast bones ; to the boys fish-hooks, and to the men knives. The elderly women particularly fancied looking-glasses for themselves and glass beads for their babies. One very unattractive woman requested me to make her an additional present of a looking-glass. A knife had been offered, which she particularly requested. She received the refusal with such a savage side-glance, that the damage was repaired at once, and the men ordered into the boat. Her sister used paint. Her forehead was besmeared with a red colour, and her lips blackened. We presented her with a large looking-glass, which she used for examining as far down her throat as possible. Pedro had a slight difficulty with one of the savages, who, he said, had stolen his knife from the boat. I replaced it, and we went on without being disturbed, though, as we afterwards learnt, these fellows not long since robbed two Brazilians on the river, who escaped down the stream in one of the bark canoes of the savages, leaving their own boat behind. At 3 p.m., thermometer, 91°; water, 85°; river, one mile wide; interspersed with islands and rocks, twenty-five fathoms depth.

On the east side a small stream of clear water flows in. The water of these small side-streams are often 6° Fahrenheit cooler than the main river water. We bottle it, as the river water is unpleasantly warm for drinking. A man fully comprehends the blessing of ice by gliding down this river. The current is fast one hour and slow the next few minutes. The men pull when they feel like it, and rest when they wish. We are moving along, more or less, all the time during the day. The river is not very winding.

SEPT. 30.—About twenty-five miles on a north-easterly course, brought us to " Doz Morrinhos" rapids. The difference of level here is slight, though the passes are difficult. A part of the baggage was handed over the rocks, which proved a prudent plan, as the boat was nearly swamped. The country is quite uneven and thickly wooded. At mid-day we had a light shower of rain, accompanied by thunder, without wind. At 3 p.m., thermometer, 87 deg.; water, 85; with a strong south-west wind. At the foot of these falls we sounded with five hundred and ten feet, and no bottom.

At a late hour in the afternoon we arrived at the head of " Teotonio " falls, the most terrific of them all. Here I was

attacked with a severe bilious fever, which brought me at once
on my back. The pain in my left breast was somewhat
like that described by those who have suffered with the
" Chagres fever." We were all worn out, thin, and haggard. I
had been kept going by excitement, as the men were careless,
brutal negroes, and Richards suffering still with the pain in his
ear.

OCT. 1.—This fall is over fifteen feet, ten of which is at an
angle of 45 degrees. The roaring made at intervals, by the
rushing of the waters over and through the rocks, sounds like
distant thunder. Our little canoe is driven for safety out of the
water to the land. The baggage was carried by a path on the
south side to the foot of the falls. Richards went along with
the first load, and remained below looking out, while I rested
to see everything sent over. The men idled their time between
us, until we were caught in a heavy rain and thunder-storm
from north-east. The boat was put upon rollers and transported
four hundred yards over a hill, and launched into the river below.
We were from daylight until dark at the work. I should not
complain, however, because men never had a more harassing
time than these have had. If alone, they would not have come
half the distance in the same length of time. They have pushed
on for me, when I least expected they would keep on.

We noticed that, at nearly all the falls in the Madeira, the
river turns as it cuts its way through the rocks, forming nearly
a semi-circle towards the eastward. After gaining the base of
the declivity, the stream returns again to its original course.
Here the path over the land describes a diameter. The storm
continued all night in squalls. The negroes took off their clothes
and laid down upon the bare rocks under a heavy rain, with
cold wind, where they actually slept, while those of the crew,
with Indian blood, built a fire and slept on the sand close by it
in their clothes. The baggage was left on the sand bank until
morning covered with raw hides. We were well drenched ;
certainly a poor remedy for bilious fever, particularly when
followed by the heat of a tropical sun.

OCT. 2.—Five miles below are "San Antonio" falls, which we passed by tow lines without disembarking our baggage. The difference of level is very small; the bed of the river much choked with rocks. The stream is divided into a great number of rapid and narrow channels. We took breakfast on the west side, at the foot of these falls, with feelings of gratitude; we had safely passed the perils of *seventeen* cataracts. Those parts of the rivers Madeira and Mamoré, between the foot of San Antonio and the head of Guajará-merim falls, are not navigable for any class of vessels whatever; nor can a road be travelled, at all seasons of the year, on either bank, to follow the course of the river, for the land bordering on the stream is semi-annually flooded. By referring to the map, it will be seen we travelled from Guajara-merim, on the Mamoré, in a due north course to the Pederneira falls on the Madeira. By the windings of the river we estimate the distance not less than one hundred miles. From the Pederneira falls to the foot of San Antonio, our direction was about east-north-east, a distance by the river of one hundred and forty miles, which makes the space not navigable two hundred and forty miles. A road cut straight through the territory of Brazil, from San Antonio falls, in a south-west direction, to the navigable point on the Mamoré, would not exceed one hundred and eighty miles. This road would pass among the hills, seen from time to time, to the eastward, where the lands, in all probability, are not overflowed. On a common mule road, such as we find in Bolivia, a cargo could be transported in about seven days from one point to the other. Don Antonio Cardoza was *five* months struggling against these numerous rapids and rocks, to make the same distance, with his cargo in small boats. We have been twelve days descending the falls, which is considered by Brazilian navigators fast travelling. The wild woods that cover the lands are unknown to the white man. Topographically considered, the lands on the east side of the Madeira are the most valuable.

Our experience with a black crew gives reason to believe the climate is more congenial to them than the white or red races. Among the half-civilised and savage aborigines, we notice very few men live to an old age; they generally pass away early; the tribes are composed usually of men under forty years. The moment we landed at Principe, there appeared before us a number of active, grey-headed old negro-women and men, grinning and bowing, with as much life in their expression of face and activity of manner as the youngest. Long after the savage has become hammock-ridden with age, the negro, born before him, is found actively employed. The physical strength of the negro is not equalled by the red man here. The Indian enjoys the *shade* of the forest trees, while our negroes rejoice in the heat of the sun.

*   *   *   *   *   *

Now that we are at the mouth of this magnificent stream (the river Madeira), we find no deeply-loaded vessels enter it. The value of the present foreign trade of South Peru and Bolivia may be worth ten millions of dollars per annum.

The distance from the foot of San Antonio falls to the mouth of the Madeira is five hundred miles by the river. A vessel drawing six feet water may navigate this distance at any season of the year. A cargo from the United States could reach the foot of the falls on the Madeira within thirty days. By a common mule road, through the territory of Brazil, the goods might be passed from the lower to the upper falls on the Mamoré in less than seven days, a distance of about one hundred and eighty miles ; thence by steamboat, on that river and the Chaparé, a distance of five hundred miles to Vinchuta, in four days. Ten days more from the base of the Andes, over the road we travelled, would make fifty-one day's passage from Baltimore to Cochabamba, or fifty-nine days to La Paz, the commercial emporium of Bolivia, where cargoes arrive generally from Baltimore in one hundred and eighteen days, by Cape Horn—often delayed on their way through the territory of Peru from the seaport of Arica. Goods by the Madeira route, sent over the Cordillera range to the Pacific coast, might get there one month before a ship could arrive from Europe on the eastern coast of the United States, by two oceans, or the old route.

THE MADEIRA

AND ITS

HEAD-WATERS.

BY

GENERAL QUINTIN QUEVEDO.

TO THE NATIONAL GOVERNMENT.

Taking, as we do, a lively interest in any great and excellent measure for the future development of Bolivia, we have read with considerable satisfaction Señor Quintin Quevedo's treatise, published in Brazil, " On the Navigation of the Madeira." The work being one of national interest, we reproduce it here.

This valuable undertaking, the result of an earnest love to his country, prepares the way for the realisation of one of the most important projects devised by our statesmen.

Various sessions of the Congress have hailed it as the most efficacious means to advance Bolivia to the prosperous destinies which are in store for her.

Cabinets of renown and progressive policy have devoted some measures to it.

The Ministry of " Mendez" (Manuel de la Cruz) took the initiative in 1842, and to it is due the administrative and social amelioration of the Beni, as well as the surest precedent to the ulterior steps relative to the navigation of its rivers.

The Ministry of " Urquidi " (Melchor) in 1851, gave to the fluvial enterprises such impulse as circumstances, political and financial, would, at the time, permit. This was done with that activity and constancy which characterise him in the public service and in the furtherance of profitable ideas.

The Ministry of Rafael Bustillo, in 1853, enthusiastic to realise such a project, spoke of the navigation of our rivers as *the prolific germ of liberty, so useful to the interests of the Republic, as well as of all mankind, and declaring, in consequence, free for the commerce of all nations of the globe, the affluents of the Amazon and the Paraguay.*

Now that the Republic enjoys peace under the auspices of a

just Government, that the national opinion has already been established with regard to the interests of our country, and that, for the honour of Bolivia, revolutions are regarded as a great calamity, and the good sense of the Bolivians recognises only work and industry; now that steam and electricity have been placed at the service of man, by which means an economy of physical exertion and monetary expenditure can be effected; and now that the resistance offered by Brazil has ceased, and that Empire manifests a willingness to enter into relations mutually beneficial; now-a-days, we say, the magnanimous President Achá and his illustrious Cabinet have resolved to give execution to that splendid project—*the navigation of our rivers.*

To you, then, General Achá, and to your worthy Minister Salinas,* we address ourselves as patriots, and we solicit from you protection for so great an enterprise; protection for the citizens who undertake it; and protection, finally, for the Bolivian who has known how to utilise his ostracism in subservience to his country, to which he dedicates himself with manly efforts and patriotic self-denial from remote foreign regions, whither an unfortunate policy has banished him!

The act of thus opening up a future for Bolivia crowns you with glory; and you immortalise your name by supporting the national prosperity.

Certain Bolivians.

Cochabamba, December 28, 1861.

To the Most Excellent Señor Angelo Thomas do Amaral,
President of the Province of Pará.

Most Excellent Sir,
        To your patronage I owe the publication of this treatise, and it is my duty to dedicate to you the work of your protection.

Impressed with the lofty sentiments of progress which you

* To Dr. Manuel Macedonio Salinas is entrusted the Department of Foreign Affairs, which has to deal with the matter in point.

cultivate, I wish you the attainment of your desires, and the deserved enjoyment of the rich results.

Now, more than ever, you are in readiness to devote yourself, with strenuous efforts, to the subject of American prosperity, as Deputy General of the Empire. I need not recommend to your attention the *Navigation of the Madeira,* nor urge upon you the importance of a reciprocally-profitable understanding between our respective countries. You are aware that two nations linked by interests of commerce, proximity and sympathy are almost unknown, and allow the advantageous elements to be wrecked, in the nullity of its reciprocal indifference.

The vivifying germ of the world, *steam,* already bears its fruit along the wide affluents of the Amazon ; and, on crossing the point where the Madeirá joins it, the last-named protests against the neglect to which it has been condemned, being, as it is, the richest tributary of the *King of Waters.*

I esteem so much the more your favourable position in this respect as I see the nullity of my own, incarcerated within the limits of *ostracism.* Unfold, therefore, Sir, all your resources in aid of so important a work, and give it, in the Tribune, the support it merits. Then, when my aspirations may be crowned with success, you will command the esteem of two nations, and the lasting gratitude of

Your Excellency's

Obedient and faithful servant,

· (Signed) Quintin Quevedo.

Belem, April 25, 1861.

THE MADEIRA AND ITS HEAD WATERS,

A SHORT ACCOUNT OF THIS RIVER, ITS AFFLUENTS AND THEIR BASINS.

From the town of San Joaquin, almost opposite to the fort Principe de Beira, I addressed, to the press of my country, Bolivia, a small sketch of the Department of the Beni, the ancient Province of Mójos, little known in the Republic, yet worthy, by reason of its features, productions and fluvial position, of the most earnest consideration. Then, also, I offered to make public the result of my journey, or exile converted into an exploration. What satisfaction attends me on the one hand, in fulfilling this promise, and what regret, on the other, through my inability to do so as I would wish !

I look back on all I have seen and passed through as a fantastic dream. Certainly, to cruise with those simple Beni Indians, in a badly-hollowed canoe, for a distance of more than five hundred leagues upon the waters of one of the principal affluents of the mighty Amazon, thus gratifying a caprice ; to cross virgin woods where Nature has lavished with a profuse hand her richest gifts, to tread under foot, on all sides, the most valuable fruits which rot in the desert ; to live in the forest amidst wild beasts, to enjoy the most picturesque views of vegetation, to drink the surrounding freshness of the green and humid woods, or to endure, benumbed, the crushing power of the tempest in the silence of the wilderness, within hearing of the cries of the bear and the tiger ; are truly scenes of infinite magnitude and variety, far beyond all conception.

I have traversed in my journey the heart of South America, and it seems that, from the centre of the Andean heights, where

I was taken, I have wandered, urged on by an ill-directed impulse, but sheltered by the hand of Providence, so that amid the rocks and precipices He has spread over me His mantle of roses, and has brought me to the mouth where the Atlantic is rolled back by the Amazon, by which, one day, the richest and most fertile regions, now deserted and uncultivated, will be brought into communication with the world, offering it exuberant produce, and receiving thousands of immigrants that Europe sends out in all directions. One of these regions of such promise is the scene of my aspirations, the fatherland of my children. How shall my heart not palpitate at the prospect of such brilliant fortunes in the future!

Providence is wont to reserve, now and then, amidst inscrutable designs, the weakest objects for the most colossal ends. I, proscribed and forcibly ejected from my country, to satiate the cravings of rancour and prejudice, on beholding myself in the solitudes of the Madeira, borne along its waters or whirled in its eddies, have often had the idea of accomplishing a mission, and my privations and sufferings and dangers have passed unnoticed, because my mind lived in its idea, and all my sentiments centred in the thought of the *navigation of the Madeira.*

My position offers a rare coincidence with that of a distinguished Peruvian, who, not many years ago, by way of the Solimões, arrived at this village, and found, as I did, a vehicle of commerce and of life for the oriental towns of Perú. Impelled with this idea, and eager for the prosperity of his country, he devoted his energies to the navigation of the Amazon, until he crowned them with the desired realisation. I, placed in the extreme east of America, and urged by a similar motive, also raise my voice before Bolivians, inviting them to an enterprise which is important to the prosperity and future development of that dear fatherland. To achieve this result, needs less effort and expense than has been supposed in Bolivia; and it may be assured that the only real impediment is inertia and want of will. I abstain from publishing the diary of my voyage, because I consider that it contains little amusement, being wearisome and scattered. I have preferred to extract from it data and more reliable accounts, so as to arrive at a trustworthy illustration of the navigability of the Madeira, its inconveniences, and the actual condition of those regions. I regret that I am not in a position to offer a worthier work and more exact data; but I am limited by my capabilities, and, through my position as a proscript, I, necessarily, lack all information requisite for earnest study and action.

Before treating of the rapids, I will give an account of the river Madeira, of its practical value, its source, and other noteworthy features.

Three principal branches form this river, which drains the eastern district of Bolivia—the Mamoré, the Iténes, and the Beni.

The Mamoré, which flows almost directly from south to north, intersecting the tracts and central towns of the Beni Department, takes its name after the confluence of the river Sara with the Chaparé. The river Sara descends from the plains of Santa Cruz de la Sierra, and is formed from the watery wealth of the river Piray, which crosses to the west of that city, and of the river Grande, which, descending from the rugged heights to the west of Cochabamba, and forming a large bend, traverses, near Chayanta, the Department of Potosí and that of Chuquisaca, makes a detour at the eastern district of Santa Cruz, and joins itself much lower down to the Piray. Still further below this confluence, the river Yapacani incorporates itself with the river Sara on its left bank. The depth of the river Sara may not perhaps, be sufficient at all times for the navigation of large vessels; but it is always available for canoes and small launches. On the Piray is a port for Santa Cruz, in a place called Cuatro Ojos, at thirty odd leagues from that city. After the Yapacani comes the Chaparé, which possesses greater volume and is more navigable than the Sara. This river has its port, named Cohoni, distant thirty odd leagues from the city of Cochabamba. It flows west-south-west, until its confluence with the Sara, and thence takes the name of Mamoré. This river, in its steady course from south to north, afterwards receives, on the west bank, the river Sécure, by which Mr. D'Orbigny descended from the heights of Cochabamba, through the forest of Totolima. This affluent, then, offers a better channel of navigation for

Cochabamba, by reason of the direction in which it flows and its proximity to that capital, thus saving the journey across the Espíritu Santo, so unhealthy and dangerous. It also receives, successively on the same side, the Tijamuchi, Apere, Rápulo, and Yacuhuma. On the east, it receives the Ibaré and the Matucaré, with other small streams. The Mamoré flows by the towns of Loreto, Trinidad (capital of the department), San Javier, San Pedro, San Ignacio, Santa Ana and Exaltacion, crossing regions and plains full of wonderful vegetation, abounding in wild horned cattle which feed on the prairies. These tracts are rich in coffee, cocoa, white and yellow cotton, almonds, tamarinds, tobacco, sugar-cane, yuca, rice, sarsaparilla, gum elastic, copaiba, vanilla, indigo, Brazil-wood, and a thousand other dyes, resins, and woods of infinite variety.

The Iténes, or Guaporé, which descends from the province of Mato Grosso, sparkling and clear as a mountain-stream, wide, deep, and majestic (even more so than the Mamoré), flows in an east-south-easterly direction, and unites with the Mamoré at about thirty leagues below Exaltacion, after passing the well-known Brazilian fort of Principe de Beira. It receives in its course the Bolivian rivers—San Simon, S. Nicolas, S. Martin, Negro, Blanco, Magdalena, and the Ipurupuro or Machupo. These rivers approach, and facilitate the communication between, the Beni towns of San Joaquin, S. Ramon, Magdalena, Guacaraje, Concepcion de Baurés, and El Cármen. After the confluence of the Iténes and Mamoré, the latter name is preserved until the reunion with the Beni. The river Iténes forms the boundary-line between the empire of Brazil and the republic of Bolivia, as far as the centre of the rapids.

The Beni, which descends from the eastern declivities of the department of La Paz, crosses, as the Mamoré, tracts covered with most surprising vegetation. In the gorges of its Andean slopes, grows the invaluable species of Peruvian bark known as *cascarilla calisaya*, so highly esteemed in Europe and North America. There, also, is cultivated the *coca*, a valuable plant, the use of which should become more general in the world, substituting, with great advantage, tobacco for chewing. The leaf is used for this purpose, possessing, as it does, many commendable qualities and no inconveniences. It supports and stimulates, dispels the tendency to sleep at " wakes," and removes feelings of weariness. In Bolivia and Perú, it is in general use among the indians, and to this is attributed the patience, frugality and hardiness of the people. On the slopes of Cochabamba, the waters of which descend to the Mamoré, there is, also, a large quantity of *cascarilla calisaya,* and coca is cultivated. Both localities produce the finest coffee in America, equal to that of Mocha; and the best kinds of cocoa. The varieties of tobacco are of the first quality with regard to flavour and strength. I do not know the true course of the Beni. I am aware that it is navigable nearly as far as the city of La Paz; that it has as great a volume of water as the Mamoré; that it includes unknown plains of wild cattle; and that, in its most fertile spots, it is peopled by various tribes of uncivilised and nomad indians. This river joins the Mamoré on the left, at the sixth rapid, called the Madeira, after which it takes the last-mentioned name.

The Madeira, formed in the manner indicated, has sufficient depth for navigation, since it does not descend below four fathoms, and generally rises to twenty more. It also has a steady width of three hundred yards, widening to eight hundred and a thousand yards in the wet season. It flows northwards, receiving, on the west, at the rapid of Pederneira, a tributary, the Abuna. With this main direction, but inclining to the east, it continues as far as its confluence with the Amazon, receiving many tributaries on the way, and amongst these the rivers Jammary, Prieto, etc. At this confluence, the Madeira takes the easterly course of the Amazon, passing by a number of towns and villages, such as Serpa, Villa Bella, Obidos, Santarem, Gurupa, Breves, etc., which are rapidly growing, in consequence of the impulse to trade given by a steamer which visits those places fortnightly, plying on the Amazon as far as the town of Manáos, and running once a month as far as Nauta, a central port in the north of Perú.

The Amazon, in its long course of 235 leagues from its junction with the Madeira as far as Belem, capital of Pará, receives many large affluents, such as the Saracá, Ramos,

Atumá, Trombeta, Tapajos, Xingu, Tocantins. The ocean-tides are felt as far as Santarem, more than fifty leagues distant from Belem. The Amazon constantly spreads over the illimitable fields included within its basin, forming numberless large and small islands, and indulging in capricious ramifications. At one time it appears as a wide ocean, immense and imposing; at another, it narrows, breaking up into a thousand picturesque channels. Sixty leagues from its sea-mouth, it parts into two great branches, of which the greater lies to the north. These branches flow into the sea together by one common mouth. Along their banks are to be seen plantations of cocoa and coffee; also neat country-houses and the modest habitations of agricultural labourers. Suddenly, beautiful hill-ranges appear, the aspect of which varies, or rather, they are succeeded by immense plains, where numerous herds of horned cattle graze, this, after Obidos, constituting an item of territorial riches. The tributaries furnish the market of Pará with gum elastic, tobacco, sarsaparilla, the clove, chestnut, copaiba oil, etc. It is a pity that the cultivation of these articles and other agricultural productions are not taken proper care of. This remark especially applies to certain products of great value; for instance: gumelastic, copaiba oil, and sarsaparilla. Those fertile plains might be made to produce articles of the highest utility in large quantities for exportation. It always occurs in new countries that there is a want of labourers, where the enticing hope of large profits on some one branch of industry absorbs attention, until commerce and the lapse of time equalise the relative importance of products, and industry is generalised.

The navigation of the Madeira is usually dangerous, as well as difficult and slow, through scarcity of traffic. On the Mamoré, on the banks of the Iténes and the Beni, there are wandering tribes of indians that attack small canoes. Midway along the rapids, I have come across various settlements of the Caripuna Indians. They are the most tractable, but, nevertheless, have not hesitated to commit acts of violence on small parties of travellers. I have seen, among them, well-arranged habitations with spacious and comfortable houses. They dwell in small scattered communities, but subject to a local chief whom they term *capitan*. There is another "capitan," superior to all these chiefs. He is called Tupi,—a serious old man, and polite in his manner towards white-faces. There are among them some cross-breeds with Europeans. Their colour and regular forms are agreeable, but this cannot be said of the indigenous portion, which exhibits little regularity, is somewhat variable, and of a coppery colour. In full dress they wear necklaces, bracelets, and earrings of long large teeth and a capivara-tusk, or fine straw, inserted through the cartilage of the nose. Generally speaking, they do not use any clothing, and when they do, it consists of a large bark-shirt. In this manufacture they trade with the Beni indians that pass by that way. The women, out of decency, wear a wide leaf suspended from the belt. Their hair is cut short on the forehead, and allowed to grow long down the back. The men wear their hair in the same manner. They stain their faces and bodies with a black shiny dye, and present a ridiculous appearance by having their sexual organ tied up with a string. One afternoon, at the rapid of Araras, I enjoyed the society of a small band, the captain of which was called Buchi—a lively and rather sensible young man. I was then able to observe something of the customs of those people. They are clean and frugal in their diet, and work in common under the orders of their leader. Men and women are employed in cultivation or rowing, fishing and the chase are left to the men, while domestic attentions are left to the care of the females. The men, in their language, have but one name. The women take Christian names, such as Maria, Catherine, etc., but they do not appear to cultivate religion. The women are honest in the midst of their simplicity, and the men are hard-working, indolent in the highest degree : they sow and cultivate only what is necessary. They display an interest in matters outside their circle. None of them are warriors, but they always carry their arrows for purposes of hunting or in case of attack. They like European articles, such as glass beads, necklaces, cheap trinkets, needles, scissors, knives, *machetes*,* small looking-glasses, etc. In exchange for these articles they give their services (which are very necessary at the rapids),

* A short sword used for chopping.

yuca,* maize, sugar-cane, fowls, and plantains. They do not know the value of the riches which exist in their mountains; but abandon their treasures to the ravages of time.

Beyond the rapids there are various other tribes, such as the Aráras, Turás, Mundurucus, and Parententins. All these are domesticated and tractable, especially the Mundurucus, who act as the merchants of the Madeira, and make war on the Parententins. These latter are wild. They are the most formidable of the tribes in Bolivia, and appear upon the highlands of the river Jammary and the Prieto, as far as Crato and Bayetas. They live there, almost among the Brazilian rubber-merchants, and I am surprised that the mishaps which, from time to time, occur are not more frequent. They have attacked some traders unawares, and feasted off their roasted bodies. A short time ago they ate two or three unfortunate persons whom they surprised in a canoe. The worst tribes of Bolivia, such as the Sirionós, Guarayos, Tovas, and Mosetenes do not practice this horrible custom, however much they may kill, lay waste, and pillage. I had entertained doubts as to the existence of cannibals until I encountered them here on the Madeira, on the most fertile banks, and almost among civilized people. The most satisfactory point about them is that, as I have been assured, they are white, and differ from the indigenous race. There are some who believe that these persons are the descendents of refugee Jesuits after their expulsion.

I have wandered somewhat from my subject for this merely descriptive account of the localities and rapids, considering that this will aid my purpose of illustrating those regions.

The rapids are nineteen in number, and lie at various distances from one another, within a space of sixty leagues, skirted by a low hill-range which appears to obstruct the course of the waters. Some of the rapids are passable; some do not offer any hindrance; some have facilities for being made more level, and only three, the worst, need heavy work and skill to render them available. I append a table showing their order, distances, nature, and obstacles.

* A root which is ground and used for making bread.

1st Rapid. *Guajará-Merim.*—Short and easy
passage, to the left, without any inconveniences. Distant from the second
rapid.. 1½ leagues.

2nd „ *Guajará-Guassú.*—Larger than the
first, but equally passable, on the same
side. Distance to the third rapid ...... 2 „

3rd „ *Bananeiras.*—Includes three bodies
of successive currents, and is divided
into various branches. The first two are
passed by the left, and the third by the
island in the middle. The current is
strong. Distance to the fourth rapid 1½ „

4th „ *Pão Grande.*—Has two branches.
The second is more impetuous and
dangerous. Passing, by the left, along
the first, one arrives at an island where
it is usual to unload and convey the
canoes empty. Distance to the fifth
rapid.. 1½ „

5th „ *Lages,* or *Jacaré,* may be passed on
the right, but cautiously. Distance to
the sixth rapid.................................... 2

6th „ *Madeira.*—Has two branches, and is
impetuous. The river is divided into
many channels. There is a passage to
the right. Distance to the seventh
rapid... 2 „

7th „ *Misericordia.*—A short but dangerous
rapid. There are many rocks and a
swift current. It is necessary to unload,
and the canoes have to be towed along
the channel to the right. Distance to
the eighth rapid .... 1 „

8th „ *Ribeirão.*—An insuperable fall, wide,
and divided into many branches. The
current forms whirlpools, sweeping
along the left bank towards the fall on

the right. Canoes are unloaded at this bank, and dragged overland for five hundred yards. Distance to the ninth rapid...................................... ....

9th   ,,   *Periquitos.*—Has two notable branches, but it can be easily passed on the right. Distance from the tenth rapid...

10th   ,,   *Aráras.*—There are foamy waves and a swift current, but the traverse may be effected towards either bank. Distance to the eleventh rapid 10½

11th   ,,   *Pederneira.*—This rapid is superable but there is a strong current in consequence of the channel passing between rocks. Distance to the twelfth rapid 3

12th   ,,   *Paredão.*—Has two branches. It is a strong rapid. Unloading is effected to the right, and the canoes are passed empty. Distance from the thirteenth rapid................................ 5½

13th   ,,   *Tres Irmãos.*—A small and easy rapid, which may be passed on the right. Distance to the fourteenth rapid ...... 7½   ,,

14th   ,,   *Girão.*—An insuperable fall. The river is divided into branches that are more or less impetuous. Boats have to be unloaded on the right, and dragged for a mile over an easy road. Distance to the fifteenth rapid 1   ,,

15th   ,,   *Caldeirão do Inferno.*—Has three channels, which may be passed, but with great danger, especially the last, which crosses from a vast rock to an island where *pascana* is found. All boats pass along the left. Some persons prefer to unload. The river is divided into many branches. Distance from the sixteenth rapid..................10   ,,

16th   „   *Morrinhos.*—Has passable canals, and although there are currents and rocks, it is passed on the left. Distance to the seventeenth rapid.................... 3   „

17th   „   *Theotonio.* — Insuperable, and the highest of all. Has two branches. Having passed the first, unloading is effected in the second, on the right, and the canoes are towed overland, over a passable road a quarter of a league in length. Distance from the eighteenth rapid............................. 1   „

18th   „   *Macacos.*—A slight rapid, which may be passed towards either bank. Distance to the nineteenth rapid............ 2   „

19th   „   *San Antonio.*—Is strong, and the river is divided into various branches. The passage is effected through the channel to the right. Some unload ....

---

19 rapids .... 60 leagues.

This account implies that the rapids may be classified in three orders, as follows:—

9 *Rapids*—Guajará-Merim, Guajara-Guassú, Lages, Periquitos, Aráras, Pederneira, Tres Irmãos, Morrinhos, and Macacos, are simply currents of varying velocity, but passable at all seasons. The passages through them are short, and even the worst are accessible. A river steamboat would surmount them.

7 *Rapids*—Bananciras, Paõ Grande, Madeira, Misericordia, Paredão, Caldeirão do Inferno, and San Antonio, are impetuous, have a dangerous current, and are, more or less, obstructed by boulders and rocks. Nevertheless, they have no falls, and are superable by canoes well-manned. In the dry season the rocks appear, and usually offer channels of access. The most difficult portions may be levelled, and none of them extend over any great distance.

3 *Rapids*—Ribeirão, Girão, and Theotonio: are insuperable falls at all seasons; the escarps rise from six to fifteen yards in

height, making a base angle of fifty degrees, and with an impetus proportionate to the immense volume of waters. These three falls are the only ones which require difficult and costly work as well as skill and time.

From this it may be seen that the Madeira is entirely navigable to steamboats, except at the rapids ; and that its three principal branches—the Iténes, Mamoré, and the Beni—are also navigable almost to the cities of Bolivia, namely, La Paz, Cochabamba, and Santa Cruz. In 1854, a large Brazilian vessel, drawing more than eight feet of water, was able to reach the capital of the Beni.

In the rainy season, the Madeira and its three affluents rise considerably, and are available for large craft.

In the dry season, from April to November, its diminution is immense. In order to ascertain the difficulties with scientific accuracy, and to calculate the cost, it is necessary to make explorations at the last-named period, in order not to be deceived by the floods. I refrain from giving my opinion with regard to cost, because it is not possible to arrive at any satisfactory calculation without having made a special voyage, and not having either instruments, resources, or the means of making observations.

However, I will say what I think regarding the systems that may be adopted.

1st.—I consider that there are three ways of facilitating the passage of the rapids, first to break through the crusts and rocks, and opening passages through that channel.

2nd. To canalize inaccessible places.

3rd.—To establish two lines of steamers to run as far as the extremes of the insuperable part of the rapids, and to act in conjunction with a railway along the Madeira.

The first plan is the most perfect, convenient, and profitable ; but I consider it very difficult, costly, and slow. I find that, in the three highest spots, the impetuosity of the waters, the surges and whirlpools, preclude all access.

The second plan is, to my mind, easier, more applicable, and less expensive than the first. Only three channels would be necessary in the three large falls, rendering the others available

by cuttings in the most convenient localities. The east bank,
along which canoes are now dragged, has suitable passages for
these channels, and although the whole of the land in the neighbourhood is of hard stone, there are neither great heights nor
long distances to be traversed. The longest canal would not be
more than one mile and a half ; the next one mile; and the
shortest, from eight hundred to one thousand yards. There is
an advantage in these falls, as well as in all the rapids, that extensive and level sand-banks intervene, which may be used for
carrying off the waters. By this means the short passages in the
river would be employed, and canals would only be cut where
unavoidable. Further, this would secure the advantage of continuing the navigation as far as the ports of Bolivia, avoiding
stations for loading and unloading, and consequently offering
greater convenience and economy for freight.

The third plan, without offering the advantages of the others,
would be the easiest and most economical. The space of sixty
leagues extending between the rapids, inclusive of the windings
of the river, would leave a shorter distance to be spanned by the
railway. The distances from the first rapids must also be
counted. They are passable, perhaps, as far as Misericordia up-
stream, and Macacos and San Antonio down stream ; since only
from there would the railway begin. On the other side, a slight
elevation of the hill-range offers an easy surface and abundance
of wood, which is almost as hard as iron. Thus the cost would
be greatly reduced.

Whichever of these plans is adopted, it is necessary to bear
in mind the abundance of labourers the towns of the Beni
afford, the small wages that these people demand, the scanty
supply of food that they require, their experience and hardiness
in work. This will render any of the works easier to execute,
and more economical.

In 1845, a great company, in order to carry out this enterprise, was started in the city of Cochabamba, by Mr. Charles
Bridoux, a Frenchman.

Its promoters collected more than five hundred (500) signatures of Bolivian subscribers. Reasons which I do not propose to
explain prevented the realisation of the project. Now-a-days, that

we possess more information about the Madeira River, now that steamboats ply on the waters of the Amazon, and commercial life and its benefits are beginning to be appreciated along its banks, and in this city, now that Bolivia, on its part, needs more imperatively an industrial action proportionate to its wealth, requirements, and the age,—I do not doubt that the dormant enthusiasm of the year 1845 will prove itself more energetic, and that the efforts of the people of Bolivia and Brazil, under the protection of their Governments, may achieve the desired success. With this object I have started a company in Bolivia, and a subscription-list has already been opened.

Approximate distances between the principal places.

From Santa Cruz, Cochabamba, and La Paz to the various
   eastern points, the distances of which have been ascertained,
   there are, by land   ...   .   40 leagues.
From Cochabamba and Santa Cruz, as far as the
   capital of the Beni, since these distances are almost
   equal ; that is to say, from its ports to Trinidad,
   there are   50
From Exaltacion to the junction of the Iténes...   34
From this junction to the first rapid   26
From the first rapid to the junction of the Beni $8\frac{1}{2}$
From that junction to San Antonio, the last rapid $51\frac{1}{2}$,
From San Antonio to Borba   60
From Borba to the junction with the Amazon   20   ,
From this junction to Serpa   5   ,,
From Serpa to Belem, the capital of Pará   ...   230   ,

                                       525   ,,

In the course of the Beni, I calculate a distance very nearly equivalent.

This route, or fluvial highway from the principal cities of Bolivia is now traversed in a most wonderful manner, being beset, as it is, with a thousand difficulties and imminent perils ; while it requires seventy days for the descent, and more than six months for the ascent. Directly steam communication is established the route will be a convenient one,—easy, economical, short, and occupying no more than twelve days to the capital of Trinidad, and twenty-two days to La Paz, Cochabamba, and Santa Cruz, including the stoppages of the steamer. What a difference between the two means of communication!

The explanations and data, which I have given with all the conciseness that a small publication like the present necessitates, confirm the justly-famous wealth of the regions watered by the Madeira, and prove its navigability. Brazil and Bolivia, so interested in this work, must shortly either push it forward, so as to meet the requirements of those two countries, and thus act up to the spirit of the age, or they must remain in an unheeding and indolent state of apathy. In the first case, they will fulfil a duty which will yield them abundant returns ; in the second, the evils, the calamities and their consequences will afflict both nations.

The present epoch is not one of expectation and irresolution. The impulse of steam pervades all. The age requires, sees, and executes. America, especially, with the vigour and exuberance of its wealth, its youth, and its climate, needs, for its existence development, locomotion, and commerce. The universal impression of advancement which permeates all corners of the earth seeks our illimitable regions for their development and expansion. The magnetic needle of human progress points towards us, as the rudder directs a vessel. Meanwhile, what are men doing ? I have seen many in my country who have devoted themselves to objects of caprice, personal negotiations of convenience, ostentation and luxury. Why do they not, for once, at least, dedicate those wasted talents to a work possessing such advantage to our country and the world ?

Brazil, notwithstanding that her political condition has been spoken against in certain publications, has, though she has not noised the fact abroad, fourteen lines of steamers which ply in nearly all parts of the country. The Government expends 4,500,000 dollars in fostering this line of communication and of commerce. Who will doubt its co-operation and efforts for the navigation of the Madeira, when it has already determined to make an exploration upon that river? With the eminently patriotic and progressional genius of the present emperor, Dom Pedro II., it cannot be doubted that the enterprise will be patronised and protected. What is now lacking is that Bolivia should make a cessation of her political squabbles, and, whatever party be in power, lend her assistance, negotiate a treaty of commerce, and make effective its right over the Madeira, as master of the sources which form this river.*

Commerce will revive the dormant riches which are now valueless for want of an exit; demand will inspire new vigour to labour, and, in view of these blessings, dissents will be subdued, and politics will be stripped of its direful attractions. Henceforth no Bolivian will fail to recognise the reality and importance of the said navigability. All, like myself, will lift up their voices, and every year, every month, every day lost will seem a crime of erring humanity, because commerce is a benefit to the whole of mankind.

The Brazilian Empire has placed in communication and contact the immense line of circumvallation, which, from the central highlands of Mato Grosso, extends by the Paraná and the La Plata, back along the Atlantic, and up the Amazon to the interior of Perú, as far as the port of Nauta and even Laguna. How little it lacks to complete the circuit of its boundaries? The *fabulous Madeira*, whose providential waters penetrate the most auriferous regions in the world, in the Brazilian province of Mato Grosso, in Chiquitos, Chayanta, Choquecamata, Tipuani, Chuquiaguillo, Los Cajones, etc., in Bolivia; descending in wide torrents from the heart of that republic, enriching its margins with the most valuable products, offering, in its enchanted valleys, the most poetical and picturesque localities, allowing the world to use the navigation of its waters, seems, by its roaring

---

* In the year 1853, the Bolivian Government offered a prize of 10,000 dollars to the first steamer that, *via* the River Plate or Amazon, should arrive at any of the free ports which had been declared available for the purpose in its waters, and to allot, from the riverine lands, plots of from one to a dozen leagues to the individuals who should have made either of the two voyages.

surges, to protest against the negligence of man. If, on the one hand, heavy expenditure and work are necessary, on the other, the realization of this plan would confer very great advantages. Moreover, the progress and prosperity of the nation—what are they not worth ?

How fair would the Empire of Brazil appear on the map of the world if it could show that all its districts were connected by navigation, conveying to its widely-extended limits an exchange of produce. And Bolivia, a central country, destined to be shut out to the eastward of the Andes chain, and endowed by Providence with the products and climates of all countries— what life and impulse will be achieved the day when, obedient to Nature, she fulfils her mercantile necessity and opens the roadway of its waters! Bolivia has no choice; she must go forward, and the navigation of the Madeira is essential to her existence. Brazil cannot fear; she has to follow the impulse of the age, and be swayed by her duty and convenience. The immunities declared on the La Plata, demand from it equal rights for the Amazon.

I rejoice if, by this feeble attempt of my patriotism, I bring about the *navigation of the Madeira*. Then would I bless my forced exile ; my privations and my sufferings as a man, and as a father. I would consider myself amply repaid.

(Signed) QUINTIN QUEVEDO.

Belem do Pará, April, 1861.

NEW FLUVIAL OUTLET

FOR

BOLIVIA.

BY SEÑOR YGNACIO ARAUZ.

TRANSLATED FROM SPANISH

BY

JAS. WM. BARRY,

SECRETARY OF THE MADEIRA AND MAMORE RAILWAY

COMPANY, LIMITED.

THE MOST EXCELLENT,

COLONEL QUINTIN QUEVEDO,

ENVOY EXTRAORDINARY AND MINISTER PLENIPOTENTIARY

OF BOLIVIA,

AT THE COURT OF BRAZIL AND THE

REPUBLICS OF LA PLATA.

———

You, Sir, who, descending from the slopes of the Andes by clear sportive currents that subside into the placid waters of the Mamoré, and, afterwards, having been whirled along by the impetuous torrents of the Madeira as far as the plateaux of the Amazon, were the first to adopt a new line of communication with the outer world, a route until then looked upon with dismay. Deign, Sir, to accept the badly-written lines that I annex, and which I have the honour of dedicating to you, requesting that you will pardon their imperfection, and accept a work, the sole merit of which exists in the importance of its subject.

Bolivia, crushed and desolated to-day by the affliction of civil war, will add to your history new insignia of glory and greater titles of gratitude, if, interposing the diplomatic character in which you represent her, and even by your personal influence, you bring about that the industrial world shall not lose sight of our Amazonian water-system, whereby they might bring us commerce and prosperity, receiving, also, a just remuneration for benefits so conferred.

Receive, Sir, the most sincere demonstrations of respect of

Your humble servant,

YGNACIO ARAUZ.

Manáos, 24th February, 1868,

NEW FLUVIAL OUTLET FOR BOLIVIA.

THE Republic of Bolivia, centrally situated, as it is, and intersected by various offshoots from the Andes, contains—although
the fact is unknown to the industrial world—the most precious
portion of its territory within its orient. Unless the unfor-
tunately-placed inhabitants obtain benefits offered by commercial enterprise, they will remain deprived of the refining
advantages of progress in all the various branches of social life.

On the one hand, while Europe cultivates ideas of exalted
progress, facilitates means of communication in all directions,
as far as its most remote confines, while its ingenuity invents
means of convenience for the most trifling wishes, and tills its
almost sterile soil, so as to produce more than might be expected from the nature of the locality; on the other hand, the
people in the eastern part of Bolivia, exuberantly rich though it
is, in all the kingdoms of nature, are obliged, in order to learn
the complicated machinery of a steamer and the simple appa-
ratus of a sail, to cross the mountain-chain of the Andes, and
traverse the wide and arid desert of Atacama, exposed to suc-
cumb through thirst, and suffering the rigour of inclement
weather, in a march of more than 320 leagues, as far as Cobija.
This is the real obstacle to the progress of Bolivia.

We, Bolivians, owe to our liberator both our country and
liberty ; but we, undoubtedly, do not owe him the benefits of a
just territorial demarcation.

The department of Santa Cruz, with its province of Chiquitos,
and that of the Beni, situated in a vast plain, the limit of
which, northwards, is yet unknown, but which extends southwards as far as 22 deg., while the longitude extends between 60

deg. and 71 deg. from the meridian of Paris. There is only a population of 200,000 souls, who participate in the same benignity of climate, and, consequently, they enjoy the same productions.

The inhabitants of the two last, which make the third part of this number, are indigenous, whose missions the Jesuits founded in the seventeenth century.

This plain is immensely fertile, and can yield all the productions of intertropical countries, where the severity of winter is unknown, where perpetual spring prevails, and where the gentle and fragrant showers freshen the air and moisten the earth ; so that, without hard work, the dibble of the agriculturist makes the hole that receives the seed, the products of which he will collect in abundance a few months afterwards.

Two-thirds of the waters of Bolivia flow through this beautiful plain into all the rivers that form the Madeira, to which they give considerable importance, seeing that it flows into the Amazon.

One of the principal branches of industry in the Department of Santa Cruz is the sugar which is produced for the consumption of the whole Republic; and the year that the crop is moderately abundant, the price goes down to less than one peso the arroba. This important production is raised by only a few cultivators, who are solicitous that their number should not be augmented, lest a greater quantity should be produced than is necessary for the national use, and a loss would, consequently, be incurred on all surplus stock. Who will deny that the cause of this evil is the lack of means of exportation?

Chiquitos,* besides the productions common in Eastern Bolivia, produces white wax, which is employed efficaciously in surgery, is also indispensable in the tributes to our religious culture, and which is cultivated, in a niggardly manner, for the same reason as sugar, in the capital.

In the town of San Javier, of this province, there are found gulches of gold, which the indians wash easily in trays, and obtain sufficient to purchase themselves the simple tools which

* A Province belonging to the Department of Santa Cruz.

they require for their work, and for which they pay exhorbitant prices, their clothes, and the earnestly-desired rum. From this source they manage to obtain the two pesos of personal tribute, which they pay the Government annually. The epoch of oblation having arrived, they retire on Monday to a distance of two leagues from the town, and on Saturday they return, so as to deliver, on Sunday, the two *adarmes* * of gold, which make up that value, bringing, at the same time, a surplus which they devote to feasting for one or two days. Let it be noted, moreover, that, during the expedition, they employ half the day in hunting, in order to sustain themselves.

It will be easily understood that the indolent life that these wild children of nature lead is owing to the abundance wherewith nature has supplied them.

The indian who has a *chaco* † of 100 yards square, sown with banana, has this fruit, without any further operation than to gather some daily, for the whole of his life, and even for his descendants ; and all that devolves on him is the necessity of occupying himself one week twice in the year in sowing equal portions with maize, mandioca, and occasionally rice, which they use but little, but which can be produced, like everything else, in superabundance. As regards meat, besides being the possessors of horned cattle that rapidly multiply, they have abundant chase, birds and quadrupeds in great variety, which they dexterously secure with the arrow.

It is excusable to speak of the exuberance with which coffee and cotton are produced, as well in this province as in its capital, Santa Cruz; this last-named having, moreover, a trade in the valuable tanned hides, and the most skilful manufacture of pure cigars, which, by reason of the excellent quality of the tobacco and its efficacious preparation, rival the best from Havana. Both articles are exported on a large scale to the shores of the Pacific; but, in consequence of the costly transport, they are little utilised by speculators. There are also manufactured *jipijapa* ‡ hats, with perfect skill and consummate taste.

---

* *Adarme* = half a drachm (16th part of an ounce) ; called in Brazil *roçado.*
† *Chaco* is an indian word, and means a *clearing.*
‡ Chilian hats in Brazil.

The Department of the Beni, or the ancient Province of Mójos, the emporium of natural wealth, situated on the north-east frontier of Brazil, is, now-a-days, the most important part of this valley, the most desirable advantages of which are accessible to the hand that attempts to develop them. Steam, the powerful centre of creation, would convert the Beni, in less than five years, into an important commercial seat. Such are its elements. And what colossal fortunes would, at the same time, be amassed!

All the waters of Bolivia that, *via* the Amazon, find their way to the sea have destined this pleasant valley as their point of meeting.

The river Grande, which rises near Cochabamba,* and directs its course towards the south-east, in order to receive the waters in the north of Sucre,† afterwards wends towards Santa Cruz, increasing its bulk immensely in the transit. It passes at ten leagues to the east of this city, with a steady north-easterly course.

The river Piray, which rises in the heights of Samaipata,‡ passes to the east, and, at less than a league from the same town, takes the same course as the river Grande, and assembles the waters of the west and north from the province of El Cercado; so that, at 32 leagues, where now exists the port of Cuatro Ojos, where, at present, Mójos has a traffic with Santa Cruz by canoes, the river is already navigable by suitable steamers. At 30 leagues from this port it unites with the river Grande, and takes the name of the Sara. Twenty leagues further down it receives, from the left, the river Yapacani, larger than the Piray, and which is also formed of various streams from the towns of El Cercado, and is navigable from Santa Rosa.||

The Coni rises in the heights of Cochabamba, and descends to the plain in a navigable body, since at eight leagues from the Cordillera lies the port of the Chaparé, which offers communication between Mójos and Cochabamba, this port being distant only 36 leagues from the capital of the last-named province; but, through the unpleasant nature of the road, this march on mule-

* Cochabamba, capital city of the Department of this name.
† Sucre, capital city of the Republic.
‡ Samaipata, a town belonging to Santa Cruz, situated to the west of the *Cuesta de Petacas* (Leather Trunk Hill).
|| Santa Rosa, a town situated at thirty leagues from Santa Cruz,

back becomes long and tedious. When the hand of industrious man shall have changed it from the condition in which it is by nature, it will be short and convenient.

Twelve leagues below this port, the river San Mateo flows in from the left. It has its head-waters in the valley of the same city of Cochabamba. Sixty leagues farther, it unites with the Chimoré, which contributes a double volume of water, and is formed of the river Surutú, which rises in the hill-chain of Petacas. It passes by the town of Buena Vista, of the province of El Cercado de Santa Cruz, and, receiving the river Pampa-Grande, called Rio Negro, comes parallel to the Coni. So close is it, that, from the port of the Chaparé to the mission called Chimoré, which is upon the river of this name, there are only two leagues.

It is very noteworthy that Mójos has its communication with Cochabamba by way of the Coni, which is smaller, and at its ebb exhibits some small cascades, and not by the Chimoré, the main channel of which is perfectly well known, and does not offer impediment.

The rivers Sara and Chimoré then unite to form the Mamoré, which is the chief fluvial channel of Bolivia, with a steady depth of more than 15 metres, and a width of more than 400, presenting to view either the most leafy woods imaginable, or beautiful plateaux, rife with the most picturesque perspective, where the man who for the first time visits these solitary plains, however unimpressionable he may be, must, at least, contemplate with sensations of wonder the magnitude of nature.

What a marked difference exists between so vast a longitude and latitude, which comprises the eastern part of Bolivia, inhabited by only 200,000 souls, and Europe, where a handbreadth of land has its value in gold; between the benignity of the climate which offers the constant spring-tide of that level tract, and the desolatory winter of Europe!

I regret profoundly my inability to furnish better data to the public respecting the riches which now lie hidden from the gaze of the world of enterprise.

Following the Mamoré, then, it will be found that, twenty leagues from its origin, it receives, on the left, the river Sécure, which rises in the province lying north of the Department of Cochabamba, and is almost equal to the former.* Facing this confluence, and on the right bank of the Mamoré, at a distance of seven leagues, is the first town of Mójos—Loreto.

Twelve leagues farther down, and at two from the river margin, lies the town of Trinidad, capital of the department. On the same right bank, at five and seven leagues, are found San Javier and San Pedro, standing as far from the river as the capital does. Descending eight leagues from this town, the river Tijamuchi flows in on the left. This offers communication with the town of San Ignacio, which is situated on this side, and at 20 leagues from the Mamoré. At twelve leagues farther, on the same side, we find the river Apere, which has its head-waters in the mountain-range towards Cochabamba. Four leagues onwards we have the river Yacuma, which has its head-waters at four leagues from the river Beni, the waters of which, not being able to flow into this by reason of the intervening hills, take a fixed direction from west to east.

It would be very feasible to unite both rivers by means of this canal; this will, no doubt, be done some day. By means of it the other towns can be communicated with, such as Reyes, which is situated at seven leagues from the river Beni, and Santa Ana, which is only distant four leagues from the Mamoré. Proceeding down the Mamoré, twelve leagues lower, lies the town of Exaltacion, which is built on the river-margin.

All the towns named, except Reyes, produce cacáo for the whole Republic, and would be capable of supplying all America if they devoted themselves, I will not say to cultivating, but merely to sowing it; since from the junction of the Chimoré with the Sára, which, as I have said, forms the Mamoré, as far almost as its confluence with the Amazon, there exist along the banks groves of wild cocoa-trees, the abundant fruit of which is positively neglected. Such indifference is not to be attributed solely to the indolent nature of the people, but to the plentiful supply. The average annual crop of cocoa in the Department

---

* This river was explored by the scientific French traveller, M. D'Orbigny, who embarked in the vicinity of Cochabamba.

of the Beni, is from twenty to twenty-five thousand *arrobas ;*
in which case the indian proprietors sell this product to export-
ers, at from ten to twelve *reals*† the arroba. If the crop reaches
thirty thousand arrobas, they sell it at one peso (eight reals) per
arroba; and, following this proportion, if the crop is doubled,
the price descends to one-half. This circumstance is exceedingly
disagreeable to those who see their labour repaid with only half
its value.

Besides the eight towns already mentioned, there are San
Joaquin, San Ramon, Magdalena, San José, Concepcion de
Baurés, and El Cármen, scattered over the immense and level
tract forming the right bank of the Mamoré. All these places
are advantageously situated upon branches of this river. The
actual productions of these towns are coffee, cotton, and tobacco
of the best quality; the first is sparingly cultivated, for the
same reason we have mentioned in the case of the cocoa, and
the cotton is limited to the demand for the manufacture of
hammocks, towels, ponchos (cloaks), and church-apparel common
to the country, which are exported to Santa Cruz, and the ponchos
as far as the Argentine Republic. The means which they use
for working the cotton are of a primitive nature. The cotton
is combed out by hand, and spun on a spindle after the same
fashion, so that an expert workman, working steadily, spins one
ounce per day, so that in sixteen days one pound of thread will
be prepared, sufficient to make one short pair of trousers, which
they weave in a frame. This takes two weeks. Afterwards they
are washed, pressed, and sold for twelve reals. When they are
woven in a loom the operation is shorter, but the value is also
considered inferior. In the neighbourhood of Concepcion de
Baurés a rich gold-mine has been discovered, in the hill of San
Simon, from which, so says tradition, the Jesuits derived a large
quantity of this metal. Some companies are now occupied in
placing machinery for working this mine, which experienced
foreigners assert to be richer than California ; and if to this
circumstance is added the large number of available labourers,
who receive the small pay of half a real per day, and the abun-

---

* An *arroba* is twenty-five pounds English weight.
† One *real* is worth nearly fivepence English money.

dance of provisions, the prices of which are in proportion to the pay, it will be seen that a favourable result cannot for a moment be doubted.

The Amazon valley, which I have just described, is pastural as well as agricultural, and possesses, in an eminent degree, facilities for the farmer. The existence of wild horned cattle, that propagate rapidly, would be enough to make the fortunes of thousands of inhabitants who might colonize these vast districts. These cattle belong to the State, and by its order any one can have a herd of two hundred head allotted to him, provided he intends to devote himself to pastural pursuits, at the cost of six pesos per year. In accordance with this law, more than two hundred allotments have been granted.

The abundance of the best kinds of building and cabinet woods, rubber, ipecacuanha, jalap, the herb *maté*,* indigo, which grows wild, constitutes a real source of wealth. The culture of the sugar-cane deserves admiration. The cultivation of maize (corn), rice, mandioca (a root from which flour is made), pine-apple, water-melon, and a variety of vegetables and other nutritive products can be easily effected; and it scarcely ever happens that the crops prove a failure. The gold and diamond gulches at Santo Corazon de Chiquitos, the immense agricultural and pastural lands, are riches which Bolivia offers to the world in exchange for colonization and industry.

Twenty leagues farther down from Exaltacion there is the mouth of the Guaporé, or Iténes, which, descending from Mato-Grosso, is enriched by all the waters of the province of Chiquitos, whence it rolls majestically onwards to join the Mamoré. This beautiful river, which waters the slopes of the hill of San Simon, will, doubtless, be the nucleus of the emigration that will be drawn thither by the presence of the rich gold-mine. Proceeding thirty leagues down the double volume of the Mamoré, Guajará-Merim is reached. It is a small rapid, and presents no difficulty, since the channel to the left would easily allow a steamer of large draught to pass. At this spot ipecacuanha, a highly valuable medicinal root, is found in abundance.

---

* In the town of Buena Vista, twenty leagues from Santa Cruz.

Two leagues more, and the first part of the rapid of Guajará-Guassú is reached. This is the point from which a lateral road will most probably be made to branch off, and thus friendly relations between Bolivia and this vast Empire will be established. Descending the rapids Bananeiras, Pão Grande, and Lages, at the distance of five leagues, the Beni flows in on the left. This river rises in the Department of La Paz; and, in the form of a stream of but little importance, flows through the city of this name. Afterwards, however, being enriched by the numerous torrents which descend from the Cordillera (mountain-range), as well as several rivers flowing down from Perú, among which may be reckoned the Madre de Dios, which, descending from the Department of Cuzco,* and known among the Indians under the name of Manu, flows into the Beni, forty leagues before its confluence with the Madeira.

It is probable that the Madre de-Dios offers free navigation from Cuzco, as far as its junction with the Beni; since, in 1861, it was explored, on behalf of the Peruvian Government, by a Señor Maldonado, who fell a victim to his patriotic daring at the rapid of the Madeira, called Caldeirão do Inferno, only two of the crew being with difficulty saved. It is much to be regretted that such an important expedition failed, since in the wreck were lost the documents which should give information regarding this new route, which interests Perú as well as Bolivia and Brazil. From what can be gathered from the ignorant boatmen, we learn that there are numerous fierce hordes of wild indians along the whole extent of the Madre de Dios.

I regret to be obliged to acknowledge that the river Beni, although the most convenient channel whereby the north of Bolivia might communicate directly with the world, has remained unknown in the most important part of its course, namely, from its confluence with the Madeira for about sixty leagues, as far as the mission of the Cabinas Indians. From here upwards, the country contains many of these missions, which, entrusted to the spiritual and temporal care of European friars, are making admirable progress.

In 1846, the Government of the Republic, in the belief that

* Cuzco, capital city of the Department of this name in Perú.

it was conferring an inestimable benefit on the Department of the Beni, ordered the exploration of the unknown portion of this important river. The task was entrusted to incompetent hands, the only result being that much money was expended, and the discovery made that there were merely three rapids below the confluence of the Madre de Dios, and that these channels are available for steamers of medium draught. According to information obtained from the Chacobos,* there is, along its banks, an abundance of rubber, ipecacuanha, sarsaparilla, copaiba, vanilla, and Peruvian bark (quina calisaya); and the missionaries assert that these missions are situated over rich lodes of various metals. A treaty, made between Brazil and Bolivia, has been announced, the bases of which are amity, commercial limits, navigation, and extradition; and it is to be hoped that this treaty, adjusted, no doubt, with clear foresight and mutual good feeling, will establish firm relations between both countries, so that, instead of leaving sanguinary strifes to our posterity, Bolivia will imperishably prolong the memory of the illustrious Minister, His Excellency General Melgarejo, and confer ceaseless praises on His Excellency Lopes Neto.

* A wild but docile tribe, inhabiting the angle between the rivers Beni and Madeira.

VOYAGE

CITY OF THE GRAM PARÁ

MOUTH OF THE RIVER MADEIRA

BY THE EXPEDITION WHICH ASCENDED THIS RIVER
TO THE

MINES OF MATO GROSSO,

BY SPECIAL ORDER OF

HIS FAITHFUL MAJESTY,

IN THE YEAR 1749;

WRITTEN BY

JOSÉ GONSALVES DA FONSECA,

IN THE SAME YEAR.

———

THE ROYAL ACADEMY OF SCIENCES OF LISBON.

1826.

———

[*Translated from Portuguese by* JAMES WILLIAM BARRY, *Secretary of the
Madeira and Mamoré Railway Company, Limited.*]

A VOYAGE

FROM

THE CITY OF GRAM PARÁ

AS FAR AS THE

MOUTH OF THE RIVER MADEIRA.

THE canoes of his Majesty started on the 14th of July from the port of the City of Gram Pará, with the intention of making a voyage on the River Amazon, and from this to enter its affluent, the Madeira, along the south bank, and to make for the military stations of Mato Grosso, in accordance with the orders of our Lord the King. .

The start was effected on the day named, taking the direction of the river named Mojú, which flows in at two hours' journey from Pará. Aided by a favourable tide, the river Mojú was entered, explored, with a S.S.W. course, shifting afterwards to the S.W. The tide beginning to run down, a pause was made for the next up-tide, with which the voyage was continued with a W.S.W. course, as far as the mouth of the Igarapémerim, where we arrived at about two o'clock on the afternoon of the 15th.

This Igarapémerim traverses the delta between the river Mojú and the Tocantins. The entrance to this valley is so narrow that it hardly affords passage to a large canoe, and continues thus for the distance of half a league, when it begins to open out into beautiful basins, the banks of which are inhabited by settlers who own plantations of *farinha* (a cereal from which a kind of flour is obtained), sugar-cane, and cacáo; and the river proceeds as far as a bay named Marapatá, which is at the mouth of the river Tocantins. Entering the said Igarapé-merim we navigated to the W.N.W., and, with this direction, arrived at the bay already mentioned.

There is also another route from the city to this bay; namely, to pass between the large island of Joannes, which lies to the west, and the coast of the delta already alluded to. But this way, although shorter, is rather unsafe on account of sand-banks and rocks, and being without shelter from storms which generally occur in the evening, and render this coast very dangerous. In order to navigate in this vicinity it is necessary to employ natives of the country as pilots. Nevertheless, the indians, in the neighbouring towns of Murtigura and Sumaúma, navigate it at all hours in small canoes; since, if the sea swells, they lighten their canoe and proceed on their way, and if they cannot do this, they always save their lives by swimming.

Having arrived at the bay of Marapatá, it is necessary to cross to the land opposite, in order to find the islands lying between that of Joannes and the mainland to the eastward. However, as it was requisite to obtain indians from the village of Parejó, adjoining the town of Camutá, we coasted along the bay to the left, ascending the Tocantins which we navigated, steering south-south-west, and then south, which is the true course. After sailing for eight hours, we entered some channels that form numerous islands, amongst which we journeyed for six hours. We then struck across for the town of Camutá, occupying three hours in crossing from the east bank to the west.

This town is situated in 2 deg. 40 min. south latitude, on the margin of the Tocantins, on a site but slightly elevated to the westward, where there is a plain of sufficient extent to build a better town than the present one, which consists of a small street of unpretending houses, and only two of them covered with tiles. There is a church, but it is a plain and poorly-built edifice. There is also a hospital of Mercenarios, which is not a better specimen of architecture than the church, but it affords greater facility for divine worship.

The inhabitants live on their cleared plantations, on which they grow the mandioca-plant, cacáo, and tobacco. They likewise manufacture a large quantity of oil, which they usually extract from a species of chestnut which they gather on the islands. This oil is used in their lamps, and is called *andiroba*. It is manufactured throughout the state. The air of the river Tocantins is salubrious; the views are delightful; the waters of the river exceedingly clear and pleasant-tasted,—they, however, produce, in those accustomed to imbibe them, the terrible disease known as "gravel," since they possess an extremely subtle power of petrifaction. The river abounds with fish, and the banks and islands with every kind of game, while the natives enjoy that fertility which nature spontaneously offers to them; and they might increase the utility of the products obtainable, if they cultivated the land with care. The soil on either margin is admirably adapted for any phase of agricultural industry.

The river Tocantins rises in a vast plateau, which extends from east to west across Brazil; and in the vicinity of its headwaters are to be found the mines of Goiaz, Maranhão, San Felis, Natividade, Corixas, and others, from all of which there are streams flowing into the Tocantins.

From the said town of Camutá, we went in search of the village Parejó, which, at a distance of nearly one league, is situated on the same tract of land towards the north. We embarked some indians from this village. They are held in the highest estimation as rowers. We navigated among the islands with a northerly course, and made for the Igarapé do Limoeiro, which is a portion of water similar in character to the Igarapémerim above mentioned. The canoes took this route, thus avoiding a journey along the bank, this being dangerous work, as at Mortigura. We entered this Igarapé with a north-easterly course. It has an outlet fronting the island of Marajó. Between the latter and the mainland, where the said Igarapé discharges itself, there is a vast number of islands, lying in different directions, while channels and bays divide one from the other. Along these canoes pass, seeking to regain their right course; steering northwards for the island of Joannes, or turning eastward for the village of Araticú. The journey to this island

—on which the said village is situated—was accomplished by the canoes of the escort in twelve hours, sailing and rowing.

Thence we turned northwards to the large island of Joannes, which we approached from the east, where that island borders on the district of Amazonas. We made observations, and found ourselves to be in 2 deg. 12 min. south latitude. We afterwards turned our course west and west-south-west, to the village of Aricurú, which is situated towards the west, and we arrived there after ten hours' sail. From this village, sailing south-west and south-south-west, we arrived, in three hours, at Aracura, a village lying to the east, in the district of Camutá, on the mainland. All these villages are missioned by fathers of the " Companhia de Jesus," and were formerly inhabited by indians called Ingahibas. They were catechised in the forests by the very celebrated Father Antonio Vieira, a member of the same religious order. Descendants of this tribe still remain. The navigation between the villages is all carried on among the islands, and in order to form an idea of the vast sea in which they are found it is necessary to be informed that the mighty river Amazon, sweeping down towards the island of Joannes, becomes divided into two branches : one, the greater, flows northwards to the ocean between the said large island and the mainland of the Cabo do Norte ; the smaller one, turning southward, mingles its waters with the rivers Xingú and Pacajaz. These waters, being thus united, form a large number of channels, bays, and estuaries, between the islands referred to ; and, passing between the mainland to the eastward and the coast of Joannes towards the west, they flow into the ocean between the said island and the district of Pará, carrying with them, at the same time, the waters of the Tocantins, Mojú, Acará, and others already mentioned. The number of islands to be found in all this space—if only a third part of them be reckoned habitable— would make an archipelago, more celebrated than any of the four so famous in the world's history. But nearly all the islands are inundated in the rainy season, by the overflowing of the rivers, and, during the whole of that period, they are uninhabitable. Nevertheless, there is in them an abundance of game and fauna, but all the flora is wild and nearly useless. However, the navigation of its canals and bays is puzzling, since the forest continues throughout the labyrinth of waters. Moreover, the boats are subject to great risks from sand-banks and rock-shoals in the said canals, except during the prevalence of thunder-storms, when the tide rises considerably.

In the villages in question, which are generally known as Bócas, the indians required as rowers for the canoes were promptly obtained, and straightway the journey was continued to Tajupurú, in a north-easterly direction. The Tajupurú comprises a number of islands, disposed in such a way that, in consequence of the smaller branch of the Amazon, combined with the waters of the Xingú, it forms various rivers, which all empty themselves into the bays and estuaries already mentioned, which were navigated in search of the villages of Bócas, as we have stated above.

We arrived at the principal mouth of the Tajupurú, and entered it with a north-north-easterly course. We ascended it, sailing and rowing, for two days, steering north for the most part of the time; and, ultimately taking the west-south-west, made for the western frontier, to some islands which lie almost north and south. Between these and the islands of Tajupurú there can be little more than a league in distance. Here the water of the Xingú flow in from the south and the smaller arm of the Amazon—which they join towards the north, being afterwards divided among the islands which form the above-mentioned Tajupurú, at the entrance of which the observation was made—is situated in 55 min. south latitude.

Three roads can be made from this part—one northwards, to Macapá and the Cabo do Norte, and round the headland to Caena, a French colony; another, returning from the west, might be made to enter the Amazon, where there is a greater abundance of water and a tolerable amount of danger; the third road would be southwards, to the fortified station of Gurupá. This is the route generally adopted by canoes that ascend into the interior.

The escort followed this route, coasting along the mainland to the left, steering south-south-west and west-south-west, following the natural indentations. After twelve hours' journey, with sail and oar, we reached the military cantonment of Santo Antonio do Gurupá, situated on the mainland referred to, in 1 deg. 46 min. south latitude. This place is the reporting-station for canoes, whether on the up or down journey to or from the interior. At this outpost there is a chief captain as commandant, provided by his Majesty. The garrison consists of eighteen men and a captain, and other subordinate officers. It takes eleven days to reach this station from Pará, which is, approximately, at a distance of ninety-one leagues.

About a cannon-shot from the said station there is a village of indians who are used for garrison work. This place is missioned by the Capuchin Friars of the Province de Piedade. There was a tolerable number of indians located there, but repeated attacks of small-pox and measles have swept down many persons of both sexes and all ages.

Navigating from the mouth of the Tajupurú towards Gurupá, the above-mentioned islands lie to the right, stretching from north to south, partly taking the directions of the rivers which traverse the mainland to the left, between which and the said islands there is a beautiful river of more than a league wide, formed by the water proceeding from the river Xingú, which flows into the Amazon among the islands in question. They are situated between the two branches, occupying more than two leagues in width, and are admired by all who visit the mouth of the Xingú, as will be gathered from this narrative.

The islands referred to, which are generally called Gurupá, are uninhabitable, because all are subject to inundations in the rainy season, and, notwithstanding so great a drawback, they contain a vast abundance of wild game, such as *pacas, corias,* deer, boars, etc. They contain, also, a vast quantity of cacáo-trees, from which the inhabitants, not only of those districts, but also of Pará, gather fruit during the season ; and, were it not for the mischievous practice of stripping the trees before the fruit has become ripe, the crop that might be collected from these islands would, in a plentiful year, be sufficient to load several canoes.

Starting from Gurupá, we sailed to the west-south-west, between the coast and the islands, and, after ten hours' sail, we

arrived at the village of Aarapejó, which stands on the mainland. It is situated on a plateau at the summit of a high bank.
This village is missioned by the Capuchins of Piedade. The
population consists of 300 persons of both sexes. We then
visited another village, at which we arrived after a voyage of
a little more than two hours, with the same course. It is
situated on the same continent as the one already alluded to,
and is called Cavianá. This place is also under the administration of the Capuchins. The population consists of 200 persons
of both sexes.

From this village we continued our voyage, with the same
course of west-south-west, between the mainland, on the left,
and the islands, which continue on the right. The height of the
locality gives command of a wide view of waters, which form
several bays of almost two leagues across, extending from portions
of the islands to the mainland. On the latter there is a settlement, called Boavista,* consisting of some seven or eight inhabitants. The locality is so pleasant and the surroundings so pretty,
that the name has been well bestowed. After five hours' journey
we reached the village named Muturú, situated on the same
margin as the aforesaid places, on a plain at the elevation of the
water-level, near the mouth of the Xingú. The missionaries are
of the Capuchin order of Piedade. Population consists of 400
persons of both sexes.

The mouth of the river presents a beautiful and pleasant
view ; latitude, 2 deg. 7 min. south. Among the great rivers which
are collateral with the Amazon, the Xingú holds a prominent
place, by reason of the vast body of water that it contributes.
The best proofs of its importance are its width, depth, and the
rapidity of its current. It flows from south to north. The
source of this river is not yet accurately known, although at only
the distance of a few days' journey. The reason of this is that
the ascent is impeded by a number of formidable *cachociras*,†
which deter the people of Pará from making the attempt, not-

---

* NOTE BY TRANSLATOR.—*Boavista* signifies *beautiful view.*
† NOTE BY TRANSLATOR.—*Cachociras* is an indian term for *rapids* and *falls* in
a river.

withstanding that several foreigners—who are located in the village with the Jesuit missionaries—venture to do so in the season, and they annually gather a large quantity of cloves along the banks, with which commodity a trade is carried on. However, the position of the outlet, and the direction in which the river flows, give reason to believe that, between the sources of the rivers Tocantins and Tapajos, which are parallel to it, rise the rivers Bacatis and Mortes, which, uniting, form an immense volume of water at its mouth ; and, between the two mentioned parallels, there is no greater outlet than that of the Xingú. It is probable that the Bocatis and Mortes have their headwaters in the same locality, to the north of the Cordilheira Geral, as the Tocantins and Tapajos, which are all tributaries of the Amazon. The waters of this river Xingú appear on the surface to be black. On being struck with the oar they exhibit very small crystals, which break away. Such is its clearness that, from its banks, one can see whatever is in it to the depth of a fathom.

It may be inferred from this narrative that the Amazon has not been properly navigated to this place, on account of the large number of islands separating its waters from those of the Xingú. With this object the voyage here described was undertaken.

From the said village of Muturú the canoes started southwards; then, turning westward, we made for the bank, to the place where a river named Akeky flows in. It is a tributary of the Amazon, which, flowing in by two mouths in the mainland, shortly afterwards blends its waters with those of the Xingú, near its outlet.

We sailed up the Akeky, steering west, west-north-west, and north-west. At a distance of four hours' journey a river called Jarauku flows in. It runs parallel with the Xingú, but is far smaller. Its whole course lies among hill-ranges, except near its outlet, where the land becomeslevel. There are some cachoeiras in this river. It abounds with fish, and along its banks, as far as the cachoeiras, clove-trees are plentiful.

After passing the mouth of this river we continued our journey, steering in the same direction, and, wearied with travel, we halted for the night, and experienced the first onslaught of an innumerable host of mosquitoes, which, like torrents of rain, pour down upon the navigators. Although we did not suffer so much as the Egyptians in the time of stubborn Pharaoh, these insects gave lots of opportunities for the exercise of patience, so that the boatmen, not being able to endure their attacks, we resolved to continue the journey, since it was not a perilous one. At seven o'clock the following morning, after twenty-one hours' journey altogether, we arrived, with a northerly course, at one of the estuaries formed by the branch of the Amazon already referred to, and then we beheld the mighty Amazon in all its vast grandeur.

Facing this outlet, towards the north, we observed a hill-range. On the slope of the hills stood the fortress of Parú, after which the hill-range and coast are named, as far as Macapá, on the road to the Cabo do Norte. The chain of hills referred to runs in this direction, that is to say, west-south-west. This name also extends further northwards.

These hills along the west bank of the Amazon appear to be of a formidable size. They stretch along, with here and there a valley interposed, even more sterile than the hills themselves, seeing that the latter contain a tolerable amount of parsley and cacáo, while the former do not even afford a sufficiency of water for the use of persons from Pará, who assemble there during the season of gathering those products. There is a wide-spread rumour that gold is to be found in these hills; but no regular or definite examination has been made. From these hills proceed the rivers which discharge themselves along the coast of Macapá, such as the Jari, Tueré, Vramucú, Cutipurú, Aranari, and others, towards the interior, from the Cabo do Norte. This range also gives rise to some rivers beyond that cape, namely, Vicente Pinson, Japoco, and others of less importance, to the east and west of Cape Orange, on the coast of Caiena.

Starting from the mouth of the Akeky to the Amazon— which, with justice, can hardly be called a river—one has to coast along the eastern part, which lies on the left when on a west-south-westerly course, and the hills on the opposite strip of land are in sight the whole way. After eight hours' journey we arrived at a place which the natives call Mauari-ajura-pára, being named after a channel which the river forms between the mainland and an island towards the south, and which can be navigated with the same course. From this channel the Amazon branches out, passing some land in the midst of that portion of water that, joined to the said Akeky, forms the river of that name, which affords an entrance to the Amazon by the mouth of the Xingú. The journey on the Amazon, as far as the mouth of the Mauari, occupied fourteen hours, during which, by sail and oar, an approximate distance of ten leagues had been traversed, after making the necessary deductions on account of the current.

In continuing the journey from Mauari it is advisable to start early. The mornings are generally clear, a very important feature, by reason of the swampy shallows, near which, even with an ordinary wind, the water is exceedingly boisterous, though the hindrance is greater to those who are navigating downstream, since, in that case, the wind proves contrary. For this reason, in order to avoid any unfortunate accident, a start is made soon after midnight, and the coast is skirted, while rowing with a west-south-westerly course, until daybreak, and after that the sails are set. After eight hours' journey an island is arrived at, lying in the same direction, and extending further to the north-east. We, therefore, coasted along this island, in order to cross the Amazon. This we effected after four hours' delay.

The canoes having been transported to the other side of the Amazon, we entered the river Vrubu-quara, in search of a village of that name, where, after three hours' journey, we found a harbour. This river is an offshoot from the Amazon to the north-west. For a distance of six hours' journey the river extends itself into various lakes while traversing a wide plateau along the slope of some hills which form part of the Cordilheira do Parú. These hills, deflecting from the river-margin, continue towards the interior in three branches. The lakes referred to become so wide and deep in the rainy season that they are navigable for larger vessels than canoes. The river abounds with alligators and water-snakes of great size, called sucurujus, giboias, and boia-ussu. Every variety of fish produced in the Amazon is plentiful here.

The above-named village is situated in 2 deg. 20 min. south latitude, on an elevated plain, the ascent to which is laborious, and occupies half an hour. Half way up, a beautiful spring flows out from a rock. The water is exceedingly clear, pleasant to the taste, and possesses diuretic qualities. From this plateau various hills can be seen towards the north and east; while southwards may be observed groves, lakes, and plains, the whole affording a delightful and entertaining panorama. This village is missioned by the Fathers of Santo Antonio da Provincia da Extremadura. There is a considerable number of indians among the inhabitants; but there was an insufficiency of house-accommodation, in consequence of a recent fire. The houses being constructed of wood and straw, the slightest spark suffices to reduce to ashes, in a few moments, a large habitation of this description.

We occupied fifteen hours, sailing and rowing, in going from the mouth of the Mauari to this village, an approximate distance of ten leagues.

From the village of Vrubu-quara we made for the Amazon, returning along the same route by which we entered; and, starting towards the parent-stream, we kept along the west bank, which lay to our right, with a west-south-westerly course. After fourteen hours' journey we arrived at a village called Gurupa-tuba, the entrance to which is by a channel similar to the foregoing in everything but the extent of the lakes, those of Gurupatuba being far greater than those of Vrubu-quara. If we are to rely on the statement of the indians, there exist in these lakes certain serpents of such an enormous size that, with the exception of the whale, no other known animal in the ocean possesses such bulk.

The village referred to stands on a height in the middle of a plain, and is skirted towards the north and west by trees. The country to the south and east is eminently beautiful, since a view is obtained of all those level tracts wherein the Amazon expands itself to form the above-mentioned lakes. The eye beholds a delightful panorama, consisting of a number of lakes bordered by trees of dense foliage, the horizon to the east being closed by lofty hills, which are sufficiently distant to produce a picturesque effect. Towards the south may be seen a portion of the Amazon sweeping throughout the whole extent of the vast plain, which seems surrounded, as if by art, with that crystalline adornment which Nature spontaneously lavishes in order to render beautiful the whole of that most pleasant country.

This village is populated by tribes of indians who are justly celebrated for their adaptability to all kinds of work. These indians are even gifted with a curious talent for making various ornaments for the use of the people, more especially painted vases, (called *cuia*), washed with a resin distilled from trees by way of varnish, which resists every kind of liquor that may be boiled in it. In order to better understand the nature of this manufacture, it is necessary to say that *cuia* is a kind of fruit like a round water-melon *(melancia globosa)*, but in quality resembling the calabash, but the husk is harder, and the interior more compact. It is the produce of a tree called *cuieira*, which is not large nor bushy. It is like the pomegranate in the ramification of the branches and the shape of the leaves. The *cuias* are gathered of various sizes. These are split diametrically, and the contents extracted until the interior is made as smooth as the convex surface outside, the husk being left of about the thickness of a piastre (pataca). These husks, when well dried, are ornamented with various paintings on both sides, and washed over with a kind of gum-varnish, which is distilled from the trees of the country. This preparation polishes and fixes paintings, so that they are used for ordinary purposes by the indians and other inhabitants.

After a stoppage of twenty-four hours at this village, which is missioned by the Fathers of the Piedade, and after furnishing ourselves with indians and provisions, we started in the canoes by night towards the Amazon, so that by dawn we might commence the navigation of the coast called Cuieiras, which is rather dangerous because it affords no shelter in case of thunder-storms, and any high wind would drive the canoes on shore, so that they might easily become a total wreck. About four o'clock in the morning we began to coast, steering south-west and south; and, after eight hours' journey, rowing, we rested for the space of one hour. After continuing to navigate for seven hours with the same points, the canoes arrived at an igarapé, or valley, the only shelter which there is in the whole of that distance. This igarapé is called Cuiciras. We proceeded twelve leagues that day.

On the following day, at daybreak, we left this place, and, proceeding to the south-west, arrived, after six hours' journey, sailing and rowing, at the islands which stretch across the river. By passing from one to another of these a passage might be effected across the river without danger, so as to enter the mouth of the river Tapajós, which we reached about five o'clock in the afternoon, and moored the canoes alongside the fort. During eleven hours' journey, we went nine leagues altogether, including four expended in crossing. From the fort of Gurupá to the Tapajós we traversed in eight days an approximate distance of seventy-five leagues.

The Tapajós is rather a large river, and is a league wide at the mouth. It flows from south to north, running parallel to the Xing. At a distance of five days' journey from its mouth there is a large number of rapids, which are very difficult to traverse when going up stream. The locality is said to be somewhat unhealthy, in consequence of the climate, and the exhalations from the water; but it has not been properly substantiated whether the cause of the many ailments which are ordinarily contracted by those who proceed thither to collect produce arises from impure air or the proximity of the rapids. Near its headwaters, it is well known there are extensive prairies, which are called Paracizes, a name given to a neighbouring tribe, which to-day is extinct, through having on one occasion put to flight and destroyed, with extraordinary inhumanity, a people called Cuiabá, which, in repeated hordes, continued to make inroads into those wide plateaux in search of gold.

These plains extend along the slope of the Cordilheira Geral, which runs from east to west between 14 deg. and 16 deg. south latitude. From this range descend the rivers Juina and Jezuena, the sources of which are in the same meridians as the Aporé and Jahur. These rivers lie on the south side of the chain, and further eastward rise the rivers Arinos, Sumidouro, and

Preto. These united form a body which, combined with that of the Juina and Jezuena, constitutes the Tapajós. It has been ascertained that the source of the Arinos is so near that of the river Cuiabá, that only three leagues along a level tract intervene.

At the entrance of the Tapajós, towards the left, upon a rocky height, stands a fortress, regularly constructed, and quadrangular in form. It has a garrison of soldiers, commanded by a captain and a lieutenant. This fortification could never be useful to command the passage of the Amazon, since between the river and the work there are several islands, among which canoes might navigate without any danger, and evade the vigilance (if it were exercised) of the above-mentioned garrison.

The only purpose for which it could be convenient would be for the missionaries in the villages on the Tapajós. These are five in number, visited by the Fathers of the Companhia de Jesus. The waters of this river are even clearer than those of the Xingú, so that if a glass be filled with some of this water, its presence cannot be detected. At a depth of almost two fathoms may be seen the sand and pebbles along the margin. Both banks abound with clove-trees. The gathering of this product is attended with danger, because the neighbourhood is infested with a number of wild indians. The mouth of the Tapajós lies in 3 deg. 4 min. south latitude.

Starting from the Tapajós on the ninth of September, steering north-west, we crossed over, amongst some islands, to the Amazon, reaching which we stood along the left bank, steering west and west-south-west; and, after traversing eight leagues in fourteen hours, passed the night on the point of an island, where we waited for the dawn in order to continue our journey.

On the west bank of the Amazon, which we had been sailing along with the last-named course, there is a lake closely resembling those of Gurupatuba and Vrubuquara—called also Surubiú, already mentioned. The Amazon empties itself through two mouths : through one branch it receives the waters which descend to that plateau from the hills, and, by means of the other, it receives the watery tribute of the various affluents which form this mighty river. The lake referred to contains several islands, as well as an abundance of fish. On the right margin stands a village named after the lake, and missioned by Capuchin Fathers of the Provincia da Piedade. The indians resident in this village enjoy considerable fertility of soil. This results not merely from the proximity of the lake, but from the fact that the wild horned cattle of the prairies have been utilised for purposes of irrigation during the last few years.

We started, sailing, next morning very early, and the flotilla coasted along to the left under sail, favoured by a strong gale, which lasted for two hours. The result was that the Amazon was broken up into waves of even greater size than those of the ocean, because the latter are borne along by the breeze, while the former encounter the resistance of the current, the effect produced being fatal to canoes going down-stream, whereas those that ascend regard it as favourable. We had the precaution to stand out amid-stream, because, when going in this direction, similar storms blow aft and from the east ; and, also, we avoided the bank to get out of the way of the vast quantities of logs which are scattered near the brink, in the bays, whither they are borne by the currents from the rivers Madeira and Vaiale, where cedars of an incredible size grow. With this gale we steered to the west-north-west, and afterwards to the north-west. We proceeded thus for ten hours, during which we traversed seven leagues. We halted at this bank for the night, within view of the fort of Panais, intending to cross over to it on the following day.

From the spot where the Amazon first becomes navigable for large canoes—which is from Jaen de Bracamoros, as far as the Cabo do Norte—there are only two narrow passages—the first, immediately before the point where it receives the river S. Tiago, as far as the town of Borja, penetrating a rocky ridge, through which it cleaves a passage merely twenty-five yards (Spanish) for a distance of three leagues. This distance, by reason of the furious rapidity of the waters, is traversed in a quarter of an hour, with considerable alarm and danger. This strait is called by the Spaniards Pongo, which, in the language of the Mainas indians, who inhabit that district, signifies " a gate," and this pass is the most dangerous throughout the whole of this vast river. The other strait is in the place of which we are now treating, in which is built the fort of Pauxiz. The Amazon is here of an astonishing width, since its volume is increased by many large rivers, which flow collaterally with it on both sides. When it reaches the hills of Pauxiz, such is the disposition of the ground in the locality that the portion which has to be traversed is converted into a sluggish abyss, so that, notwithstanding a depth of over three hundred fathoms, for the distance of almost a league, no current is perceptible. In this passage it is remarkably strong. The banks are one thousand fathoms apart. This channel, compared with the expanse above and below it, appears like a Bosphorus, for it forms a narrow link between two sea-like portions of the river.

The time now arrived to cross over to the fort. It was already broad daylight, and we proceeded along the east bank for a distance sufficient to allow for the swiftness of the current. We then turned inwards, and, bending to our oars with a north-easterly course, brought to, in a quarter of an hour, by the opposite bank, near the landing-stage for the said fort.

This is situated on a slight declivity, in latitude 2 deg. 40 min. south. It is built of stockade-work, irregular in form, in addition to which the walls are breached in the angles, and the curtain skirting the river needs considerable repair. This fort is supposed to command the passage ; but, through the elevation of the site, the guns of the artillery would be no more able to prevent contraband traffic than in the case of the castle of Almada, should the defence of the Tagus be entrusted to it. Smuggling-craft could navigate on the Amazon with security. This is the only height which stretches across the river sufficiently to serve as a key to the navigation, not only for the purpose of suppressing any attempt to transgress the laws, but also to serve as a stronghold in case of foreign invasion.

The garrison of this fort consists of a captain, one lieutenant, a sergeant, and a company of soldiers, which is rarely complete. This detachment is furnished from the troops quartered at Pará.

The hill-range, on which the above-mentioned fort stands, is distant rather more than a musket-shot from the water-side, and, with this interval, follows the course of the Amazon as far as the mouth of the river Trombetas, already alluded to, a distance of three leagues. On the mainland-heights is a small indian village, missioned by the Capuchin fathers of the Provincia da Piedade. They have succeeded in converting some of the indians employed at the fort. Not long ago a number of these escaped to the wilds of the Trombetas, after having committed various enormities and crimes, for which they deserved punishment.

It was necessary for the flotilla to stop two days at this fort, waiting for a supply of provisions which had been sent for to the village of Surubiú, obtaining which, we left that port on the 15th of September, about five o'clock in the afternoon; and coasting to the right, with a west-south-westerly course, we arrived in seven hours at the mouth of the river Trombetas, a distance of three leagues. Here we halted.

This river flows to the north-east, which is probably its main direction. It is not a large river, but abounds in *cachoeiras.* Hordes of wild indians people its banks. With these there is but slight communication because of the difficulty in ascending the river. In the interior there is an abundance of excellent specimens of the finer kinds of wood, especially that named *borapenima,* which is capable of receiving as high and permanent a polish as tortoiseshell. The action of water tends to greatly enhance the lustre.

Very early on the morning of the 16th, we continued our journey, coasting to the west-south-west and south-west; and, after twelve hours' journey, during which we travelled six leagues, we moored off a sandbank in the middle of the Amazon; the only one available for those navigating up stream at low water. From this point upwards, in the dry season, other such banks are observable, which, at the lowest level of the ebb, are of tremendous extent.

On these sandbanks turtles are taken very plentifully, when, leaving the water, they sally forth to deposit their eggs in the sand in such vast numbers that they cover a large portion of the localities. Persons not only from Pará, but also from all parts of the Amazon, encamp here with a double object in view: first, to catch turtles for their use; and second, to collect from the sand the buried eggs, from which an immense quantity of butter is manufactured and used throughout the whole of that vast country. This article is used as a substitute in all cases where butter or oil is employed in Europe.

The turtle-gathering is most extensively effected every year in the month of October; and it is credible that if this destruction were not made both upon the Amazon and its tributaries, so numerous would these animals become that they would not only render that vast continent more fertile, but, in some parts, they would render navigation more difficult than might be imagined.

On the 17th we sailed along the same west bank, with a west-south-westerly course, for rather more than an hour. We then struck across for the east bank, passing close by some islands which lie extended in the river, having their diameters parallel to its course. During that portion of the day which still remained we coasted to the left, sailing in the same direction we have mentioned, and to the south-west. In eight hours we had proceeded five leagues with difficulty, because there was no wind, and the current was strong.

In this latitude, on the west bank, is a lake called Jamundas, closely resembling those of Gurupatuba, Vrubuquara, etc., to which we have already drawn attention. At the mouth of this lake there is a formidable whirlpool; and it is necessary to exercise great caution in passing to the village, also called Jamundaz, on the margin of the lake. This place is missioned by the Capuchin fathers of Piedade. This lake communicates with the mouth of the river Trombetas by an arm which is navigable, but not frequently used, through fear of falling into the said abyss, called by the natives *caldeirão* (cauldron), and Jupiaz by the inland inhabitants of Minas.

On the following day we continued our journey, coasting to the left, sailing to the south-west, south, and again to the south-west, until we came to our proper course of west-south-west; and, after journeying for ten hours, we had traversed seven leagues, the wind having been favourable, and by nightfall the flotilla arrived at a hill which rises on the margin of the Amazon to a greater height than that of Pauxis, and for a considerable distance does not present any declivity, rising abruptly to a great height, the gorges abounding in shrubs and thick vegetation, while the summit is crowned with a close forest of virgin growth.

On the 19th, scudding along the same coast with a good wind, with the points of west-south-west, west, and west-north-west, within sight of the peaks which skirt the river on that side, the squadron arrived, in seven hours, during which time we traversed five leagues, at the mouth called Abacaxis. Here the flotilla halted. This mouth of the Abacaxis is more properly a smaller opening of the river Madeira. Since it is a branch which the latter throws off to receive the contribution of various rivers, streams, and a large number of lakes, as it proceeds on its way to the Amazon, into which it falls with a width of four hundred fathoms at its mouth. Along this the flotilla did not pass, because the settlers of the village called Abacaxis which used to be on that branch have removed to the main body of the Madeira. Here it was necessary to resort, and, for this purpose, we were obliged to continue the navigation of the Amazon, in search of the chief mouth of the Madeira.

Thus, on the 20th, coasting to the left in the directions of west-south-west and west-north-west, up to four o'clock in the afternoon, it was then necessary to cross to the west margin, in order to avoid the navigation of the very wide bays which lay along the east. The passage across lies between islands, and the journey is attended with great difficulty, because the channels formed by the river among these islands are not convenient, while the river at that spot is exceedingly wide. After an hour of our transverse journey, with a northerly course, we reached the first island, and, passing through the channel, we stopped at the second, since night was already drawing on. That day, after thirteen hours' journey, we accomplished seven leagues.

At daybreak on the 21st, we continued on our way, proceeding from island to island, until nine o'clock in the morning. We then began to coast along the west bank, steering north-north-west, west-south-west, and, finally, south; and, after ten hours' journey, during which we had gone six leagues, the flotilla drew up in a bay which lies at the mouth of a little river, which flows into the Amazon between two rocky promontories, where the waters are considerably broken.

On the 22nd, continuing the navigation with the same southerly course, we passed shortly afterwards to the south-west and south-south-west, until we arrived among some islands, where we halted, having accomplished seven leagues in twelve hours.

On the 23rd we found it necessary to wait for a canoe which had been delayed, and as we did not wish to lose the time in idleness, orders were given to fish in the several large lakes that exist in these islands. The result was that we obtained an abundance of fish, sufficient to supply the whole escort. The above-mentioned canoe having at length arrived, we proceeded for four hours, steering west and west north-west, during which we traversed a league and a-half.

On the 24th we continued our journey, coasting to the left, from daybreak till noon. Soon afterwards we began to cross, passing between some islands to the eastern margin, to the principal mouth of the river Madeira; and, navigating until nightfall with the points of south and south-west, the flotilla stopped at a large island, opposite the mouth of the said river, at more than a league's distance. This day, having proceeded for fourteen hours, we accomplished eight leagues.

At dawn on the 25th we doubled the point of the island, towards the south; and shortly after daybreak we had the entrance of the river Madeira full in view, towards which we rowed in a south-westerly direction, and with the same course effected our entrance, and continued our journey.

NAVIGATION OF THE RIVER MADEIRA.

Begun on the 25th of September, 1749.

BEFORE entering the river Madeira we stopped, about day-break, on the 25th of September, at a very wide sandbank which stretches from one of the many islands in the river Amazon, fronting the said river Madeira.

Daylight disclosed to us the whole horizon, which terminated towards the east and west with the immense waters of the Amazon, and towards the south-west with those of the Madeira at its mouth, which is eight hundred fathoms in width, flowing into the Amazon between two low promontories, clothed with vegetation like that generally found on the latter river.

From this sandbank we struck across towards the Madeira, about seven o'clock in the morning; and, after an hour's journey, rowing to the south-west, we entered the mouth, where we did not perceive any greater swiftness in the current than that of the Amazon up to this place. Before arriving at the first turn which the river makes, it was necessary to wait until mid-day to make an observation. This was done, and by this means we ascertained the entrance to be in 4 deg. 14 min. southlatitude.

In this first bend of the river we remarked that there was no firm land eastward—that is to say, to the left—or westwards, capable of habitation, because it is all inundated at high tide. It only appears when the limit of low water is reached, when it is covered with an alluvial deposit, for which cause it is called, in the country, Alagadiço (muddy); and the same nature of soil characterises the Amazon to the east and west, where the Madeira empties itself.

Having taken the observation, we continued our journey to the south-west, and presently, having doubled the second bend, we navigated to the south-south-west, which soon proved to be the proper course to take.

Before passing the first bend there is, to the right, a lake which fills at high-water and diminishes at low. In it are fish in abundance, of which fact travellers take advantage. Rounding the second headland, there lies, to the westward, a sandbank, on which an islet is forming; and, on this bank, there is abundance of turtles at the proper season; that is to say, at the new moon in October, when the water is low.

Arriving at the third bend, we observed it to be rocky, this being the nature of the margin to the right, which, although not very high, does not become inundated when the river is flooded. At the entrance to the bay, which commences at this bend to be of the same nature as the river-bank, there is the site of what was a Gentile village, and there exist remains of former habitations, in the shape of fruit-trees, which are there preserved. In the interior there are cocoa-plantations, and these were the origin of a large number of others that are to be found along this river. The locality is inhabited by some settlers, who occupy themselves in curing fish.

Continuing our journey under sail (on this day and the following the usual wind having prevailed), in the directions referred to, we arrived, about seven o'clock in the evening, at a halting-place between an island and the land on the right, called Paranámirim, (that is to say, in the native idiom, *little river*, not because there is really a river there, but that the indians thus name that portion of water dividing the island from the mainland). During this stoppage we tried the experiment of angling, and the scaly wealth of the river was evinced by our securing, in a short time, as much fish as we required for the occasion.

In the seven hours that we travelled this day we traversed four leagues.

On the 26th, about six o'clock in the morning, we began to row in the directions of the south-west and south-south-east, and, after proceeding for a little more than an hour, we crossed over to the left side, by passing through a channel that there is between the mainland and the other island, which stretches almost across the river. In the part directly opposite this island, there is a lake called Do Padre S. Paio.* In it there is an

* This was Father João de S. Paio, a Jesuit.

immense quantity of turtles and various kinds of fish, suitable for curing.

We proceeded on our way towards the right, coasting along a wide bay which communicates with another lake, but of less extent and utility than the preceding. From here there extend some low spur of hills, which are occasionally rocky, as far as the village called Dos Abacaxis, where we arrived about ten o'clock in the evening, having gone two leagues in four hours, which, with the four leagues of the previous day, made a total of six, and this is the distance from the mouth of the river to this first village.

Is is situated on the margin of the bay referred to, fronting an island parallel to the course of the river. This island is subject to inundations, like those already mentioned. The village was located on a branch of the river Madeira, which flows into the river Amazon, with the name of Dos Abacaxis, or Topinambas. From here it was changed to the locality mentioned, because the former site contained a number of lagoons, from which resulted a considerable amount of sickness and mortality among the inhabitants, who, even in their present situation, are not safe from such a calamity, because the settlement stands on a small strip of land between the river and the lake, which, in the dry season, causes illness. For this reason few persons inhabit the village, and the greater portion live in dwellings scattered here and there among the rocks on the main land in the vicinity.

In consequence of the mortality arising not only from the malignity of the climate, but also from the two epidemics of small-pox and measles, which ravaged the country from 1743 up to present year, 1749, less than a third part of the population remains. During the administration of Father João de S. Paio, of the order Da Companhia, there were among the indians more than a thousand warriors and able-bodied men before the two plagues alluded to. The land is sufficiently fertile; turtle and various kinds of fish abound; but *farinha* (flour) is so scarce that the missionaries had to send to the city for whatever was needed in the way of food for themselves and also some indians who did not possess plantations.

It was absolutely necessary to stop two days at this village, in order to obtain fifteen indians to take charge of the large canoes after embarking the staff in the small ones, which have to be constructed before encountering the *cachoeiras*. This was accordingly done in the manner which will be related farther on.

Having procured these indians, about eleven o'clock on the morning of the 28th, we continued our voyage about four o'clock that afternoon with a south-westerly course, coasting to the left; and after rounding the angle of the bay on which the said village is situated we continued to the south-south-west and west-south-west, and last to the south, towards the Abacaxis estuary, where the Madeira falls, by that way, into the Amazon.

About eight o'clock at night the canoes halted, after six hours' journey, during which we had gone two leagues, at the entrance of the northern part, known as the mouth of the Topinambas, notwithstanding that the river Madeira forms the Abacaxis branch, dividing itself into two portions, with an island in the middle, and traversing the mainland as far as the Amazon. Beyond the twenty-three lagoons on both sides, there flows into the said branch a river called Canomá. It flows from the mainland, and is tolerably large. The banks are inhabited by various Gentile tribes, not of a very hostile disposition, but they do not accept the privilege of civilisation, although several amicable attempts have been made to persuade them to do so.

This explanation clearly shows that the river Madeira enters the Amazon by two mouths, making the principal outlet the matrix of the river, and inferior to the Abacaxis, which receives the waters of the Conomá, forming an island of that large piece of land which is washed by the Amazon, the said Madeira, and the Abacaxis. Some inexperienced authors of geographical maps place this large island in the middle of the Amazon, opposite the mouth of the Madeira, giving it the name of Topinambas. The most recent description we have of this locality is from the pen of Monsieur Lacondamine, in his map of the Amazon and its confluents, printed in Amsterdam in the year 1745, an explanation of which was furnished in his diary. This map was printed after he had navigated the Amazon from the province of Quito to Pará; and, notwithstanding that this mathematician did not enter the Madeira, nor examine practically the communication existing with the Amazon by way of the Abacaxis, he has recorded accurate information furnished to him by persons of experience in Pará, who had navigated the entire length of the Madeira; and, had he not made this investigation, he would have fallen into the error commonly made by geographers with regard to this region, as did Condamine himself in the map referred to, with regard to the large island of Joannes, at the mouth of the Amazon, and the size of the islands of Tajupurú, persuading himself, perhaps, that no further water intervened between the mainland to the east and the said island than the channel he traversed in proceeding to Pará, and afterwards along the coast of the Cabo do Norte.

About three o'clock on the morning of the 29th we began to look out for a suitable place to say mass, and at daybreak the canoes halted at another island, larger than the preceding, opposite the mouth of the Topinambas, where there was a splendid sandbank whereon to raise the portable altar. We named that island St. Michael, it being the day of that glorious archangel when we celebrated mass there. This spot is situated at the distance of a league from the northern mouth. We started thence at dawn towards the tract lying southwards, which we reached by the time the sun rose. St. Michael island, as well as the foregoing one, and the mainland formed by the mouth of the Topinambas, are all covered in the rainy season. After celebrating the holy sacrament of mass we continued our journey, sailing and rowing, to the south-west, and afterwards to the west-south-west, coasting to the left the low lands on both banks, and passing three lagoons to the east, we stopped, after travelling for six hours, during which we had proceeded five leagues. During this day we noticed three islands in the middle of the river, which are covered at high water. There is a strong current to the left, at the end of a bay, where the banks are steep and rocky. The canoes halted at the margin of the river, a little above the said current, beyond the promontory of the bay.

On the 30th, we began our journey at six o'clock in the morning, proceeding to the south-west, and soon afterwards to the west-south-west and south-south-west, and subsequently to the south-west, and after doubling a headland which stands out from the middle of a bay, we found a village called

Trocano, opposite to an island lying lengthways to the course of the river; we had rowed for four hours, during which we went two leagues, and came to the conclusion that this village is about nine leagues distant from that of Abacaxis.

From the entrance of the Madeira, as far as the headland to the north of the bay in which the town above mentioned is situated, the river preserves the width of from three hundred and fifty to four hundred fathoms; but after arriving near the said bay, and passing the two islands on the right, the width becomes contracted for the space of half a league, where it extends but little more than one hundred fathoms, until passing the headland of the bay referred to. Here there is a sandbank stretching almost across the river. The eastern bank is rather rocky, and thence as far as the village named the margin attains a slight elevation, thus placing it out of the reach of inundation in the rainy season.

This village of Trocano, established under the auspices of St. Anthony, was built between the river Jamari and the first rapid of the Madeira, and is composed of persons who, in the year 1722, formed part of an expedition to the Madeira under Francisco de Mello Palheta.

The missionary of this settlement was Father João de S. Paio, of the Companhia de Jesus. After passing some years there, and seeing that the spot was not suitable for the health of the indians, while they were also harassed by the neighbouring tribes, he adopted the expedient of removing to the site of the present village of Trocano.

It is built on a plateau, which extends along the barrier of hills at the bay above mentioned, to the east of the Madeira. The breezes are pleasant, and healthier than those of the Abacaxis, and the construction of the village is superior to that of the foregoing. It is missioned by the members of the Companhia, the chief of which was not in the village on the occasion of our visit, he having ascended the river Negro with the intention of preaching to the forest-tribes, so as to win converts for this village, and not only with this object, but also to avoid some disagreement among the indians. We halted the canoes at the sandy shores of an island lying lengthways up-stream, to the right, at the distance of a league and a-half from the village, from which its lower point is visible. We then proceeded in a light canoe to arrange whatever was convenient for the service of the escort.

At less than a day's journey up-stream from this village, there are several habitations of unconverted natives, who have had the boldness to attack the people of the said village. As a precaution against such inconvenience, the missionary lives in a house fortified with stockade-work, so that, with the assistance of two acolytes, who aid him in his duties, a better resistance might be offered in case of an assault. These officials were administering to the religious welfare of the village during the absence of the chief, at the time when the canoes arrived there. So little attention was paid by the population in carrying out the instructions of this missionary to aid the escort with indians, that one of them hid in the forest with the greater part of the indians, and some who were taken from there in order to bring back the large canoes were obtained by tact.

Thus, we could not there obtain a supply of provisions, because even if they raised sufficient for themselves they will have none to sell. This was also the case with the *farinha* (mandioca-meal), which was the most essential article of food.

At about seven o'clock on the morning of the 2nd of October, leaving at the village a light canoe with an official and two soldiers, with a view to purchase some farinha, the canoes started with a fresh wind, in a north-north-westerly direction, and after passing a bend to the right, where there was a large sandbank, we navigated to the south-south-west. After travelling for three hours, during which we went three leagues, we halted at the sandy beach of an island situated to the right, where we awaited the small canoe already mentioned during the remainder of that day and the next, which was the third.

When we steered to the south-west, towards the right of the Madeira, we passed close to the mouth of a river which flows into a small bay called Coaota. This river is not large. The land near its mouth is low enough to be flooded. As we did not enter it, we were not able to examine the direction of its current. Coasting along the same western part in the bay which succeeds, there is found an island with a very wide sand-beach, which extends to the bend of the other bay, in the middle of which, on an island similar to the preceding, we made a halt, mooring the canoe to the bank, as usual.

Facing this island, on the right, is seen the entrance to a lake which abounds in fish. This is the first place on the river where wild indians are to be met with. Although none were to be seen, we took every precaution in case of an attack by them.

The water of the river Madeira, from its entrance as far as this place, is clear and nice to the taste, but from this headland it becomes turbid in parts where the banks are of clay, and where lakes flow in; and only where there are reefs or rocks does it become less faulty in this respect; and, up to this spot, the winds known as *ventos geraes* prevail, but in a diminished form, so that only in the case of thunderstorms is it necessary to have recourse to ply oars against the current.

On the 4th day of October, the little canoe having arrived which was left at the village, without having effected any business of importance, we celebrated mass on the said sandbank, the day being that dedicated to S. Francisco de Assis, out of respect for which we called the island after the name of this saint, and, starting at seven o'clock with the course of south-south-west, we left, on the right, a large sand-beach, and with the course of south, we proceeded in search of two islands which were on the left, on whose margin are numerous cacao-trees, but the land on the bank is washed away beneath, so that it continually falls in with the weight of the enormous trees that are there produced. This passage (and there are many such on both margins of this river) is one of the greatest danger, and the most formidable that can be imagined; and, finally, up to the *cachoeiras* (rapids and falls) we did not, in our expedition, meet with another river of greater importance. The islands referred to, which are situated close by each other, are called in the indian language Carapanátuba; their Portuguese name signifies "land of many mosquitoes." The same name is applied to the lake of which mention has already been made.

Passing the islands referred to in a south-westerly direction —which is the direction of both—coasting to the left, in the middle of a bay there is a great bend of rocks, between which the river runs very rapidly, and in this place, about four o'clock in the afternoon, some uncivilised indians were seen seated on the rocks. Seeing a light canoe in which were a soldier and two indians as escorts, they slunk into the forest to watch, but, a signal being made to acquaint them that more canoes were coming, as soon as these made their appearance the indians in ambush fled into the interior of the forest.

Having passed the currents at the rocks without experiencing any of the mishaps that we might have encountered, the canoes were crossed over to the other side of the river, in search of a channel which there is between the mainland and the island of Jacaré, because there is a strong current to the east, and at high-water the river forms, midstream, a whirlpool (called in the country *Caldeirão*, or cauldron), which it takes a deal of work to avoid; and coasting along the bay on the west, the bank of which is composed of shelving earth, we made for the beach of the island, and there brought up for the night, taking necessary precautions. A lake, also called Jacaré, flows into the channel on the east. On this day, after going for five hours and a half on our way, we accomplished three leagues.

On the 5th, continuing our journey, we coasted to the right, with the points of south and south-south-west, and after three hours' journey by oar, we found an island called José João (it took this name after a native of Pará of that name, who for many years attended to the cultivation of the cacao-tree, with which the island abounds), and, in the immediate vicinity, succeeded another island, smaller, but with a long beach, extending in the direction of the river. We continued to navigate towards the south and south-west, and soon, to the west, a little above the said beach, the water's edge of which is skirted with a strip of sand, we saw a place which appeared to be frequented by wild indians, but we saw none. When one has passed a rocky bend, on the left, where the river is narrower, there is a lake to which the indians give the name of Matámatá. Merely turtles are to be found in it, of which the wild tribes make use. We crossed over to the right side of the river, and when it was almost nightfall we made a halt at a little distance from the bank, as there was no open landing-place or island where we could pass the night in greater security. This day, after eight hours' journey, the greater part under canvas, we went some four leagues. Meanwhile, there was nothing of importance to notice, since the lands are floodable; the trees are all wild along the margins, and do not give any indication of possessing value.

On the 6th we began our journey with the course of east, south-west and south-south-west, coasting the right bank, and, after ten hours' journey, rowing and sailing, we accomplished about six leagues, not observing anything noteworthy on the way.

On the 7th we continued our journey, navigating to the south, coasting to the right; and, after two hours' journey, rowing, we saw, to the left, the river called Aripoaná, and, crossing to examine its mouth, we found that it was approximately eighty fathoms wide, with a direction almost from east to west. The water of this river is clearer and of better taste than that of the Madeira, into which it disembogues in front of a small island, almost oval in shape, lying north and south in the said Madeira, at a slight distance from the bank where the said Aripoaná flows in. The locality is inhabited by some wild tribes, consequently no one has yet navigated to ascertain its source and the nature of the land in the vicinity.

Leaving the mouth of the Aripoaná, we proceeded on our journey in a south-westerly direction, coasting to the left, on which there was some tree-bark eighteen feet long and three feet wide, fastened at the extremities with ipecacuanha-sinnet in such a manner as to form the stern and bow of a boat, leaving in the middle a concavity of little more than two spans. This is the kind of boat used by the wild tribes on the whole river. The skiff we saw must have conveyed to this spot some of these persons, who might have been in the forest busy preparing a dwelling; because, with such boats, the wild indians never leave them moored by the landing-stages, but guard them under water. When these craft are required, the owners dive to unfasten the rope, and then ascend with them to the surface by swimming. This boat we came across would hold four persons, allowing sufficient space for rowing and steering, but this would be in calm weather, because if there should be a swell on the water, however slight, the boat might be swamped. From such accidents these indians extricate themselves admirably by swimming, some bailing the water out of the boat, others taking care of the arrows, which they esteem as the chief articles in their possession.

We continued our journey, steering west in search of an island called Araras, which is seen from the mouth of the Aripoaná, already referred to; and, arriving at it, we found the land to be firm, and inhabited by a tribe of wild indians called Araras, whence that island takes its name. It is formed by the Madeira, an almost equal breadth of water lying on either side, the wider portion being about two hundred fathoms across. This island follows the sinuosities of the river for the space of two leagues, such being its length. We could not judge as to its width with any accuracy.

We passed between the said island and the margin to the left, steering south-west, coasting along a bay, which terminated in a promontory of rocks, where the current ran very swiftly. Before reaching this point a small lake flows in, and there is another, which is also about the same size, after passing the current. We halted the canoes at a short distance from this spot, along the same bank, at a little rocky island a short way off the land. This island is submerged at high water, its elevation being less than that of the bank, which also becomes flooded. On this day, after eight hours' journey, we had traversed a little more than three leagues, because there was no wind.

We started, at 8 a.m., from the little rocky island, steering south-south-east, coasting to the left; and, before arriving at the point of the bay, a lake of small importance flows in. Following the southerly course, after passing a sandbank which there is at the point referred to, we encountered the strongest current which, thus far, we had met with. There being an island* with a wide beach in the middle of the river, close by the point mentioned, the water rushes with great force by reason of the shallowness in all these parts, and it was necessary to take the canoes to the island, and tow them until dinner-time, when we rested.

* On the margin of the right side facing the point of this island, there being one Antonio Correia, an inhabitant of Pará, on his cacáo-plantation, the Muras attacked him there and murdered him and five indian servants, with arrows.

Soon continuing our journey, in a south-south-westerly direction, towing, as before, along the beach of the island, it happened that a light canoe of scouts became separated from the larger boats, and after passing the left side of the bank, some ten or a dozen wild Muras indians rushed down to the brink and fired a volley of arrows at the skiff, without, however, injuring any one. Those in the little canoe having recourse' to arms, the aggressors soon made off, and did not make their appearance the rest of the afternoon, which was passed in towing the canoes as far as the island extended. On it we passed the night, using great vigilance and caution. On this day, after eight hours' journey, we accomplished three leagues, being impeded by the current, which we had to overcome until stopping for the night.

On the 9th, when it was already clear daylight, we started, seeking the left margin, near where the indians had appeared, and coasting to the south-south-west, we found, on a beach there, a nailed arrow, which the pilots told us was a signal of defiance by the indians, who were probably awaiting us on the headland of the bay, where there was a current. We, therefore, proceeded with necessary caution.

Nothing fresh of any importance occurred on this passage; but, on passing round a bend which next presented itself, the two skiffs in the van discerned five canoes* of indians, steering down stream, very near the bank. Without evincing any fear, they began to show fight, but no sooner did they perceive the first large canoe than they leaped to earth with incredible agility, and sank the bark-canoes in which they had been journeying in such a manner as to leave no trace behind.

From the middle of the bay, after having passed the said bend, we crossed to the right, where the bank was higher, and the interior abounds with cacáo-trees. We coasted along, with the course of south-south-west and west-south-west, and lastly to the south-west, until arriving at the island of Mautará, in front of which we halted the canoes at a point of low land, where there was no fear of an attack; nevertheless, we still passed the night with our usual precautions.

* These canoes were of the same kind as already described.

Almost in front of this island a stream flows in, on the left from which the island takes its name, so that both are named Mataurá. There was not an opportunity to examine its mouth, because night came on as we approached, but the pilots say that it is larger than the river Aripoaná, already mentioned on the 7th instant, and that several wild tribes dwell on the Mataurá. The bay into which it flows, near the headland whence the little canoes started, is above the river at its flood, and the soil seems to be of a good quality for tilling, because the trees, besides being high and leafy, were devoid of the usual thickness which prevails in parts inundated by the river. This day, after seven hours and a half of journeying, we accomplished three leagues.

At ten o'clock in the morning, crossing to the right, we navigated with a southerly course, which we soon changed to south-west, the current running rather swiftly. Then, after three hours' journey, we saw, to the left, a rather reddish hill-range, which, in the chief language of the indians, is called Guarapiranga. Extending for a quarter of a league, with its flora of virgin forest, there flows in, at the end of this range, a stream called Matapí. It is not very important. Facing its mouth is an island with a beach stretching from one end to the other. After passing this island, one meets with a reddish hill-range, similar to the one on the left, but less extensive in length. We continued our journey until six o'clock in the afternoon, and the canoes drew up for the night by a projecting portion of the left bank, after having traversed three leagues and a half in nine hours. During the greater part of this time we steered south-west.

On the 11th, at daybreak, we continued our journey, going south and south-west, coasting to the left, when we soon met with a rocky bank, in front of which stretched an island parallel to the direction of the river. Between this and the margin we made our way, so as to avoid the current between the rocks to the left. After the said bank followed a wide bay, skirted by low land, in the middle of which flows a large lake. called Manicoré, and the further promontory of this bay is clothed with a cacao-plantation. In front of this begins a large beach, or sand-bank, which extends from the middle of the river, which it divides into two channels. In the part to the right two small lakes empty themselves, and the banks are wholly composed of fallen land. After coasting along this, we arrived, at seven o'clock in the evening, opposite the mouth of the river Unicoré, which flows in on the left bank, almost in front of the point of an island, at the beach of which we halted the canoes.

This river Unicoré is of a tolerable size; it flows to the south-east, and is inhabited by wild indians.

On this day, after ten hours' journey, we went five leagues under canvas, being assisted by a gale of wind blowing from the stern. Our principal course was south-west.

On the 12th, we continued our journey, coasting to the left, through a channel which the river forms between a shoal and the island named, the beach of which, on the south, spreads out for the space of more than a league, with a direction from south to south-west throughout its extension. After this, trying the course of south-west, we found, at the beginning of another bay, three islands, and, by way of the channels between them, where there was less current, we made our way until four o'clock. We left the island straight astern, and steering south to the bay, we brought up the canoes at a beach, which, from the point of the bay, to the left, extends to the middle of the river. On this day, after nine hours' journey, we went four leagues, principally with a south-west course.

13th.—On this day, we did not go further than the entire length of the bay, steering towards the south and round by the east to the north. Terrific currents exist throughout the bay.

When proceeding with an easterly course, we met with a bank of red earth on the right, and a rocky headland, where the water ran with great fury. This bank extends for a distance of three quarters of a league along the river, and contains chestnut and cacáo-trees. It terminates in a lake called by the indians Capaná. This portion of the continent is the best adapted for the foundation of a town, which we had hitherto met with ; but, besides the land being suitable for cultivation, the creek of the river abounds in fish and game—both bird and beast. The bank on the left is almost entirely of fallen earth, and great quantities of timber had drifted in heaps along the margin.

The canoes were halted, at six o'clock in the evening, at a beach which extends from the further point of the creek on the right, almost to the middle of the river.

On the 14th, we commenced our journey at six o'clock in the morning, coasting to the right, by a channel between the bank and an island, steering south-east, south, south-west, and up to west. At half-a-league from the bay, the mouth of an *igarapé* was seen, which, striking across to the centre, falls into the lake called Capaná, of which mention has been made on the previous day; and, continuing the course of west, about four o'clock in the afternoon, a storm aided us in rowing, and the canoes were brought up, at six o'clock, at a new island. During ten hours' journey we had gone about four leagues.

After this island was another of tolerable extent, and between the mouth which both form, a bank was seen on the left, where Captain João de Barros da Guerra pitched his camp, when, in the year 1719, he was sent, as the commander of a troop, to destroy the tribe of wild Indians' named Torazes that inhabited these districts. They were so daring that they would paddle downstream to the Amazon and plunder the canoes which proceeded upwards from Pará to the cocoa-tract on the Solimões, and kill the crews. The war waged by the said Captain exterminated them.

15th day.—About six o'clock in the morning, we proceeded on our journey with the course of west, by a channel between the land to the right and an island which stretched downstream, and, almost at one of its terminal points on the left margin, a river called Araxiá flows in, on the left bank, apparently coming rom the east. Continuing along this same channel, after four hours' journey we found land more elevated and red, after passing which a lake called Macoapi was seen to flow in, in which there is a serpent of enormous size, called by the Indians in their principal dialect *boya asu*, which signifies in our language " big snake." These Indians relate incredible things of these animals, and we, therefore, draw attention to them as a curious feature.

Having passed the mouth of the lake referred to, at the distance of three hours' journey from the right side, and leaving the said island straight astern, we saw some smoke issuing from the bank, near the brink of the water, and, on the vanguard canoe arriving at the spot whence the evidence of fire issued, some Indians who had lit it were now seen, and no sooner were they observed than they fled, and hid themselves in the thick foliage. We continued on our way without any fresh incident, and at about an hour's journey from there brought up the canoes at the point of an extensive beach which an island near the left bank presented. On this day, in eight hours' journey, we went about three leagues.

16th day.—Proceeding through the channel between the mainland on the right and the island mentioned, we continued still with the course of west for the space of two hours' journey, and afterwards arrived at the point of a bay, in the middle of which (after having passed the island) a rocky promontory, where the river, at high-water, has a very considerable current. On entering the bay which follows, we steered west-south-west, and, after a while, arrived at the general course of south-west, and with this we proceeded on our journey, coasting to the right, where the bank is high and red, and approximately half-a-league distant, lying along the river, but, in the centre, this land spreads out. On it, Captain Francisco de Mello Palheta made a halt when he was sent, in the year 1723, by the Government of Pará to explore this river Madeira; and in this place they built light canoes, in which they proceeded on their way as far as Santa Cruz de los Cajubabas, or, as it is otherwise called, Exaltação, a village situated on the right margin of the river Mamoré, by which we entered, leaving the Madeira to the left, into which the Mamoré flows. Having passed the mainland referred to, where we stopped to dine, we continued our journey, skirting the shelvy bank which followed. It used continually to fall in, and with it some trees. In consequence of the fear caused by this danger, which we could not avoid, as at the other margin there was not sufficient depth for navigating, we finished this work at six o'clock in the evening, when the river offered us an island with its beach, where we drew up the canoes, the passengers taking only the ordinary precaution of keeping one as sentinel to look out for indians. This day, the current being stronger, we went about three leagues in nine hours' journey.

On the 17th day, we began our journey at dawn, coasting along the right margin, with a southerly course, between it and an island which runs in the direction of the river, for a while, and, after half-an-hour's journey, we began to pass another island, still steering south-west. This island on reaching the left bank expands at the various creeks along the river for the distance of five hours' journey. In the smaller channel which the river forms between this island and the land to the east, a river called Vrupuni flows in, the direction of which we could not with certainty ascertain, because the river is full of shoals and the current would not permit the passage of a large canoe. The banks of this river are peopled by a tribe called Mura, and the flames of the fires they cause in this district can be seen by night, and we observed them on that and the following day. Having passed the island referred to, at an hour's journey there follows another much larger, also lying along the coast to the left, and following the indentation of the river, with a main direction of south-west. Almost at the terminal angle of this island, we halted the canoes, after ten hours' journey, during which we had gone four leagues, with the said course, because, at this height, the current is very rapid and almost insuperable to large canoes.

18th day.—On this day, we began our journey, coasting along the island referred to by the principal branch of the river between the island and the right margin, and, during this day's travel, we did not go further than the two creeks, that is about three leagues, by reason of the powerful current. The principal course was south-west. At a beach which extends from the point of the island where it terminated we halted the canoes, and we remained at this beach on the 19th.

As, in this district, the currents were strong and the shoals very numerous, all of which impeded the navigation of large canoes, we expended the whole of the 19th in exploring the forests of the said island and of the land to the east, with the view of finding woods adapted for making light boats, in which we might continue our journey, as much to overcome the obstacles alluded to as to pass the rapids, for we were now approaching that difficult passage on our route.

We did not find upon that island trunks of a capacity suitable for the work aimed at, and only upon the right bank of the main land did we meet with anything that would answer our purpose. But there was the objection of having to make proper arrangements for the construction of canoes upon the main land, where necessarily there was greater risk of attacks from the indians than would have occurred upon the island, where they were not accustomed to molest travellers, as they would not in that case have an easy retreat after making their onslaught. This does not happen on the main land, where their safety is favoured by the nature of the soil and the forests in which they conceal themselves, as we afterwards had occasion to experience.

As we had no other choice, for we could not stop at the land to the right of the river, the level surface of which was, for the time being, free from floods, and as we wished to proceed with the operation of making the new canoes, we transferred the large ones at daybreak, on the 20th day, to that spot in the locality referred to which was found most proper for forming an encampment. For this purpose we reserved a sufficient portion of the indians, and the rest we appointed to cut down the trees that were necessary, in order to make from them the new embarkations. Luckily we found the requisite kinds of wood in the neighbourhood of our halting-place, and, in addition to this advantage, we also found a suitable amount of eatables for our sustenance, not only in the way of game but of fish. Merely in a lake, which we discovered near the encampment, there was such an abundance of turtles, that, besides being sufficient to support the whole party, enough of them might be captured on leaving to last for several days on the voyage.

Having constructed, on this day and the 21st, our camp. surrounded by a trench, stockades, and fascines, as a suitable precaution for any invasion on the part of the indians, we garrisoned it all round at night with three sentries, who were to give the signal of any movement which might occur, and we adopted the military style of making the officials of the escort visit these sentries, so as to avoid any lack of alertness.

Twelve days passed in our encampment without any alarm

of indians, until, at dawn on the 3rd of November, some of our indians being occupied in hollowing out the canoes by burning (a process which up to that time had proved very successful, as there was no wind, which is prejudicial to the work), suddenly they were attacked by members of the Mura tribe, who, aided by the darkness and the thickness of the forest, discharged a number of arrows upon the said indians who flew to arms, and were officered by the sergeant-major, commandant, and adjutant of the troop.

In this advance we were not able to make any response till one of our indians was bold enough to fix an arrow in one of the enemies; this we saw in a part where the glare of the fire-light was shining. The alarm was sounded, and the whole garrison got under arms (leaving a sufficient guard over the canoes which were at the landing-stage), and it happened, when one of the indians rushing out of his hut at the sound of alarm, he was struck by an arrow between the furcula bone and the left side of the neck, of which wound he soon died.

We extended a cordon of defenders outside the trench towards the part attacked, and another by the forest, so that if they should find any of the assailants in the vicinity at daybreak they might punish them for their boldness. This project was not effected, because the Muras, taking charge of their wounded man, who was uttering loud cries, were far from that spot when the day broke. There was not an opportunity on the occasion of this attack to use our guns, because we could not see where to aim. The arrows fell but we could not distinguish the hand that impelled them. After a while the fire of arrows ceased, four shots having been previously discharged in that direction.

In order to carry on the work of construction it was requisite that the sentries at the trench should be doubled, and when it was necessary to go to the forest for ipecacuanha-sinnet, leaves, tow, and other materials, the indians always went escorted, and only on one occasion did the Muras appear while our men were cutting wood, but, being seen, they expected that firearms would be brought to bear upon them, and, giving vent to their cowardice, they retired without making any attempt. Only, by night, they would skulk round the camp, and would not any longer venture near the trench. Nevertheless, most of the canoes were burnt hollow within the gates.

As the indian labourers found it tiresome to do wood-work during the day and keep watch at night, we thought it advisable, after hollowing out five canoes that were required, to change our camp to some more suitable spot, where the night-watches would be less irksome. Carrying out this resolution, we abandoned that place on the 19th day of November, and, after four hours' journey up-stream, we selected a little island with a beach, where, on the 20th, we made arrangements to finish the five canoes. This work we carried on without the least disturbance from our enemies until the 1st day of December, when, taking charge of our new boats, the next morning we despatched the large canoes downstream supplied with provisions, arms, and wood, so as to avoid going on shore to cut it, and we gave orders to the soldiers who steered that they should halt at the village of Abacaxis, and there to await the return of the squadron.

The flotilla continued on its way up-stream on the 3rd day of December, about eight o'clock in the morning, steering west, and soon with the usual course of south-west. We coasted along the part to the left, where the bank almost entirely consists of fallen earth ; and, on the right, a lake flows in which is very abundant in fish. These should be dressed with seasoning, or salted, as they have no flavour ; for this reason, the indians, in their language, call the lake *Lago de Jerupari-pirá*, which signifies *Devil-fish Lake*. It is not known with certainty to what to attribute this extraordinary insipidity.

Night having already come on, we halted the canoes at an island called Santo Antonio ; and, after eight hours' journey we might have gone about three leagues, steering with the points referred to.

On the 4th of December, we continued our journey coasting, on the left, a large creek, the low flooded banks of which stretch down from the reddish hill-range. At the end of the said creek, a lake called Pirá Jacaré flows in, opposite the mouth of which begins an island, extending along the coast to the right, and which reaches to half the succeeding creek. Opposite the point of this creek there was a shoal of rocks, where the current ran very swiftly ; after surmounting which, as it was already night, we brought up the canoes at the bank on the left, after thirteen hours' journey, during which we had gone about four leagues, steering, at first, south ; then, for the greater part of the journey, south-west ; and east when we halted.

On the 5th, we continued our journey, coasting to the left, with the course of south-west and south, along low flooded land. Two lakes flow in opposite, the land about which does not flood. After this creek, there follow three islands, lying to the south-east. As regards coast, we followed the windings of the river, and, passing the creek, we coasted to the right, and entered, while it was already night, a channel between the main-land and an island, and halted for the night at a bank at about 8 p.m., after having travelled five leagues in fifteen hours.

On the 6th day, which happened to be very stormy, with wind and rain, we continued to navigate the said channel with the points of south-west and south, two more islands, besides the fore-going, making their appearance. Issuing from the creek and coasting to the right, we saw a man on the bank preparing to dispatch an arrow on the crew of a canoe, but the musket of a soldier was quicker in dispatching him. We do not know whether he was struck, but one thing is certain—we saw no more of him.

Shortly afterwards followed a shoal of rocks, which we passed with some labour, towing the canoes. This operation finished, as it was almost night, we drew up the canoes at the beach of an island that there was in the middle of the river. On this day, in six hours' journey, we went about two leagues, with the said points.

7th day. We continued our journey coasting, to the left, with the course of south-south-west and south, and, on the right bank, a lake flows in opposite the point of the creek, and, a little further on, a stream also, of slight importance ; and, before arriving at some red hill-ranges, two lakes, likewise in-significant, flow in. Having passed the said ranges, a river called Marani disembogues itself, having a width of (to all appearance) about fifty fathoms at its mouth, and its direction seems to be to the west.

Continuing on our way to the left with the course of east-south-east, we found the land on the bank higher, and, within view of the water, we saw several guard-houses (after the style of sentry-boxes) covered with straw, in which the watchers are accustomed to make notes as to whatever passes along the river, and, when the banks are left bare of floods, they proceed to fish. At the end of this margin, higher up, a lake called Piranhas discharges itself. From this place, even until evening, some islands were seen, between which the river divides itself into various channels in such a manner that the green of the trees represented in the still waters offers to the eyes the most agreeable object which we had hitherto enjoyed. By this time, it being night, we drew up the canoes at the beach of one of the islands, after travelling for a dozen hours, during which we went about five leagues.

On the 8th day of December, being sanctified to the sacred mystery of the most pure conception of Our Lady the Virgin, to whose sovereign protection we had dedicated the camp for the construction of canoes, with common consent, we decided on the morning of this day to merely hear mass, and, having done this, to resume our journey only in the afternoon, with the points of south-east and south, and, after six hours' travelling, we had gone about two leagues.

On the 9th, setting out about three o'clock in the morning, we passed to the left side of the river, steering south. Having passed the first creek, we proceeded to the south-west in the second, where the banks were high and of red earth. From the midst of a ravine gushed a rill of clear sweet water, which pre-cipitated itself as far as the river. At the end of this land a large lake flows in, and opposite to it an island begins, between which and the mainland a narrow channel intervenes, and passing to the largest, which lies to the right, we navigated with the course of south-east and soon of south. It was already dark, and we drew up the canoes at the beach of an island which lay in the middle of the river. On this day, after travelling twelve hours, with the points referred to, we went about four leagues.

On the 10th, starting from the said island with course of

south, which was soon changed for south-west, and after three hours' journey, in which we went about a league, we arrived with this course at the mouth of the river called Gi-paraná, in the language of the indians, which signifies *axe of the sea*. This name was bestowed upon it by the indians from their finding in this river some molluscs similar to oysters, the shells of which served them for cutting new bread.

This river being the largest we found flowing into the Madeira, it was necessary to make some examination in order to ascertain, so far as possible, its direction, size, and the latitude in which it disembogues itself.

This river pays watery tribute to the Madeira between high banks. It divides itself into two branches, forming an island of small width, but of considerable length, and running in the direction of the said Gi-paraná, needs, so they say, two days to pass. The channel to the east has a width at its mouth between the mainland and the point of the island 257 Portuguese yards (*varas*), and that to the west has a width of 177, making together a total of 434, which is the entire width of the estuary of this river. Proceeding up it for an hour, we found that its direction lies to the south-east and east immediately at the mouth. We took the latitude of the mouth, which we found to be 9° south. As regards the source of this river, attention will be drawn to it when we treat of the river Jamari, which is next in succession.

About three o'clock of the same day we left the mouth of the Gi-paraná, and continued on our way with a south-westerly course, coasting along the left bank, where the red hill ranges were the highest we had hitherto met with. It was necessary to cross the river to arrive at the beach of an island near the west bank. When proceeding with this course, we noticed, from the middle of the river, a number of hills which seemed to be a long way off, and they proved to be those where the rapids begin, which we had, of necessity, to encounter. Accordingly, we stopped at the said beach, it being already night, having, in four hours, with the said course, gone two leagues.

On the 11th, of the same month of December, starting from the beach referred to at dawn, with the course of south-west, we navigated to the right, and, after three hours' journey, we found that the bank along which we were coasting was of schistose rock ; and soon after reaching the end of it, a small stream flows in, which was once the habitation of the Torazos, who have removed to the neighbourhood of the Muras, where they wage war on them, of which mention has already been made.

In front of the rocky bank, on the east margin, a lake of small consideration discharges itself. The rest of the day we made way to the west-south-west and west, and brought up the canoes at the west bank, after twelve hours' travel, in which we had gone some four leagues, having encountered formidable currents on that day.

12th day. We began to navigate to the right, with a westerly course, and, after going an hour, we arrived at a rocky bank by which there was a powerful current, so much so that it was necessary to quit it for the left margin, where we proceeded with less effort, and following the course of south, we found flowing in on that side a stream of small importance, and facing its mouth was a large beach on which trees are already growing, so that the place will, in a few years, become an island. Between it and the land on the right we continued our way with the same southerly course, and issuing from the channel we observed a stream flow in, on which was situated the village of Santo Antonio, that is now called Trocano, of which mention has already been made.

This stream is called Aponiao, and, it is said, the lands on it are good for cultivation. It is not large, and flows towards the west. Soon afterwards, we passed four small islands with beaches, between which the river is divided into various branches, and by the chief one, which is to the right, the canoes entered the bay, and, doubling the point of it, we halted the canoes as it was already night. After eleven hours' journey, we had gone four leagues, with the said points of west and south.

On the 13th day, with the course of west, we began to pass between the mainland on the right and an island called Tucunaré, fronting which, on the left, was a bank of red sandstone, and, at the end of this, the lake also called Tucunaré flows

in. Having passed the island of this name, another soon follows, which begins to form in a bank there is in the middle of
the river, where, navigating to the south, was seen the mouth of
the river Jamari, for which we crossed with the course of south-
south-west, and, about ten o'clock in the morning, we halted,
after five hours' journey, during which we went three leagues
without a current.

This river Jamari is reckoned, in Pará, to be more important
than any other of those which flow into the Madeira, and the
reason is that this river possesses a greater abundance of wild
cacáo, which the inhabitants of Pará come every year to gather
at the proper season ; the crews of four or five boats combine, so
that they may in a body resist the attacks of the Muras ; and,
when this junction cannot be effected, the gatherers are exposed
to the cruelty of the indians and the fury of animals.

We entered this river with a south-east course, and afterwards
east, and thus round to the north. We again proceeded to the
south-east, which is probably the right course of this river.
This conjecture was made after two hours' journey, and, proceeding a little further up-stream, we halted the canoes. This river
is not without a strong current. Its waters are crystalline and
pleasant to the palate, especially to those who have been accustomed to the turbid waters of the Madeira, after which those
of the Gi-paraná and now of the Jamari offer an agreeable
change.

The Jamari flows into the Madeira, having a width of two
hundred and forty Portuguese yards (*varas*) at its mouth, which
lies in 9 deg. 20 min. south latitude, as shown by an observation
made on that day with a quadrant, the horizon being good.

On this occasion we made investigations as to the sources of
this river and the Gi-paraná, acting on the information given by
the inhabitants of Mato Grosso, who travelled with the squadron, saying that by the course which the Madeira took to the
west, they presumed that one of these rivers was one which, with
the name of Candeas, has its head waters to the north of those of
the river Galera, which opens into the Aporé, with the difference
that the Candeas has necessarily to cut across from the east to
the west to join the Madeira to the east of the main hill-range

and the west of the same cordilheira. The Galera flows almost from north to south to enter the Aporé. The way to settle this controversy is to navigate the whole river, and the Candeas only to its headwaters. In this doubt, we called in experienced aid, in attentively examining the entire cast margin, until arriving at the other mouth of the river, which would equal that of the two foregoing, so as to give more opponents to the sources of the Candeas. Otherwise, to continue on the Gi-paraná and Jamari is the contingency referred to. At the proper time this matter will be treated with greater individuality and clearness.

14th day. We continued our journey on the same day, starting from the mouth of the Jamari, and, with the course of south-west, at about one o'clock in the afternoon, we coasted to the left, and at four o'clock in the afternoon, when going with a southerly course, a tremendous thunder-storm came upon us, which obliged us to take shelter at the beach of an island lying near the left bank, and there we bivouacked for the night. The canoes went a little more than a league. The reason why we halted nearly a whole day at the Jamari was that, besides the explorations that made, our indians wanted to wash their clothes, as there were no hopes of quickly finding clean water for the purpose.

On the 15th day, at daybreak, starting from the island mentioned, with the course of south-west, we went on our way, coasting to the left, and having passed a small bay, we once more steered south and south-east, between an island and the mainland, the bank of which was of rock, finishing with a reef, on which the current was strong, to avoid which we had to pass to the right; and, with the course of south-west, leaving the rocks astern, we bore once more to the left at midday, and in the middle of a small bay, with firm banks, a stream flows in, in which place, at the entrance, was founded the small village already referred to of Trocano, when first the forest-people descended, near the second rapid, as will subsequently be mentioned. In this place were still found a number of limes, sour oranges, and other fruits which are produced in that place since the time the said habitation was there. Thence, with the course of south-west, we sought a large island, having passed which, we halted the canoes, it being already dark, at the beach of the last headland. On this day, after eight hours' journey, we went three leagues.

16th day. With the course of west, we began our journey coasting to the left. In front of the beach, on the right, a stream of little note flows in, and soon a lake in the same creek. We prosecuted our journey with the course of south-west, in a wide bay, and passing to the right to get out of the very powerful current we proceeded with the course of west, and with it we halted at a beach that there was in the middle of the river, on which there was an immense abundance of turtles, and of these we took a good stock. It was the last opportunity the Madeira offered of this species of fish, since there are no more up-stream, where the rapids begin. On this day, in six hours' travel, we went about two leagues-and-a-half with the points referred to, the principal being south-west.

On the 17th, we began our journey with a south-west course, coasting, to the right, a bay, at the end of which, on turning for another, we encountered a current so violent (arising from a rocky shoal located there) that we could not overcome it by any means. We crossed to the left side, and, passing some sand-shoals, we navigated for a short distance to the west, and soon to the south-west and south along a wide bay, at the end of which, entering another to the south-west, we saw when on this course the first rapid, and with the same course we arrived at our halting-place close by, about five o'clock in the afternoon, having in eight hours' travel gone about three leagues.

The bank of the bay which stretches down to the said rapid is of high land, which begins from a rocky point on the left. It is well timbered, and shows throughout its extent that it is capable of being settled upon and brought under cultivation.

SHORT NOTICE OF THE HILL RANGES

WHENCE PROCEED THE

RAPIDS OF THE RIVER MADEIRA.

As it is necessary in this diary to advert frequently to the hill range whence the rapids proceed, it appears advisable, before entering on a description of them, to offer, in this place, an idea of the extent of those hill-ranges, as much for the reason that they can be plainly seen in places on the journey—either from their vicinity to the river or their great elevation, they are better observed in the centre—as also for the exploration which was made with the courses which the range takes from its origin, as recounted by trustworthy persons who travelled a great part of the wide district which they occupy.

Although some expositors of the Scriptures have raised the question whether, or not, hills and mountains existed before the universal deluge, nevertheless, whether they were a part of the original creation (which is most probable), or they were a safe-guard thrown up by Nature when the world was oppressed with that great inundation of waters, these giants of the earth, many of which soar among the clouds, are objects worthy of being admired as displaying the greatness of the Supreme Architect, God, in his marvellous creatures ; and every one knows pretty well from books of history, ancient and modern, sacred and profane, how much the world is taken up with these prodigies of Nature. Passing from plain-spoken terms to figurative ones, in all geographical maps, whether of the world or of certain countries, the authors always represent the principal hills and mountains, which in divers positions spread over the old and new world. With this assurance, we will here only treat of those which have any bearing on the subject above referred to.

The most reliable historians and exact geographers in describing and delineating the celebrated mountain-chain of the Andes, assign to it the very great extent which is intercepted between the Straits of Magellan and New Spain, traversing the whole of the coast of Chile, Perú, and the Isthmus of Panamá, comprising a vast line of more than a thousand leagues of ground. Its elevation in parts is so excessive that, it is said, the most active birds cannot wing their flight across.

To this wall—raised by Nature as if to defend the land from · the inroads of the Pacific Ocean, and which, by its immense height and profound solitude, cuts off the wealth of gold and silver which form the opulence of the West Indies, and enriches the Castilian Monarchy—there corresponds another range, not less extensive in size nor less abundant in precious metals and the finest stones, with which brilliant adornment the august royal crown of Portugal sparkles. This range fronts the Atlantic, stretching along the coast of Brazil from the Capitania do Seará as far southward as Cape St. Mary, the promontory which terminates the northern peninsula, where the large river La Plata flows into the Atlantic.

The range takes this direction when in 23 deg. south latitude. Close to the south of the town of Santos another chain strikes off from the place called Serra do Mar, which, penetrating the country in various directions, ramifies through the districts of Minas Geráes and Goyazes. In the vicinity of the town of San Paulo, this range begins to fertilise the earth with copious waters, giving (among many others) rise to the large river designated on foreign maps the Paraná, which flowing westward joins itself with the Paraguay, and not without some claim to rival its immense volume of waters. Nevertheless it loses its name in that of the Paraguay, in . . . deg. south latitude. In the same chain rises the famous river San Francisco, which, collecting in itself the waters of many other rivers that also have their sources in this range, flows in a semi-circle northward and falls into the Atlantic between Cape St. Augustin and the town of Bahia, in latitude . . .

Finally, after that, from this range, in the district of Goyazes, descends the celebrated river Tocantins, which is enriched by a

R

large number of streams flowing down the slopes of the same chain, this chain proceeds with the direction of west, and as if to form a boundary of the dominions which pay watery tribute to the ocean, it divides, at the distance of more than two hundred leagues, the waters into great rivers, some of which seek, with a northerly tendency, to terminate their course in the celebrated and mighty river Amazon, and others go southward to form the river La Plata or Paraguay, of which individual notice will be given in a more opportune place.

With the west course referred to, making various semi-circles of mountains; and throwing off many branches to the south (and some also to the north) the range goes widening out as far as the end of the Campos Parcizes, which it leaves on the north, parallel to the sources of the rivers Madeira and Jahurû, whereof more distinct notice will be taken further on; and seeking the margin of the same Madeira, leaving the course of west, it accompanies this river with a direction of west-north-west for the space of more than one hundred and eighty leagues; then striking off to the north-east with the said river for the space of sixty leagues it forms the cachoeiras (rapids and falls), which are described further on, as far as 9 deg. south latitude, where it leaves the cachoeiras and turns off with the course of west, in which direction it is lost to view. It probably unites itself with the mountain-ranges of Perú, which form the mentioned cordilheira of the Andes.

The extent of the unconverted tribes that inhabit, above and below, the rugged slope of these hill-chains, and possess a greater tendency to associate with wild beasts than men, requires particular treatment. It would need a large volume to inform the world about a great part of the unfortunate paganism that still exists. In this diary a short notice of it will, where suitable, be given, as, also, of the abundance of riches which this range has disclosed, and the search for which is still hopefully carried on in parts of the same chain, further to the west, in the district of which we are treating.

This range is called by the name of the Cordilheira das Geráes, or Chapada Grande, by the inhabitants of Cuyubá and Mato Grosso, and we will employ one or other of these names as we proceed with this diary.

CACHOEIRAS OF THE RIVER MADEIRA,

WHICH WE BEGAN TO PASS

ON THE

18th DECEMBER, 1749.

FIRST CACHOEIRA.*

ARRIVING on the 17th of December, about four o'clock in the afternoon, in the vicinity of the cachoeira, called by the indians Aroaya, and by the Portuguese S. João, we sent the most experienced indians, in a canoe, to examine which of the channels the river makes between masses of rocks would be capable of being passed with less danger ; and, by the information that we obtained, it was shown to be advisable to proceed with our canoes along the left margin, and not by the right, because the middle was impracticable ; and, in the same way, it was equally dangerous on the right, since the river had already begun to fill and to rush along that bank with a greater impetuosity of current than on the left. With this view, we undertook, on the following day, the passage, which was accomplished with work, but happy success.

After entering upon the navigation of the river Madeira to this place, the first rapid, we found that both banks were low, and during the flood-month they are inundated for a distance of one or two leagues from the centre on each side according to the quantity of rain that falls in the wet season, which sometimes is more copious than at others. Hence result the numerous lakes which, at low water, the river forms along both

* NOTE BY TRANSLATOR.—Now called San Antonio.

banks, so that, it being rare that the land rises to any great height—the bank never continues high for more than a league —the small portions that rise are around the lakes, and seem to be islands rather than the mainland. However, on approaching the first rapid the high bank corresponds with the centre, and is not subject to inundations, and from this part the hills of the Cordilheira Geral begin when going up stream, and terminate when coming down. These were the same hills that we had seen.

These hills extend themselves along both banks in various directions, and between them the river Madeira makes its way ; and as they are composed of rocky masses, so the current of the river flows swiftly between them, so that at the first rapid the land was observed to make a small bay, to the east, composed of masses of rock, which, traversing the river, form in it two islands : the larger contains lofty trees, and is 200 fathoms from the mainland or the left bank, the smaller, which stands opposed to the current in its midst, is almost bare. On the west begins a wide bay, and on the point opposite to the right there is a similar series of rocks of the same kind and position as that already referred to.

The current of the river bursts through between the two promontories of firm land, offering to view a spectacle equally formidable and pleasant; because by reason of the impetuosity with which the water rushes to overcome the impediments which in parts are opposed to it, it has precipitated two blocks of rock, and issuing between others already torn off, it goes on whirling, dashing, and seething until it subsides into quiet water in the creeks, where in the calm water are reflected the ever-leafy trees on the margins. Everything combined conduces to contemplation and amusement, although enough horrible objects are presented to travellers to cause fear.

At that season, there were three channels to be found in that rapid. By that of the middle no one has yet passed, nor can anyone help infallibly falling a victim in the attempt. By that on the right, in the dry weather, any canoe can go without danger; but directly the river begins to swell, there only remains the channel on the left, which in the dry season has no water in it, and during the entire period of high water is the best for navigating. At the season of our visit—the beginning of the first rains—there was enough water in to pass the canoes. This we did on the 18th, after performing mass in memory of Mrs. O., whose birthday was celebrated by the church.

Two passages appeared in the part referred to, both among rocks, but with the difference that by the first the canoes could go loaded, being towed with great care. We passed this after four hours' labour ; and arriving at a calm piece of water near the other passage, we unloaded the vessels, and, carrying the cargoes over the very numerous rocks for the distance of 200 fathoms, they were left in a convenient place, where, without danger, we had already embarked. This being effected, we continued to convey the canoes through the remainder of the channel, which was formed by a broken portion from a boulder, whence the water issues from a height of two palms with little current for a distance of some thirty fathoms. For this purpose, it was necessary to make keels of wood to save the canoes from harm, which they would have suffered against the rocks that were not yet covered. With this precaution, we successfully pushed forward the canoes, employing the whole afternoon in transporting the cargoes to a safe place, quite free from the dangers of this first impediment ; and each one carried the things that belonged to him, so that on the following day we could proceed on our journey.

On the 19th, when already it was day, we began our journey with a west-north-west course upon the smooth water, which there was between the rest of the island and the firm land on the left ; and, in less than half an hour, coasting to the west, we saw the cachoeira ahead, where could be observed the sources of the water, which descended between two islands and the channel on the right. The scene was, in truth, more sombre than that already referred to.

Coasting along the same left margin, we found, after an hour's journey, going with the course of west-south-west, two masses of rock close by the shore, one of which formed an island, and from it rocks stretched out to the middle of the river, where there

was rather a strong current, which we overcome by oar; and had
it not been that the rocks in the middle were well covered with
water, there would have been fresh work for us to do, as if at a
rapid.

From this part we crossed to the right, coasting along which,
while yet the preceding mass of rock was in sight, we found
another very similar. It was less like an island, but the current
was greater. This we could only pass with some danger by towing. At this place, there were on the mainland many cacáo-trees
with fruit which was already nearly ripe, and plenty of trees
bearing chestnuts and other wild fruits which the Indians eat.

Having accomplished this passage, we began to hear the roar
of the waters of the celebrated cachoeira called Gamon, and
coasting along the same right side with the points of south-west
and south, taking forthwith as a guide the noise of the waters
referred to, we saw, on turning from the south-south-east, that
barrier of water which extends across the whole width of the
river, and with the same course we arrived at it about four
o'clock in the afternoon, and, in six hours' journey, we went
three leagues, from the first cachoeira as far as the second.

SECOND CACHOEIRA.[*]

On the east and west the river forms two similar bays, so that
it appears to make a circle, uniting together the opposite headlands upstream, with a direction of north-west and south-east.
Both these headlands are formed each of one solid rock, and
both communicate, forming, as it were, a dismantled wall, the
ruins of which, being precipitated into the water of the river
with tremendous violence, make a terrific noise, which, were
there any inhabitants on the banks, would probably cause them
to suffer from deafness. It is said that this happens to those
who live near the cataracts of the Nile.

[*] NOTE BY TRANSLATOR.—Now called Macacos.

At this impediment the river breaks with such noise and fury that there is no means for persons to effect this passage by channels or smooth water, because there is no such thing near the rocks, and these are not perceived, for between the breaks of the rocks all is a waste of seething water. Scarcely does the trunk of a tree, however large it may be, reach here, than it is absorbed in an instant, and quickly ejected, to be again soon drawn in, and hence at times is shot out in foamy billows, and, thus detained, remains gyrating for many days until the river swells, and thus opens up a way of escape. From one headland to the other, the length and breadth of the precipitated water would be about two hundred and fifty fathoms. The height of the leap in parts showed, on this occasion, when the river had already increased, no more than sixteen fathoms, estimated at the utmost.

As the furious current which issues from the leaps which the waters give among the rocks flows near the right side of the river, because it encounters an island and bank which oppose it in the middle of the bay, and gives passage to the greater quantity of water between the land on the same right side and the island, the canoes took their way by the left margin, and we brought them to in a small bay, where, as it was useless to look for channels, everything was unloaded, and, having transported the provisions and utensils by land, winding over the rock for the space of 600 fathoms, we dragged the canoes overland on wooden rollers, in which labour we expended two days; and some of the canoes becoming disjoined in the passage where there was an elevation of land, where greater effort was required to move them, we expended another day to repair the damage, the neighbouring forest supplying, in the place of oakum, a fibre taken from between the bark and wood of a tree called Jacepocaya, and, only with the small attention of stripping it off as a membrane and drying it, the fibre becomes capable of being used for this purpose; and from another tree, called Cumaá, we drew the sap, which served as tar. It is even better for caulking seams than if they were stopped with oakum.

The land contiguous to the rocky mass on the left, whence branches off the series of rocks from which the cachoeira is formed, is elevated like a hill-range, which runs towards the centre, and the extremity of it finishes in the promontory on the north-west, where the rapid originates. On the bank of the bay on the right, before and after passing the cachoeira, there is a kind of earth of such surpassing quality that on it alone are sustained animals—quadrupeds and birds—which denizen those woods, such as tapirs, wild boars, deer, and other animals of this kind, also parrots, macaws, mutuns, and others of this species, which are caught for the support of travellers; for these animals do not find anything in the bushes and thickets that would supply them with greater nutrition than the said earth, in eating which they often find insects, and this quality of earth is known by the dens which those that are maintained by it leave. The taste of this kind of game is more insipid than of those that feed on plants and wild fruit. After entering upon the cachoeiras, the fish were of better flavour than those taken before arriving at this district; only the water is still muddier by the rapids than before reaching them, and, in order to drink it without scruple that the intestines may be clogged with mud, it is necessary to throw some alum into the utensils in which the water is. This has the virtue of precipitating all the mud, however fine it may be, and leaves the water clear, which, if drunk thus, is of good taste, although it has a slightly diuretic quality.

This rapid is found in 9 deg. 40 min. south of the Equator. We did not take the latitude of the first rapid, through not having a horizon capable of making an observation with the quadrant.

On the 23rd, we began our journey, coasting to the left with the course of south-west, and with it, after little more than an hour's journey, we found a cachoeira, the rocks of which it was composed being already almost covered, for which reason it was easy to pass it, and channels were found to the right and to the left, whereby, with little labour, we towed the canoes through. From the left bank of the river at this place issues a rocky headland, which expands, forming various rocks, reaching across the river to the right, which has three small islands formed of the same stone, with rather a considerable quantity of wild timber, and between these islands and the mainland we navigated in the manner referred to for the space of half an hour with the course of south and south-south-east; and proceeding by oar, coasting along the bay, we turned to the south and south-south-west; and at the end of the bay lies an island surrounded by rocks almost in a circle in the middle of the river, and offers a passage between it and the mainland on one side, and the other without current or labour.

Having passed the island, we continued our journey, veering round the bay with the course of west-south-west, and coasting to the west it was found to be a bank which began in the headland referred to, a vertical wall of rock of a good height. Soon afterwards, a strong current proceeds from some rocks which stretch out from the same bank towards the middle of the river, and it was passed by towing with but little labour.

Following the same course of west, and passing to the south-west, coasting along the same right part, we came to a cachoeira similar to the preceding, composed of various little islands detached from the rocks, which spread across the river almost from north-west to south-east; and as the water would already cover a great part of the rocks, it afforded us a passage between the mainland on the right and an island, towing with little effort; and the same was the case with the other canoes of the squadron which went to the left. Thus we ascertained that this cachoeira and the previous one are easy of passage on either side of the river between the bank and the rocks.

Thence, coasting with the same course as far as a small bay, in which we halted the canoes on the same right side, and had, in nine hours' travel, accomplished three leagues.

THIRD CACHOEIRA.*

On the 24th, we began to travel with the course of west-south-west, crossing to the left so as to escape a current which there was after passing the headland of the little bay, where we stopped the night. This current was produced by some rocks which at a little distance seemed to be out of water ; but coasting with the course of south-west half the distance round, we found in the middle of it, after little more than an hour and a-half's journey, a rocky promontory which stretched almost to the middle of the river, where the current was strong. This we passed by towing, and coasting with the same course for the space of an hour and a-half we arrived at the cachoeira called, in the language of the indians . . . . and presently, out of respect to the celebration of the birth of Christ our Lord, we called the cachoeira " Natal." In three hours' journey we went one league.

This cachoeira consists of two islands of rock densely timbered. Both cross the river in the direction of north-west to south-east. Lots of rocks extended off the east and west banks adjacent to these islands. These rocks occupy both channels between the islands and the mainland. This is the reason why a great difficulty is offered in the passage of any of these channels in the river when not full. The space that intervenes between the two islands is also interspersed with rocks, between which the river rushes with the greatest force of its current ; and, for this cause, the ascent or descent by this way is impracticable.

Therefore, having in view an object which, on all sides, appeared formidable, the pilots noticed that to the left there used to be a more favourable passage, and with this consideration the guides were despatched to examine the channel, which, in fact, was found such that it could be passed by the canoes half-loaded, as the rocks were not yet well covered where a route was offered.

* NOTE BY TRANSLATOR.—Now called Theotonio.

The canoes being unloaded to the extent aforesaid, we effected their transport, and for the whole of that day we remained on the other side of the cachoeira, each one with that portion of the cargo which he carried. We thus overcame a powerful current which there was in the bay; and, this done, we halted the canoes at the bank on the left. As soon as it broke day, we looked for a suitable place in order to celebrate mass, and no other which was more so offered than a small beach which there was in the island on the north-west, just clear of the cachoeira; and for this place we crossed the river on the morning of the 25th, and there we heard the three masses permitted in the celebration of the sacred birth of our Lord Jesus Christ.

On this 25th day, about three o'clock in the afternoon, we continued our journey, crossing to the left, and coasting with the course of west, west-south-west, and south-west. It was necessary, after three hours' going to cross to the right side, to halt at a small beach. This was done when it was already dusk, and we went, in the said three hours, a league and a-half.

On the 26th, after hearing mass, we began our journey, at seven o'clock with the course of south-south-west, crossing to the left, and coasting to the south between a large island and the mainland. We proceeded once more to the south-west until reaching a promontory, where began a wide bay, in which were found three islands lying in the direction of the river, between these there is a great current, which was augmented in parts where the rocks touched the points of the said islands. Notwithstanding these embarrassments, we drew up, while it was already night, at a beach in the middle of the river adjoining the last of the said three islands, and in ten hours' travel we progressed four leagues.

On the 27th, we continued on our way, crossing to the right of the river, and with the course of west were coasting for the space of an hour, and afterwards we navigated to the west-north-west, and soon again to the west, and from this course we changed in a short time to the south-west and south-south-west. At this bend, we saw some high hills, which apparently ran from east to west. At the point of one of these lies the cachoeira, whither we directed our way, passing two islands, one on the right of the river, which prolonged itself as far as midstream, and the other on the left, close to the land, and taking the direction of the bay. Continuing thus with the same course of south-west, we saw the fourth cachoeira, and moored the canoes to the left, at nightfall, at a small beach close by the bank. In ten hours' journey we had gone, on this day, four leagues.

˷ FOURTH CACHOEIRA.*

THE 28th day arrived, and, after celebrating mass, we coasted, on the right side, a small bay to the south-west. The point in which it terminates is composed of rocks, which advance as far as the middle of the river, between which we steered with some difficulty against a great current which prevailed in that place ; and as soon as we doubled the promontory we halted the canoes close to the cachoeira called, in the language of the indians, Guarâassû, that is to say, great guarâ (a bird the size of a gull, with feathers of a bright scarlet colour), and there we waited the examination of the channels so as to know which was the most convenient to adopt.

This cachoeira is composed of a labyrinth of islands surrounded by a series of rocks, which crossed the river from one side to the other, with the course of north-west and south-east, at the distance of almost half a league, which width the river attains in that place, because it forms creeks or bays on both parts corresponding in the concavity. Four islands are located with the direction of the river, and fronting a vast number of smaller ones which fill up the gaps, following the same course at a distance of almost three-quarters of a league in latitude, all of solid rock, on the heights of which are produced trees of serrated foliage and extremely picturesque. Rocks branch out copiously on all sides from them, leaving no part of the river

* NOTE BY TRANSLATOR.—Now called Morrinhos.

free, added to which, between the numerous rocks there are various precipices, and from these result whirlpools of water, which engulph everything that comes near, whether it be a canoe, a log of wood, or any other body capable of suffering the misfortune of being whirled to the bottom with incredible violence.

When the river is half full, and until high water, when there are many rocks, or all covered in the channel between the land to the south-west and the island adjoining, there is a sufficient passage at this place; but on the present occasion, when the water began to increase, everything here consisted of vortices and whirlpools of water. By the channels of the middle it was necessary to take care, since the exit of the waters was more formidable with the violence of the whole current. For this reason we resorted to the expedient of exploring the channel on the north-west, and in it, coasting the contiguous island, we found a passage, but the most troublesome we had hitherto encountered.

There are three channels which, between the four islands, make a passage in the river with the violence mentioned. But at a distance of fifty fathoms, more or less, there stands opposed a rocky island which crosses the river with the same course from north-west to south-east, where the waters break with the fury with which they dash between the rocks of the channels referred to; and at the lower part of the same island the water forms a quiet pool, by which the canoes crossed to the north-west side; and coasting the island which runs in the direction of the river, two promontories of a series of rocks were found on this island. Between these rocks we took the canoes, on which task we expended the afternoon of the 28th, and there waited for the following day to continue the passage of greater danger.

On the 29th, we proceeded by the quiet pools which intervened between the currents of the mainland and of the island, and we contemplated the passage of a rocky headland which there was on another island fronting the one mentioned, whence we could convey the canoes when half-loaded, and by this method of transport we passed to a very wide pool which intervened between this place and other islands of the portion above, and by it one could pass without danger to the other margin of the river to overcome the last fall offered in it ; but, on drawing up the first canoe half-loaded, there being confusion among the indians in towing unless under supervision, while we were not yet clear of a rock which caused the greatest ebullition of water, we made them, with a vigorous effort at the bow of the boat, to force it on the top of the rock, and soon inclining the whole towards the current, the water rose so that we had not power to resist.

We loosened the cords, and, in an instant, the current bore away the canoe to a pool behind the other island, where we found it safely floating. We drew it on land, emptied and bailed it out, it having sustained no further damage than wetting the cargo contained in her, and this event was less unfortunate than if we had lost all, for the canoe (as were, also, most of them) was made of wood which does not go to the bottom in such cases.

In consequence of this confusion, it was necessary to halt on the 30th and 31st, so as to dry our clothes and look after the provisions which had got wet, and on the 1st day of January of the new year, 1750, we varied our route, passing again to the island which on the 29th we had coasted, and, unloading everything, the canoes were towed past two rocky headlands in which work we were engaged until four o'clock in the afternoon, at which time, by the pool which there was between the main portion of the cachoeira and the last islands of rock, we crossed to the east part of the river, and caused the canoes to be halted near the last impediment of this cachoeira, which was a canal between the series of rocks which issues from the mainland and another similar promontory which communicated with one of the islands in which ends the labyrinth of those rocks which compose this troublesome cachoeira.

The 2nd having arrived, we unloaded everything from the canoes, and with good success we conveyed them over the badly-covered rocks of the channel, where it was necessary to make thick wooden keels to avoid the damage which might result from the rocks, and to render easy a slight elevation which they attained in a distance of five fathoms. The canoes having been transported, and already laden on the other side of the cachoeira, we effected a start about two o'clock on the afternoon of the same day, coasting the land on the same east side. From the first passage as far as this last one there is half a league of longitude and one-third of latitude.

FIFTH CACHOEIRA.*

ACCORDINGLY, on the same day, about three o'clock, we continued our journey, coasting to the left with the course of west, and soon, to the south-west, we saw, on the portion of the river to the right, some hills, which form the margin of it and run towards the west. They were of tolerable height, and timbered with rough trees exceedingly serrated. Arriving at a bay, still with the latitude of south-south-west, we encountered a very furious current occasioned by a shoal of rocks, which, with great effort, we passed, towing the canoes, and, taking the course of south, we saw the cachoeira called . . which we reached after four hours' journey.

This rapid is the most terrible we had hitherto encountered. The foregoing hills continue along the right bank, and consequently leave in this place so great and disordered a lot of rocks across the whole width of the river that they do not yield a passage to the waters other than they can cleave for themselves, bursting through that deranged machinery for more than 800 fathoms along the stream, without any channel by which a canoe could be passed notwithstanding every effort be made, so that there is no other choice except to unload the canoes and haul them overland along the left margin, until clearing the whole distance where any impediment exists.

Two days were expended in laying a roadway of logs so as to roll the canoes on them, and, as the land was of tolerable elevation and scattered with rocks for a distance of a third of a

* NOTE BY TRANSLATOR.—Now called Caldeirão do Inferno.

league, we occupied two more days in transporting the canoes and their cargoes, so that, on the 7th, we resumed our journey, expending four days of immense work on this troublesome cachoeira.

Accordingly, on the 7th, at six o'clock in the morning, we began our journey with the course of south, coasting to the left, and soon we came across a rocky shoal (a remaining relic of the preceding cachoeira) rather difficult to pass, but, having overcome it, we coasted to the west a wide bay where, on the right, the same hill-range continued, but with less height than the previous, and follows the same course of west. After passing this bay, we entered to coast another with the course of south, at the beginning of which, on the same left side, there was on the bank the celebrated earth which the birds are accustomed to eat. Presently we passed to the south-west, and with the course of south-east, with ten hours' travel, we halted the canoes already by night, and had gone three leagues.

SIXTH CACHOEIRA.[*]

ON the 8th day, at about six o'clock in the morning, we began our journey with the course of south-south-east, and presently south-east and south-west for two bays, at the end of which began an island which divided the river into two channels; that on the right side is greatly embarrassed with a series of rocks, where the water rushes with such violence that, encountering that which issues from the channel on the right, a rapid and furious current was produced, which from the headland of the island extended downstream a tolerable distance.

Coasting along the left, we entered, with some trouble, which was the result of taking a south-south-west course, and in ten hours' journey we went two leagues, and brought up the canoes at the island referred to, it being already dark.

* NOTE BY TRANSLATOR.—Now called Girão.

From this island and the mainland on both margins issue the rocks that make the cachoeira called Arapacoá,* so that on the following day, that is to say on the 9th, we continued by the same channel with the course of south-south-west and south-west, coasting the island until arriving at the last head-land, where there is the main body of the cachoeira, in which place the waters divide for the two channels referred to, and, on issuing from where we were navigating, we passed the rapid, rowing, overcoming only the currents which resulted from the stones already covered with water, for which reason that passage is facilitated, without further trouble than the said one of having to use the oar.

From the point of the island, on the west side, there ran along the river the highest hill-range we had noticed up to that place. These hills followed the same direction as the previous ones, from east to west. Leaving them on the bank on our right, we coasted to the left on the same 9th day, with the course of west, west-north-west, and north-west, and with ten hours' travel we went two leagues and a-half.

On the 10th we began our journey coasting, to the left with course of west-south-west, the hills continuing on the right in the direction mentioned. Down a gorge in them ran a stream, at the mouth of which we found six small canoes of bark placed as a safeguard. This was a sure sign that in the interior from that spot there were indians, who, when necessary, travel in such canoes. We allowed them to remain in the same spot, and continuing our journey still to the south-east for four hours, during which we went a league and a-half, the hills disappeared towards the centre in their course of west.

We proceeded with our journey to the west, leaving, in the middle of the river, after passing the hills, a great quantity of rocks still badly covered, which occasioned a great current, and, having overcome this, we began to pass the first currents arising from the rapid which happened to be near.

* The name of a bird of small estimation.

SEVENTH CACHOEIRA.*

IT was not possible, on that day or afternoon, to proceed further than to tow the canoes past two rocky headlands with great difficulty, and we halted at a small bay after ten hours' travelling that day, during which we progressed two leagues and a-half.

At the place in which stood the first rocky headland, mentioned above, on the left margin, issues a wide channel between the mainland and an island extending in the direction of the river, which has a bed of solid rock. Quantities of rocks of enormous size are hurled down into both channels, leaving the navigation impracticable, because the water in it has no other outlet than precipices in such a confused position, that an examination of that intricate pass cannot be effected at sight. The channel on the left, which we followed, is interspersed with rocks whence the water, being thrown off with great fury, goes to encounter that which runs close to the bank, that is also of rock with shoals, so that, from the encounter of these two furious currents, results a continuous series of whirlpools, which every moment engulph in their seething depths the waters and everything else that they can absorb.

For the space of one-third of a league does this terrible way extend, and, having passed it, we soon met with three small rocky islands, against which dash the waters of the main body of the cachoeira called Paricá, and between the little islands they run so furiously that, being intensified by the encounter with each other, they occasion terrible currents and whirlpools, like the preceding ones, which, when in sight of the precipices of the cachoeira and the roar resulting from them, make that watery spectacle more formidable.

At the end of the bay, which begins at the headland of rock already mentioned, rises a promontory of gigantic boulders, which, extending as far as the middle of the river, give rise to some gorges where the water rushes precipitously. In the same

* NOTE BY TRANSLATOR.—Now called Tres Irmãos.

direction of these rocks follows a rock-island thrown length-ways with the river-course. Between this and the rocks, at the distance of three hundred yards, is the largest channel, but it was impracticable on that occasion, when the waters had risen almost half-way up the bank. Between the island and the mainland on the side of the rocks, there was another outlet of waters, but having precipices of equal extent to the channel on the left. The distance here from one bank to the other is about nine hundred fathoms, with little difference, and its course is from east to west, its latitude north and south, and here were the islands already mentioned.

On the 11th, the journey was continued, hauling the canoes close to the bank, which being of rock, this passage gave incredible trouble. It took us six hours to arrive close to the rocks on the left where we coasted, and we went half a league on our way.

Having assembled the canoes at a slight hill which issued behind the rocks, we transported the cargoes to the other side of the cachoeira. This we did overland, a distance of some four hundred fathoms, and it occupied us the rest of the day referred to.

On the 12th, we laid a wooden roadway across a ravine in the rocks, nearer the mainland from the same eastern part, where some water flowed, which, being small in quantity, did not offer violence of any account. This gave us an opportunity to push forward the canoes that had remained all the morning above the rapid proper. We effected our object with some considerable effort and great care to avoid any danger.

About two o'clock in the afternoon of the same day, we continued our journey with the course of north-west, coasting to the left, and we saw to the right a hill which had the same direction as the preceding ones, still in sight of the cachoeira. We passed to the course of west, and after a great current that resulted from rocks in the middle of the river, we halted the canoes after three hours' journey, in which we went a league and a-half.

EIGHTH CACHOEIRA.*

On the 13th day, we prosecuted our journey, coasting to the left with the course of west, and with a little more than four hours' journey, going still with the course of south-west, we found a cachoeira called Maiarí, which consists of a great portion of rocks that cross the river from one margin to the other with a direction from east to west, and, as the waters had already covered the rocks further to the left, by this way we towed the canoes so as to surmount the great current which prevailed for the space of 300 fathoms, more or less, which is about the distance this cachoeira extends along the river, while its width is more than 1,000 fathoms.

Having passed this cachoeira without more trouble than stated, we coasted a little to the south, and soon to the south-west, and, at the end of the bay which has this course, there lay in the middle of the river a large rock in the form of an island, from which proceeded a furious current from both margins, and as the lesser was on the right, the canoes went by this. Steering to the south, we arrived at the mouth of the river called Aboná, where the canoes were halted, after nine hours' journey, in which we went four leagues.

The river called Aboná flows in on the left bank of the Madeira, with little force at its mouth, which is 300 fathoms wide. We navigated in a little light canoe for a day to ascertain its direction, and we found it to be from west to east. We did not penetrate further than to meet with a cachoeira which the river forms from one part to the other, with a tolerable height of rocks, between which the water rushes. It is rather abundant in fish, and the banks in game. Its waters are clear and of good taste. The bank is high on both sides, and in few parts flooded at high tide. It is rumoured that a peaceful tribe of indians called Ferreirús dwells on this river. They are capable of living in a civilised state by being settled in mission-villages. It was necessary to halt on the 14th day, during

* Note by Translator.—Now called Paredão.

which we conducted an examination of the river, and on the 15th day, in which we finished executing on the canoes some repairs they needed.

On the 16th we began to journey, coasting to the right, with the course of east and east-south-east, and, having passed the bay which took these two directions, we continued with that of the south-east. This gave place to the east, and soon we turned to the south-east, and finally to the south-west. Nothing memorable occurred on this day, and we halted after ten hours' journey, during which we went about four leagues.

On the 17th day we continued on our way with the direction of south-south-west, and soon south, going thus for half an hour, and going to the south-east and east we passed the sharp point to the south, and with this course at nine o'clock in the morning we encountered the cachoeira called Tamanduá.

NINTH CACHOEIRA.*

WITH the course of south runs a bay from the east part, beginning at a point of rocks, standing opposite another bay intervening between an island almost triangular in shape and of tolerable size. Two channels result from the position of this island. The wider one on the right, whereby the river conveys away the greater part of its waters which thread their way between the vast quantity of rocks to be met with at the mouth of the same channel, for which reason that transit is impracticable. The narrower one is the channel on the left, which is supposed to contain rather a swift current. Nevertheless, it can be surmounted by rowing. By this means the canoes were transported, with two hours' journey, to the part above the cachoeira, without further work than to surmount by oar two currents which, at the mouth of the channel, proceed from rocks

* NOTE BY TRANSLATOR.—Now called Pederneira.

yet badly covered with water. These cross that passage to form a junction with those of the other channel with the course of almost from east to west. Thus are formed various islands of rocks crowned with forest-foliage, presenting to the view a pleasant scene.

TENTH CACHOEIRA.*

HAVING passed, in the manner indicated, the cachoeira of Tamanduá, and starting from it towards the south, we went coasting to the left side with that of south-south-east and south-east, and with this, in two hours' journey, we encountered the cachoeira called Mamorini, which is composed only of rocks which cross the river from one side to the other with the direction of north-west and south-east, and, as they are not very large they are almost all covered. Currents arise hence. These we overcame by towing, and by oar at the left margin. Afterwards we left this astern, and for little more than an hour we continued our journey with the course of south-south-east, and, lastly, to the south, and halted the canoes on the right margin after ten hours journey, in which we went five leagues.

ELEVENTH CACHOEIRA.†

ON the 18th day, following the course of south with two hours' journey, we came up with the first rocks of the most toilsome, wearying and dangerous cachoeira we had hitherto encountered. For more than a league and a-half extends this

---

cachoeira called Mamorini. Islands of rocks covered with cheerful vegetation and bare boulders cross the river from one margin to the other, between which obstacles the water breaks, producing a variety of phenomena, because rushing precipitously in parts it continues to form among other smaller rocks continuous whirlpools, which become intensified and attract whatever approaches them, the waves dashing with the fury of a tempetuous ocean. They break down everything that opposes them, which they engulph without any chance of saving. This terrible spectacle continues upstream for the space of a league before arriving at the body of the fall, which consists of various precipices of water which cross the river, and do not offer a passage anywhere, not with all the energy and skill in the world. The only way is to drag the canoes on land so as to launch them on the other side of the falls. Attempting the passage in any other way would certainly end in failure.

This done, the canoes were drawn up at the left bank, and with the course of south, the first currents of the three points of rock were passed, which issue from the land to join those which cross the river, and, in order to surmount this first passage, it was necessary to go with great care, making haste with the oars, with which we aided the towing, until we came to a little fall, which there was between the land and an island of rocks, having passed which with great effort, there soon followed another, which, with equal work, we managed to surmount; and it being already six o'clock in the evening, we halted the canoes after six hours' journey, during which we went a quarter of a league.

On the 19th, we continued our journey, coasting the same left margin with the course of south and south-west, between a bank of the river and rock-islands and shoals, which issue· from the mainland of the same sort as on the previous day, and, with equal towing-work among rocks and light shrubs, we proceeded all day until perceiving the monstrous fall of water which there was in the highest part of the cachoeira. This was not of hollowed rock, but a declivity of more than five hundred fathoms' distance. In seven hours' journey we went a league.

20th day. We began our journey coasting, with the course of south-south-west, a bay, in which dashed the waters that issue from among the rocks of the cachoeira, which, going back to the same part where they run with fury, raise along the whole bay crested waves, which represent a stormy gulf; and thus, with great danger, we navigated this bay until the canoes took shelter in a stream which flows in at the centre, where the cachoeira makes its last leap. The passage of the bay took two hours, being a distance of about a quarter of a league from the stream, where we halted about nine in the morning.

The remainder of the same 20th day was expended in laying a wooden keelway along which to haul the canoes overland, this we did on the 20th and 21st, and on this day the canoes were already transported from the other part of the cachoeira with the cargoes that belonged to each. This land-transit extends for six hundred yards, where there is an elevation on both sides, making it more troublesome. The cachoeira runs across the river with a course from north-east to south-west.

On the 22nd day, starting from Mamorini, with the course of south, we coasted to the left, and passing in a short time to the south-west, we found a contorted block of bare rock, which extended almost across one-third of the width of the river. This water issued from between an island and the rock opposite, as also between this and the mainland on the left side and in the middle. From the encounter of all these currents there results such a great one, issuing from the point of the rock and extending over the width of the river, that it was not possible to overcome it by rowing, not only by the impetuous violence of the waters and back-driven waves which there arise, but also because along that terrific shoot are vortices of water which seem to rise more than six palms, and soon are formed into the most formidable whirlpools that have hitherto been mentioned.

Accordingly, there was no other choice than to introduce the canoes by the small channel exceedingly embarrassed with rocks and light shrubs, which intervened between the rock and the east bank which we were navigating; and with hard work towing we accomplished our project, which took four hours' of troublesome passage. We continued our journey with the same course of south-south-west, and brought up the canoes within sight of the twelfth cachoeira, having gone, on this day, three leagues.

TWELFTH CACHOEIRA.*

This cachoeira is composed of various masses of rocks in the form of little islands, with wild foliage, and others bare, almost similar in quality and position to those which there are in the preceding cachoeira, before arriving at the larger falls in it. They also cross the river from one part to the other from the circumstance that, on the right, the river does not offer a passage when it is already increased, because the greater weight of water inclines to it, so that, breaking between some rocks and rushing over others, it forms strange currents insurmountable even by force and toil. On the left margin, it shows itself more tolerable between the worn rocks, badly covered with water, and the mainland, which also has its bank of friable rocks. At the last headland of such a rough series of rocks there are two islands, stretching from one to the other side of the river with the course of almost east and west, between which the river forms three channels, that on the west being the most impetuous and most terrible, as has been declared above. The formation of these islands at their base is entirely of rocks, upon which there is some sort of a level surface, giving rise to the opinion that the earth was introduced by the inundation of the river to produce in them leafy trees of a pleasant appearance.

On the 23rd, we journeyed with the course of south-south-west, coasting to the left, and after half-an-hour we encountered the first rocks of the cachoeira above mentioned, called Vainumú; and, by working at the tow-rope, the canoes being half loaded, that first impediment was overcome, and also the second which soon followed; and lastly, the third, which together made one day of troublesome work, during which we went a quarter of a league, which is about the distance this cachoeira extends along-stream, because it finishes actually at the two islands above mentioned.

24th day. Having passed the islands with the course of south-

---

* NOTE BY TRANSLATOR.—Now called Ribeirão.

south-west, with two hours' journey we sighted the mouth of the river Beni, for which we crossed, and halted within it, in order to ascertain accurately what we possibly could regarding its direction.

Having made in that place the observation as to latitude, we, found that the river Beni falls into the Madeira, in 12 deg. south latitude. Its entrance is with the course of south-south-west, and navigating by the same river five hours we found that direction to be the most frequent, which appears to be its original one. It is somewhat extensive, and almost equal in current to the Madeira, into which it effects its entrance on the west bank, with apparently 800 fathoms of mouth. Its waters are muddy, by reason of the quantity of mud which at flood-times falls from its banks, which are very similar in height and foliage to those of the Madeira. With this the Beni could dispute the maternity of waters, if the Madeira did not show that it continues in its course with its islands and rapids in the same direction which it takes as far as that place, and, conjointly, the Madeira being of greater width and bringing more water to that part in which it receives the Beni, which, in fact, there loses its name and existence as a river.

As regards this river Beni, there is no document or tradition showing that it was navigated by Portuguese or Spaniards because these, in whatever tracts the Beni rises, were ignorant until the year 1713 as to the true course of this river, because they supposed that it flowed into the Amazon without mingling with any other waters; and the Portuguese who ascended in 1723 as far as Santa Cruz de los Cajubabas, and others who previously had gone to traffic with the indians, did not enter this river with any careful observation. Accordingly, it will be appropriate that some account, abstract though it may be, as to the sources of this river, should be related here. The second consists of a map printed in the said year 1713, and it is supposed that the graduation of it cannot be very exact, because, besides not conforming to the general geographical maps, it labours under a great error in the degrees of latitude, which was made, on this occasion, because the author of the said map, describing in it the river Beni supposes it still to be in 11 deg. south latitude, although by the observation on the said 24th day of January, 1750, it was found that it falls into the Madeira in 12 deg., as already expressed.

From the serras of Perú, parallel to the city of La Paz, in latitude 12 deg. 20 min. south, the river Beni is shown to have its source, and running between the same hill-range is joined with the Chuquiabo, which comes from the said city standing at the head of the Beni to the west. The two rivers together form the Beni, more extensive among the same hills as far as 15 deg. latitude, where running along level ground it falls into the Madeira at the part above mentioned.

Along the whole margin of this river on both sides there does not appear to have been, up to that year, 1713, more population than that of a village called Os Reis, inhabited by three thousand persons of both sexes and all ages. More to the centre, to the right of Os Reis there is another dedicated to Saint Paul, populated by two thousand seven hundred persons. The wild tribes that dwell in the district of the Beni are called Romanos, Chumanos, Chriribas, and Toromonas. Towards the west, some streams descend from the hills and fall into the Beni, where it already flows through the plain, these are the Apioana and the Amantala, and there are in their districts among the mountains three peoples, namely, Apolobamba, San João, and Pelechuco, from 15° 30 min. to 14° 40 min. south latitude. It receives also the waters of the stream called Enin, which has its origin in the neighbourhood of the city of Cuzco, situated in 13 deg. 20 min. latitude and 301 longitude. The observation being made, and navigating the said Beni in a light canoe five hours up stream, we started from it on the 25th, about one o'clock in the afternoon, pursuing our way coasting along the right margin with the course of south-south-east and south, and with this course we found ourselves at the

THIRTEENTH CACHOEIRA.*

THIS cachoeira is known by the denomination of Tejuca. Its composition consists of a number of not very large rocks, which cross the river from one side to the other with the course of north-east and south-west ; and as it was, already, almost covered with water it offered an easy passage, by towing, along the left margin where we navigated. Accordingly, after five hours' journey with the directions mentioned, we halted the canoes near the cachoeira called Dos Javalis, having gone about a league and a-half from the mouth of the Beni as far as this place.

FOURTEENTH CACHOEIRA.†

ON the 26th day, we did no more than to pass the cachoeira mentioned until mid-day. This consists of two islands surrounded by blocks of rock, a creek at the margin, and another to the left with the same course as the preceding, north-east and south-west. These two islands are united by various rocks, and throw out others down-stream at a distance of more than 800 athoms, in such a manner that the river attempting its passage between these impediments forms three channels, one between the bank on the right side and the island, narrow and of terrific current, the other in the middle, in which the water in various precipices makes the transit impracticable, and another between the island on the left side and that portion of the land that is most favourable for the purpose of a passage, because it only offers two currents near the land, where the canoes could, when empty, be towed. Thus was it effected on the day referred to until two o'clock in the afternoon, and, about three, our journey

* NOTE BY TRANSLATOR.—Now called Misericordia.
† NOTE BY TRANSLATOR.—Now called Madeira.

was continued with the course of south-west, and in it, after going half a league on our way, we encountered the fifteenth cachoeira, near which we halted the canoes so as to undertake the passage on the following day.

FIFTEENTH CACHOEIRA.*

THIS cachoeira called Dos Papagaios, was found to be somewhat intricate, because it was formed by natural projections from both sides of the river, with the course of west-north-west and east-south-east, composed of islands surrounded by boulders and enormous rocks, which extend at irregular intervals from some of the islands to others. For this reason, the river is forced to break through these impediments, in a manner almost strange for the maintenance of its course, since in places, with various directions and conflicting with different currents, it issues between the last rocks with noisy fury and foaming rage.

It barely offers a channel close against the right bank, between the bank formed by lofty rocks and an island lying close off the same part, between which the water ran with less fury, as there was then not much flowing on that side, for which reason the rocks were badly covered, so that it was necessary to make a wooden roadway there, so that on it the canoes might be hauled empty.

On the 27th day, we performed the above-named work with very great labour, and we passed the entire day on that toilsome service. In fact, on that same day the canoes were transported to the part above the cachoeira, which extends for the distance of a quarter of a league alongstream.

On the 28th day, we prosecuted our journey, coasting to the right, with the course of south-east, and with it, after a little more than an hour's journey, we found a cachoeira, called Das

* NOTE BY TRANSLATOR.—Now called Lages.

Cordas, which crosses the river from one bank to the other with the course of north-west and south-west ; and as, already, almost all the rocks were covered, it was passed by our surmounting only the two currents between the west margin and one of the islands that form the same cachoeira.

SIXTEENTH CACHOEIRA.*

FROM the two islands formed of rocks is formed this cachoeira called Das Cordas. Both are adorned with picturesque flora, not of a lofty kind, but leafy and pleasant. They cross the river with the course named. One of them approaches closely to the left margin, so that even at half-flood it does not offer a passage. The other situated to the right permits a passage at all times except at the end of the dry season, in the months of September and October. The river then only offers a channel to travellers between one and the other island, where it forms a declivity of rocks with good direction, wherein, for the space of sixty fathoms, the water is discharged with tolerable violence.

This cachoeira is passed in the manner above mentioned, and continuing our journey with the same course of south-east, there was found to flow in on the right margin a stream of clear water called Tiahoam. It is about 100 fathoms wide at the mouth, and its small torrent flows almost into the midst of the rapid we had passed, for which reason the water in it appears clear, and directly on reaching the stream the cause of the phenomenon is explained. Facing the place in which that brook empties itself, there is on the east margin an elevation of earth which forms a small hill, clothed with very high and thick foliage, and does not reach the extent of half a league along-stream. Towards the centre, there was not an occasion to make an examination of its direction.

---

* NOTE BY TRANSLATOR.—Now called Púo Grande.

Having passed the stream Tiahoam, we continued our journey still with the course of east-south-east, when on veering round a promontory of rocks, where there was a strong current on the same right margin, the oar-blade broke which served as the rudder of a canoe, and it was necessary to make another, so that on this day we did not go more than two hours, during which we proceeded little more than half a league.

29th day. On this day we began our journey coasting to the right with the course of east-south-east, and soon to the south in which we passed a bay. We continued to the south-east, and in this course we found a rapid called Panela, which is one of the largest and most embarrassing in the river.

SEVENTEENTH CACHOEIRA.*

FROM an intricate labyrinth of islands all founded on boulders and enormous rocks this cachoeira is formed, in such a disordered position for the space of almost a league along-stream, that the river, to effect an outlet from such peculiar impediments, splits itself into a variety of channels between the islands and rocks, so that in half a league of width from one margin to the other, and in the whole extent alongstream there is nothing else but whirlpools of water in every part, so that with noise and fury the currents encounter each other, irritated by the rocks that they pass and others over which they rush. Many channels from which the water issues in whirlpools and with great impetus are scattered over with islands or rocks which act as a barrier. This obliges the waters to make a way at the sides. In consequence of the stout resistance that proud element experiences from the rocks of the margin, the water is driven back, and forms tempestuous crested waves as of the sea, so that it is very difficult to wade across. Finally, so as to expose one's self as

* NOTE BY TRANSLATOR.— Now called Bananeiras,

little as practicable to all the subterfuges with which the water here endeavours to pass this cachoeira, I confess that, not finding fit and expressive terms whereby to signify such a very embarrassing passage, a great digression that might be used in description would be out of place; and, for this cause, I have recourse to engraving. This might be so effected as to give an idea of the phases which are difficult to express here.

Within sight of such a formidable impediment, which, on all sides, threatened fatal consequences in the attempt to surmount it without dragging the canoes overland, it was fortunate to find a channel between the bank on the right and the islands of the cachoeira, where, avoiding a not very large space of waves that resulted from the conflict of a current with the series of rocks on the mainland, the canoes were able to arrive at a suitable place, in which, over the rocks, still badly covered with water, they could effect a passage. And, in fact, we kept on in this channel, surmounting by oar its successive currents, until arriving at the place mentioned, where there was an ebb and the flow of the water. Furious waves were raised along the whole channel for a distance along-stream of more than the range of a musket-shot. With great care and force of rowing that perilous portion was passed, and already by night the canoes were halted near the great boulders, between the openings in which, on the following day, we had to make our way.

The 30th day arrived, during which with incredible diligence, a keelway was laid along the rocks, the canoes being unloaded, and these were conveyed to the part above the cachoeira, until the hour of midday, as soon as we returned from collecting the keelway that belonged to each of the canoes, and placing it therein, we continued our journey about three o'clock in the afternoon, with the course of south-south-west, and in it we found the cachoeira called Quatí, after an hour's journey, during which we went the distance of half a league.

EIGHTEENTH CACHOEIRA.*

THIS cachoeira is of a better appearance than that which is
offered in the previous one. It consists of an island of tolerable
extent, launched out with the course of east-north-east and west-
south-west, which has for a base an immense rock. On all sides
there is distributed a large quantity of rocks and boulders, but
regularly disposed by nature. Thus the river is parted into two
channels by reason of the island indicated. The water flows
without a decline for the greater part of the unimportant
channels proceeding from the scattered rocks on both sides
without opposition from either. Accordingly, on that same
afternoon of the day mentioned, we passed the part on the right
surmounting its channel by rowing without further embarrassment, than some currents without danger, which we successfully
passed as far as a bay, in which the waters existed in a state of
tranquillity even in the midst of the cachoeira, where we halted
the canoes, having gone a league with the course mentioned.

On the 31st day, we continued to pass what remained of the
cachoeira Quatí, which only in the mouth of the channel between
the point of the island and the land requires any trouble, because
it was necessary to use the towing-cable between rocks and
light shrubs for the space of two musket-shots. From the bay
we steered to the south-east, and on issuing from the channel it
was to the east. This cachoeira extends for a league along the
river, and half in width.

The canoes being freed from that transit, until nine o'clock in
the morning we crossed to the left side of the river, coasting to
the south-south-east, and with this same course we halted the
canoes at the right margin after six hours' journey, during which
we went three leagues.

In this place is obtained the first view of the hills called the
Cordilheira Geral, or a large plateau to the left of the river.
They run from north-east and south-west, the general direction
of the river, and from here these hills begin to cross the river

* NOTE BY TRANSLATOR.—Now called Guajará-Guassú.

or the latter to break its extremities and gorges, in which are
formed the cachoeiras of which we have treated, and now we will
do the same for the last one one meets with in ascending and
the first in descending.

February 1st. On this day, starting from the bay, where we
had passed the night, with the course of south-south-east, we
soon observed a cachoeira called Tapioca, which in this place is
the last that is offered.

NINETEENTH CACHOEIRA.*

In the bay on the east side is situated an island almost oval
in figure founded on rocks. On the west margin the river forms
another very wide creek, and in the bight of which is formed
another island of similar nature to the preceding, but of much
greater size, because it occupies all the vacant space of the bay
referred to, to the concave part of which the island corresponds almost in perfect proportion with the convex. From
these islands spread out infinite important rocks and other
large ones. Some of these cross the river, and others stretch
along stream for the space of a quarter of a league. There
are three channels whereby the river effects its passage
between the islands indicated and the mainland on both sides
producing a furious current, insuperable along the east margin'
and about the middle, since by both these channels not even at
the flood of the river (as it then was) do they succeed in covering
the rocks which oppose these two currents.

Accordingly, having examined the western branch, we found
it practicable, since it was only requisite to surmount a tolerable
current, which was passed with the cable ; and coasting to the
right between the large island and the mainland with the course
of west, and without further hindrance, we finished the same

* Note by Translator.—Now called Guajará-Merim.

circuit of the bay, with that of east, and without further impediment than some currents which were passed rowing. We halted the canoes already at the other part of the cachoeira at the right margin, after six hours' journey, in which we went two leagues.

This cachoeira, Tapioca, crosses the river with the course of east-north-east, and west-south-west, and is the last that offers before the waters of the river Mamoré incorporate themselves with those of the Madeira. This passage we were effecting on

February 2nd. On this day we continued our journey, coasting to the right, with the course of east-south-east, leaving on the same margin the first marshes that there are in this river. These are produced by the river overflowing its bed and flooding the land for a distance of more than two leagues to the centre, which forms, as it were, a lake along the direction of the river many leagues in length. A small bank intervenes amidst the marsh, and is covered with lofty vegetation. The land that succeeds towards the centre, which is flooded, is like a plain, having only a few firm points or islands clothed with vegetation distant from one another; thus these wide plains are very picturesque.

At these marshes, fish are taken at the time when the inundation prevails and the river has emerged from its bed. At this time, by chance or a marvel, not a fish was to be found. This was also the case with the game—both bird and quadruped—which retreated to the mainland, a reason why already from this place upwards, a scarcity of provisions began to be felt. This want generally places travellers in great consternation.

With the course referred to, towards the east margin, appeared the Cordilheira Geral, following the said river at a distance with an east-south-easterly direction, a course which here begins to give a new direction to the river, as might be judged by the frequency with which the needle now sought the east and south-east. Thus, with six hours' journey, we went four leagues and a-half, halting the canoes, when already dark, at the east bank, having started at midday from the place in which we halted the canoes the preceding day. We occupied ourselves on the morning of the present day in hearing mass and resting from the hard work of the cachoeiras, the transit of which had been effected with the success indicated, not without the admiration of the experienced, since in so numerous and toilsome passages not one person experienced danger or hurt out of the one hundred whites and indians that constituted our staff, though, on many occasions, all worked, without any exception.

On the third day of the same month we began our journey, coasting to the left with the course of south-west, and having recurred once more to that of west-south-west, we crossed to the right, where we coasted to the south-west and south-east, and lastly to the south; and with twelve hours' journey we went eight leagues.

In the course of this day, there was nothing to note except that the river has not a current of any consequence. This is attributable to the impediment which the waters experience in the cachoeiras, and through this cause the waters are restrained and bereft of force throughout the navigation effected above the cachoeiras.

In the place in which the canoes were halted on the 3rd day, which was at the east margin of the river, there being in that spot that which was necessary for repairing the damage the canoes had received in the passage of the cachoeiras, we passed the fourth day in the place referred to. There we repaired whatever was damaged in the embarkations, so as to continue our journey with greater convenience.

On the 5th, about four o'clock in the morning, we began our journey, coasting the left with the direction of south-east and south, and at the break of day we passed to the right, and coasted to the south-west, south, south-south-east, and south-east, and with this course we found an island which followed the bend of the river, and between it and the right bank we navigated. Rounding this bend, we steered all points from south to east, and after navigating to the south-west as far as the place where we halted at the east margin, after eleven hours' travelling, in which we went eight leagues. As far as this place there was no current.

On the 6th, at daybreak, we continued our journey with the course of south and south-east, coasting to the left a large bay, and afterwards to the south-east, and lastly to the south ; and with eleven hours' journey we traversed seven leagues. On this day there was still some current, which we overcame without great effort.

7th day. We began our journey very early in the morning, coasting to the right with the course of south-west and soon to the south and south-east, and whilst steering east-south-east we passed between an extensive flooded island and the mainland to the east. Issuing by this channel to the east, at the end of the island on the left side, we navigated to the north-east, north, and once again to north-east, veering thence up to east. With this course we halted the canoes opposite the point of another small island, after ten hours' journey, during which we proceeded seven leagues.

On the 8th day, after mass, the canoes started at eight o'clock in the morning, coasting to the right with the course of south-east and south for the space of an hour, and, crossing to the left, we coasted to the east, south-east, and south ; and with this course, it being already four o'clock in the afternoon, we began to find the waters of the river less muddy, which we observed to be the case during the rest of that afternoon, until, when already dark, we judged by what light there was that the water near the left margin was all clear. Directing our course thither we found that there were three mouths in the river in that locality : two on the south, and one to the east, which circumstance, with the diversity of water that is found, renders it certain that one of the two mouths to the south is that of the river Mamoré.

The 9th day broke, and with its light the delusion which the previous night had not permitted to be perceived was dispelled. In fact, with the course of south-west, the river Madeira flows in, having a mouth of more than five hundred fathoms, and towards it the canoes were steered, crossing that almost gulf of water formed by this river and the Aporé in the junction that both bodies of water effect, those of the Oaporé being very clear, and those of the Mamoré having the turbid appearance exhibited by those of the Beni, which continues for the extended space whereof mention has been made in the Diary on the 2nd day of October of the preceding year. From the concourse, in this place, of both waters these are extended along the east bank, and form various lakes. The mouth on the east side is that which was noticed on the previous night.

From the mountains of Perú, which form the main chain of the Andes, descends the river Mamoré in latitude 18 deg. 30 min. south, and its direction being almost from south to north it meets with the Guapaix, which has its sources from the same mountains mentioned, and passes by Chuquisaca, or the City of La Plata, and by the new Santa Cruz de la Sierra until, in latitude 16 deg., it effects, with a northerly course, a junction with the Mamoré, and, both incorporated, receiving various torrents which from the western and eastern parts flow to it ; traverses the Province of the Mojos, a level district, slightly fertile, and so sterile of riches that it is evident there is no species of metal in its wide tracts ; continues along the same plain, and, passing the lands of the indians called Cajubabas, blends its waters with those of the Aporé in latitude 12 deg. 40 min. south.

In the year 1723, the Governor of the State of Pará, João da Gama da Maia, received information from some men who went to the river Madeira to contract for labour that, above its cachoeiras, there were habitations of Europeans, but it was not known for certain whether they were Portuguese or Spaniards. In order to explore the said river he sent a troop, of which Francisco de Mello Palheta was commander, who, after passing the cachoeiras, while navigating, met near the mouth of the Mamoré a canoe of Spanish indians steered by a mestizo, who guided the said Palheta as far as the villiage of Exaltação de Santa Cruz dos Cajubabas, and after holding conference with the missionaries who rule the place, he returned to Pará with the account of what he found, without further data concerning the rivers, not only of the Beni which flows in among the cachoeiras whereof mention has already been made, but even of the Aporé, which was so clearly shown to him, and which could not but be seen on entering or quitting the said Mamoré.

Time running on, and Mato Grosso being stocked by the inhabitants of the Cuiabá, in the years 1736-7, there resulted from them new settlements in the year 1742. . . . These men had a desire to traffic with the neighbouring Castilian priests in order to keep up the supply of cattle and horses, with which object, navigating the Aporé downwards, they effected an entrance by the Mamoré, and, after a journey of ——— extent, they halted the canoes at the same village of Santa Cruz dos Cajubabas, where they were well received, but without realising the project they had in view. Three members of this party detached themselves from it, and, proceeding down-stream and, safely crossing the cachoeiras, they arrived at Pará, where, being apprehended, two of them were sent as prisoners to His Majesty for being transgressors of the law of. , . , and the third companion called Joaquim Ferreira Chaves, was placed under surveillance of a guard of soldiers. Effecting his escape, he proceeded by the Maranhão in search of Goiazes, and from these mines he passed to the Cuiabá, and lastly to Mato Grosso, where by means of this inhabitant the first intimation was obtained that by the Madeira communication could be held with Pará.

Meanwhile., some inhabitants returning to make the journey from Mato Grosso to the villiage of Exaltação, without effecting any negotiation, until, in the year 1747, those settlements finding themselves in great want of salt, a surgeon called Francisco Rodrigues da Costa went to trade some of this article, and, in fact, he effected an exchange of dry goods, and also traded in wax and cotton cloth, all of which were useful to him. Induced by his success on this occasion, he established, as it were, a compact with the missionary of the village referred to, the latter giving a list of articles the above-mentioned people desired to exchange, referring for the exchange to the missionary of Santa Rosa, a newly formed settlement on the east margin of the Aporé.

Indeed, in the following year, 1748, the same Francisco Rodrigues making a purchase in Mato Grosso of assorted goods of which he took a list in writing, and making a journey as far as the said village of Santa Rosa, he found the population changed in such a manner that not only would they not make an exchange of goods, but even the missionary declined acceptance of a present, with which, as a matter of policy, the said Francisco Rodrigues endeavoured to flatter him; giving, as a reason for breaking the engagement, that they had received most stringent orders from their Superior, resident in Santo Cruz de la Sierra that the missionaries of that province should not have any commerce with the Portuguese of Mato · Grosso. The surgeon referred to returned by those settlements with the same wares which he had taken to Santa Rosa, and on the occasion of the squadron from Pará arriving at the settlement of S. Francisco Xavier, they were there for sale in a public store.

These were, up to that time, the Portuguese voyages on the river Mamoré as well from Pará as from Mato Grosso. This being premised, it remains for us to give some notice of the villages that there are on the Mamoré referred to, and whatever is memorable in connection therewith.

Navigating the river Mamoré up stream, it was found, after half a day's journey made in a light canoe belonging to the squadron, that its entrance is to the south-west, free from cachoeiras, and has not any islands.

It is evident, from the information given by the Portuguese of Mato Grosso, who navigated it as far as Exaltação, and from that which has been written and printed in the Spanish language, that this village is the first which occurs in its navigation during eight days' journey up-stream. It has been founded more than fifty years on the west bank of the river, on a flat piece of land almost square in figure that juts out. The houses are built of mud and thatched with straw. The church has a tiled roof, and is tolerably ornamented. It has missionary-fathers and a coadjutor of the order of Santo Ignacio. It appears that the indian inhabitants are of the people called Cajubabas, and number nearly three thousand persons, all told, and of these 460 are capable of bearing arms.

These indians occupy themselves in making clearings for maize as food and pasturing some horned cattle. Their possessions consist of wax, which is extracted from the trunks of trees where the bees naturally deposit it. They weave cotton into clothing, for their wear and for trade purposes, also some sugar, for which they have a mill. They go to Santa Cruz de la Sierra to bargain these wares for goods, which they need for their use and allowance to the missionary.

A few days' journey from this village, on the west margin, a stream called Aporé flows in. In the part where it is divided into various branches, is founded the village of Santo Ignacio, composed of almost three thousand souls, all told; including three hundred and twenty-eight catechumens. Of those already baptised, 570 are able to use arms.

Having passed the mouth of this river, at a slight distance from it, to the east, is found the village of S. Pedro, which consists of more than two thousand, among them 926 catechumens and 640 of the neophytes capable of taking arms. Continuing up-stream there flows in, on the west bank, the stream called Tiamachu, and soon, above its mouth to the left side, is situated the village of S. Xavier, with nearly four thousand souls, and 560 of these indians capable of being soldiers.

Above the mouth, on the east margin, is founded the village of the Santissima Trindade, the inhabitants of which are called Mojos. It has 1,700 souls all told, of those baptised, and 1,106 catechumens, and of the neophytes 750 capable of using their bows.

Where the river Mamoré unites with the Guapaix to the east is situated the village called Loreto, also of Mojos, with 2,900 souls, including 923 catechumens, and those baptised 660 capable of bearing arms.

Following the Guapaix upwards, quite near the city of Santa Cruz de la Sierra, a stream called Palometa flows in on the right. On it is founded the city of S. José, almost on the slope of the Andes chain. It contains 2,105 souls, all told; and of these 700 are capable of using their darts. Proceeding from the centre to the western part of the Mojos, is founded another village of S. José with 3,177 souls, and among them 1,717 catechumens, and of those baptised 500 capable of bearing arms.

By the same centre, to the western part of Exaltação, is located the tribe of the Mobimas, that, in 1709, becoming apostates to the faith, martyred the Venerable Father Balthasar de Espinosa, who had instructed them in a Christian mode of life. By this people were afterwards founded two villages almost at the head waters of the stream Maniqui, which flows into the Mamoré a little below that of Exaltação. These two settlements are named S. Luiz and S. Borja; the first consists of 1,630 souls, of whom five hundred persons are capable of bearing arms; and the second of 1,300 indians, of whom 400 are able to use bow and arrow, arms employed by all the indians above mentioned.

All the villages here indicated are missioned by the members of Santo Ignacio, of which sacred order the Venerable Father Espinosa, martyred by the Mobimas, was an alumnus. The superior of these missionaries assists in the city of Santa Cruz de la Sierra, and by the direction of this superior is governed all this province, which is called Dos Mojos, through this tribe being the first to relinquish idolatry, and permit the Evangelical promulgation under the monarchs of Castile.

The greater part of the land, which the river Mamoré traverses and where are founded the villages referred to, is so flat that in the rainy season the rivers become swollen to such an extent that the plains are flooded and rendered navigable, the crops receiving great damage. This also impedes the propagation of cattle, and as far as these settlements they suffer great losses, not only from the danger which might result to them from the inundations, but to the towns themselves when the floods are very immoderate. The climate of this region is extremely variable which, besides occasioning terrible maladies, contributes very little to the production of fruits and provisions, so that in some of the towns here mentioned they suffer from a considerable scarcity of all the necessaries of life.

NAVIGATION OF THE RIVER APORÉ

UNTIL ARRIVING AT THE

MINES ON MATO GROSSO.

BEFORE pursuing the narrative of this diary, from where we began to navigate the river, which the inhabitants of Mato Grosso call Aporé, and the Spaniards Iténes, it is necessary to give in this place a true notice of the confluence of the waters of this river with those of the Mamoré, in such terms that an opinion may be formed as to which of these two belongs the maternity of the waters that form the river Madeira but as, for want of instruments, we did not make a regular examination of these rivers, we cannot with assurance affirm which of them is the base.

It is without doubt, that on crossing both rivers at their mouths one cannot venture to decide, by appearance, with indisputable certainty which is the greater in this part, because they there appear equal. Nevertheless, applying all attention to this examination, all those who are experienced judge that the Aporé showed the greater width, and soundings having been taken at the months of both, that of the Aporé was found to have six and a-half fathoms' depth, and the Mamoré seven ; but, notwithstanding this excess, it must be considered that the Aporé has in the part where the trial was made a fathom of sand which crosses its mouth almost from one margin to the other, and, this difference being admitted, no great advantage can be given in this part to the Mamoré.

On the east side, opposite the mouth of the Mamoré, the Aporé forms a bay, in which it shows itself to be the recipient, since on the west side the land forms a convex figure where the Mamoré flows in, and there intervenes between these two bodies of water a small peninsula, as it were, which separates the two rivers, which for three hours' journey run parallel with the course of south-south-west, with which the Mamoré keeps on, as already mentioned, and the Aporé, after the two said hours' journey turns to the east-south-east, which direction it preserves from its source, as will be shown in the course of this Diary.

The banks on both margins of the Mamoré were, in the distance which was traversed up to midday, similar in quality of earth and production of vegetation to those banks which we had observed since we entered the mouth of the Amazon. But it has not any islands, nor does it appear to have any cachoeiras as far as the hills where it has its source. The margins of the Aporé, after it takes the direction of east-south-east, are totally different in the nature of the banks and trees, but it continues with its islands in the same way which it had done from the beginning of the navigation. Its waters, as already mentioned, are clear, and it only remains to conclude that they are similar in taste and colour to those which are found at the entrance of the river, until they began to become billowy, and to contract impurities that are supposed to proceed from the Beni and Mamoré, as has been already related.

In view of all the circumstances here briefly referred to, it will not be sufficient to consider the Aporé as the principal trunk of the Madeira. This opinion has the general support of the Spaniards, who navigated it as far as the long island (whereof notice will be given in its place); and they judge it to be the parent-stream, after having navigated almost the whole of the Mamoré, which, without doubt, they found to be smaller.

After four days' delay at the mouth of the Aporé, we continued our journey on the 14th up the same river with the said course of south-south-west, for the space of three hours, and passing to the south, south-west, east, and with greater frequency east-south-east, in twelve hours' journey ten leagues were traversed, without any diminution, within that space, of the width it exhibited before receiving the waters of the Mamoré, and there was nothing to note further than the diversity of courses that were adopted, and which, when in sight, the cordilheira was also observed to follow.

On the 15th, we continued our journey with the course of south-east and east, and with two hours' journey there was found to flow in on the left margin a lake of considerable size, and continuing with the same course the needle passed at one bend of the river to the north, and at another it turned to the east, in which there was found on the same left margin another lake, by which, at the time of flood, the river introduces such an abundance of water that it forms a wide swamp. This it is customary to navigate in search of big game, which withdraw to the firm land in the centre. From the course referred to we turned to the south at another bend, and with it we halted the canoes at the west bank, which was not flooded, having gone seven leagues in nine hours' travelling.

On the 16th day, we continued on our way with the course of south-east, coasting to the left, and passing to that of east, we found an island lying lengthways along-stream which we had no sooner passed, with the course of east-south-east, than, with this same course, we found another island of less extent than the preceding; and, navigating by the channel on the left, at a little distance, with the same course, we found that the river effected an entrance by the land on the same left bank, expanding itself into a very wide swamp, the mouth of which differs little from the width of the river, which from this place begins already to grow narrower, and only in the parts where there are islands does it divide itself into large bays. After seven hours' journey we halted the canoes, and went, on that day, about six leagues, through having encountered little current.

On the 17th, we started on our journey at four o'clock in the morning, steering to the east, and soon we passed to the south-east, and daybreak found us steering south-south-east, with which course and that of south we went two hours, and on doubling, with this course, the promontory of a bay, we saw in the middle of it a canoe close by the bank on the right, and soon another pushed off from the left margin towards the one we first saw, for which place we were making our way, and on arriving near we found the canoe that crossed, and which was drawn up a distance of two boat-lengths off the shore, waiting the novelty which it offers to the indians to see embarkations different to their own, not only in build, but also in mode of navigating. Not making any attempt to halt, we continued on our way, without altering or diminishing the ordinary speed with which we were accustomed to travel. As soon as the distance would permit their voices to be heard, the indians who were waiting called out loudly, saying these few words in Spanish, " Friends, friends Christian by the grace of God,' accompanying this protest with affirmative actions, one of them drawing from his head a cap, and frequently raising his hand to his breast in token of goodwill and sincerity.

Recognising these indians as being domesticated, as well as several of them that were on the bank in front of the two canoes, and that they uttered the same sounds as the first, our embarkations passed among them without our giving them any further recognition than that our indians replied with the same words, " Friends, friends," etc., and, as no intercourse with them was admitted, they remained apparently occupied with admiration, following us only with their gaze, and resting immovably on their oars.

The canoes in which they navigate are similar to those that our (Portuguese) Paulistas use, and are called *ubás*. They are formed from a bole fifty, sixty, and more palms in length, and from seven to ten in width, without further embellishment than hollowing out the interior with an axe, and cutting out the prow of it from three to five palms with the same implement. In these the indians row standing, as many of them as the length of the canoe will allow. Two persons steer, one at each end, each with his oar supplying the place of a rudder.

The dress of these indians consists of some cotton tunics, well woven, and without sleeves or an opening of any kind at either side, which cover the body half way down the leg, with the hair down long in plaits, or tied at the nape of the neck, and all of them with rosaries of beads or medals hanging to the neck. Only one indian, who, among those in the canoe, showed distinction, was he who mosts tremuously made amicable demonstrations by words and gestures. He wore a shirt, though not of linen; nor was it machine-woven, for it was of a membrane that grows between the bark and trunk of some trees; over this a jacket of blue flannel, cut in the Spanish style ; his hair was gathered into plaits, and his head covered with a turban, also blue, which he removed with great promptitude when he gave the first announcement of his politeness and Christianity.

That bay finishes with the course of south-east, and soon there follows another in that of east, where we saw some islands, which experienced persons said were a quarter of a league distant from the village of Santa Rosa, and, as it was necessary to pass out of sight of the said village, in accordance with the orders that the squadron should not be forestalled by the missionary, we took the expedient of collecting all together at one of the said islands, and navigating by night in front of the said village with the requisite silence, so as to avoid any encounter of questions and replies ; and accordingly it was ten o'clock in the morning when all the canoes were halted at an island by the right margin, between which and the mainland there intervened a channel of, apparently, little more than two hundred fathoms. Here we halted for a convenient length of time.

It was three o'clock in the afternoon when from between the forest-clumps on the banks ·in front where the canoes were moored, two small *ubás* issued, each of them steered by two indians, who no sooner saw our canoes than they made for them, and when they had already arrived at a little distance off, they droned out the cry of the preceding ones, laying stress on " Santa Rosa, Santa Rosa ; " and, notwithstanding that we gave them to understand that they should retire and go their way, they did not attend to that intimation until, with great emotion, they drew nearer to the canoes, and, amidst apparent smiles, in fact, reached them, repeating the expressions of friendliness they proffered before, and, as they gave us to understand that they were, going from their settlement to the village of Santa Rosa, the chief of the squadron deliberated so as to devise a means of deterring them from doing so, in order that they might not go and give an account of the meeting, and thus enable the missionary to make inquiries about our canoes.

Accordingly, we began to traffic with them, and by means of the same Castillian phraseology, together with great courtesy, we effected the purchase of a little maize which they carried with them, and of some fruits, which they willingly bartered for fish-hooks and Flemish knives. We asked them if their plantations were a long way off, and if they could go and fetch some more maize and fruits, since what they had just sold was so small a quantity for such a number of people, and we showed them needles and glass trinkets to excite their cupidity. They conferred among themselves about the matter; and, asking if we would wait there until early the following morning, and hearing us reply in the affirmative, they turned with all haste towards the spot whence they had issued, and plunged into the forest; and for a time that embarrassment was overcome.

Meanwhile, the chaplain of the squadron made a requisition to the commandant thereof, seeking permission to go to that first village to confess to the missionary of it, as he felt indisposed to say mass without that sacrament, which he needed. He was answered that, according to the orders which he was obliged to obey, he had not the power to grant such a permission; that the reverend gentleman was well aware of the precautions and subterfuges he was then adopting in order to conceal our voyage from the Castillian priest, and it was not right that all this should be set at nought, and in disregard of the orders of His Majesty conferring on himself that power which the reverend gentleman requested him to exert.

This reply, though it would have been conclusive to any sincere and rational mind, was not so to the said priest, friar João de S. Tiago, a Capuchin of the Provincia da Conceição da Beira, because he pressed his case with the commander with unusual petulancy, concluding that in case of having orders from His Majesty, they did not apply to him, because he, the said priest, had no obligation to obey them, and that if he (the commandant) did not wish to grant the permission that he (the priest) sought, he would not say mass any more. To this, finally, the chief replied that it would be easier for him to die than to grant such permission. So that, ill satisfied, the said priest retired to his canoe, showing that he did not pay attention to the refusal of the permission he requested.

Night came on, and orders were given to all the canoes that followed their leaders in that of the vanguard, in which the commandant travelled; and, in particular, the two indians who steered the boat of the chaplain were warned not, under any circumstances, to withdraw from the protection of the other canoes, although the priest should wish differently, under pain of severe castigation.

We continued our journey, coasting to the right with the course of east-south-east, and soon the hills on the right margin were seen, on the slope of which is built the said village of Santa Rosa, in the bay in the direction referred to, wherein we passed all the bay and the cachoeira which there is in the river, arising from the various shoals of rocks which issue from the hills mentioned, and cross from one bank to the other along the land on which is built the said village, the river there having the width of the range of a cannon-shot. No other trace of the cachoeira was found than that of a greater current than ordinary, which was easily overcome by oar, as the rocks were already entirely covered through the rise of the river; and so, with two hours' journey, after passing the village indicated and the cachoeira the canoes were halted at nine o'clock in the night at the margin, of the other bay to the right, having, in eight hours' journey during the whole of that day, gone seven leagues.

The sentries were doubled that night to avoid any escape of the indians, but they were not able to avoid the desertion of the said priest, who, that same night, favoured by the darkness of it, hid in one of the two little fishing-canoes; and, when these started very early in the morning (as was usual) in search of game, the priest persuaded the indians to take and land him at the village and straightway return. This was easy for him to effect, because the indians are of a very fickle disposition, and they have no repugnance to being persuaded to any misdemeanour, and the vile interest of a little spirits suffices to buy them over as slaves to any purpose.

At five o'clock on the 18th, we began our journey without having the least knowledge of the assistance related, until, the day breaking, those who managed the canoe of the chaplain gave notice that he was not in the canoe; neither had anything issued from it, nor that of an indian fisher in the service of the priest, who was also missing. Then was supposed that which was mentioned above, after verifying from the sentinels that only

the little fishing canoes had started early that morning. Such
an unlooked-for event caused a very great commotion; but as
the religious gentleman had only taken his breviary with him,
for that was what was missing in the canoe, we resolved to
wait that day, but in a distant part, so as not to expose the
squadron to new desertions of some soldiers of the indians.
Accordingly, continuing our journey to the east-south-east and
east, we found, with this course, that on the west side another
river called Itonomas flows in, which by appearance has a width
at the mouth of 400 fathoms. Its name is derived from a tribe
of indians who have settled there in two villages, the one
called S. Martinho, and the other Santa Maria Magdalena, missioned by the friars of Santo Ignacio, subject, like the preceding
ones, to the superior of Santa Cruz de la Sierra. In this village
of Santa Maria Magdalena, which is eight days' journey up-stream
from the mouth of the Itonomas, Francisco Leme and his companions stopped when they made the first navigation by the
Aporé downwards, and of which mention has already been made.
They say that on a stream which flows into the Itonomas there
is another mission dedicated to St. Peter. It was not possible
to ascertain the number of indians in this or the two referred
to, but the contents and trade resemble those of the two villages
of the Mojos and their neighbours. This river Itonomas
has its entrance to the south-south-west, and its waters are
clear and similar in taste to those of the Aporé. We continued
our journey, leaving astern a great current arising from rocks
which there are on the left margin in front of the mouth of the
Itonomas, and of an island which extends along-stream, at the
point of which the same Itonomas flows in, and having navigated
two hours with the course of east we found that it flowed into
the Aporé. We continued our journey, already leaving astern
a great current arising from rocks which there are on the same
west bank, another stream almost equal in width at its mouth to
the preceding. It is called Baures, the name of a tribe of indians
that dwell there, and are to be found congregated in mission villages like the Itonomas. Its direction appears to be to the south-
west and its waters are very similar to those of the Mamoré and
Beni, but as they are less in bulk, their turbid nature is not

apparent in the Aporé. It is reported that there are four vil
lages on this river, namely, N. Senhora da Conceição,
S. Martinho, Trindade, and S. Gabriel, all presided over by
members of the religious order already referred to. Their communication with the city of Santa Cruz de la Sierra, to the
Superior of which they are subject, is by navigating down-stream,
towards the Aporé, and descending by this to enter the Mamoré,
ascending which, after a two months' voyage on the Mamoré
alone, they reach the city referred to. The same thing happens
to the members of the mission of Itonomas for their annual
supply of necessaries. These consist of the same kinds of
cotton, wax, and sugar, which the missions of the Mojos use for
bargaining. The tract which these two rivers traverse towards
the Aporé is level and floodable, and does not produce metals of
the same sort as the land of which mention has already been
made, when in this Diary we treated of the Mojos. The climate
throughout this region is so irregular, that there is no perceiving
any change of seasons throughout the year, since a perpetual
summer-heat is maintained, while suddenly a very cold wind
from the south might spring up. This change throws the human
system into disorder, and occasions in the inhabitants many and
very perilous diseases ; and from this variableness, which is
almost continual, those plains are incapable of producing
the fruits of Europe, and there is barely a moderate crop of some
vegetables and rice, as well as plantations of sugar-cane.

These missions of the Baures were founded in the last years
of the past century and the first of the present, the apostle of
this gentile mission being the Venerable Father Cypriano Baraso,
a native of Navarre, and illustrious son of Saint Ignatius, who
after twenty-seven years of immense labour in the conversion of
many barbarous nations, and having founded the village of
Santissima Trindade, was martyred by the said Baures in the
year 1702, at the age of sixty-one.

Having passed that district in which the Baures flows in, we
pursued our journey with the same course of east, and halted
the canoes at eleven o'clock in the day at the east margin, where
we were still more than five leagues distant from the village of
Sta. Rosa ; and on this day, in 5 hours' journey, we went 4 leagues.

In this place the commandant determined to wait and see what might turn up about the chaplain, and, it being already five o'clock in the afternoon, the little fishing-canoe of the squadron arrived. Having asked the soldier, who went with authority, regarding the absence of the priest, he did not give any information ; and, being questioned respecting the other little canoe in which two indians and a white Paulista went out on a hunting-trip for the staff of the two priests José Leme do Prado and Francisco Xavier, who accompanied the squadron, replied that this little canoe started at the same time in the morning, and proceeded up-stream as far as the first island. They then put the canoe about and came down-stream along the other side of the river, and did not turn to look, nor did anyone quit the canoe but the usual persons. He was perfectly sure that the priest had effected his flight in the little canoe of the missionaries, this view being supported further by the fact that there was in the nature of the said Paulista every disposition to allow himself to be persuaded by the imprudent suggestions of the said priest.

In a little while this statement was verified, because the Paulista arrived at seven o'clock in the evening in his canoe, so confident and satisfied with himself as if he had achieved some feat of eternal renown, for being asked regarding what had happened, he replied that he entered the little canoe of the priest over-night, and when it started with the other at dawn, he took it to the village, and returning soon with the said priest, looking from the point of the headland, and seeing the canoes in the same place the priest presumed that his absence was already known, and returned with him to the village, and from there he sent to fetch his luggage, as was set forth in a letter the same priest wrote to the chief and which he presented to him.

This Paulista, who was called João Leme, was placed in irons as a recompense for the great action which he thought he had accomplished. The letter of the priest having been read, it was found to contain what the said Leme had asserted, stating, moreover, that he wished to take in that place a cure for some ailments which vexed him, and after having recovered, he would follow on his journey as far as the large island or Mato Grosso in a canoe of that mission. This unfortunate resolution completely disturbed the repose and quiet with which for so many months we had journeyed up to this place.

A great perplexity was offered in the resolution which had to be taken on account of this embarrassment, without prejudice to the royal service, because, in order to go and fetch the priest by force, it was necessary to show the authority which, perhaps, was not yet known to the Castillian missionary. To continue our journey without the chaplain would result in far more serious consequences, because if any of the indians should die without being confessed it would not be practicable to get any of them to navigate our boats. This difficulty we had already begun to experience. The same thing occurred with the whites, as was presently found ; and, moreover, if the father chaplain had not already revealed anything about the details of the flotilla, he would be able, after being irritated through our not having sent him his luggage (which the fierceness of his temper, or continued communication and familiarity with the Castillian might induce him to carry off, after committing the greatest excesses), to relate everything about it, and to give every information as to altitude and courses steered to the same missionary, from whom it was so necessary to conceal all acquaintance with the operations of the escort, since that was the most stringent of the orders given it on starting.

Accordingly, it was necessary to use some judgment to extricate ourselves from the difficulty without embarrassment, which would produce disagreeable results to his Majesty the King. With this object the commandant conferred on the subject with José Gonsalves da Fonseca, who was taken with the escort with the object of furnishing the necessary data for this Diary The conclusion arrived at was that, in any case, the chaplain ought not to be left behind, for the reasons above stated ; while it would be equally unadvisable to repair thither and remove him by force. The medium course which offered itself was that, as there were in the squadron two missionaries, José Leme do Prado and his brother Paulo Leme, who in coming from Mato Grosso had stopped at that village and conversed with the missionary of it, these two could go with the pretext of introducing themselves to the village, and when they should find an opportune occasion they should use all diligence to persuade the chaplain to return ; and so that these two men should not in any way overstep the commission given to them, they were accompanied by a person of confidence in disguise, who was charged to keep near when they were talking to the Castillian priest, in order that they might not have an opportunity of explaining any matter which it was not desirable they should communicate.

The Chief concurred in this view, which he stated he would put into execution provided the said José Gonsalves should be the one to accompany the Lemes, and take upon himself also the duty of persuading the priest to return. There was no hesitatation on the part of the person invited in accepting the commission, as soon as he could get ready the canoe, men, and arms that he thought proper as a safeguard against anything that might occur. To this decision the captain assented, and fixed the following day for the task.

At daybreak on the 19th of February, José Gonsalves sent to unload his canoe, and delivered to the Chief papers which he carried respecting the voyage. He chose four soldiers and a sergeant, and assigned places to two indian rowers and two mestizoes, their slaves, and a negro, and with three other black slaves of the Lemes they were to work the canoe, taking only one dozen indians qualified as pilots, and two picked men for the bow, with arms proportionate to the number of whites and slaves, who were embarked with José Leme and Paulo Leme. At eight o'clock in the morning they started on their way down-stream.

On the way there was sufficient time to instruct the Lemes in the course which they should pursue with regard to the missionary, and they were advised that in his presence they were to treat the said José Gonsalves as their servant, so as to adapt he treatment with the humble dress in which he was attired, without admitting him to any share in the conversation ; and when, whilst still a long way off, the village was seen, the soldiers were assembled under the awnings, with the order not to appear unless called.

At eleven o'clock in the morning the canoe stopped at the

village, and the three persons selected having disembarked, the canoe was made to stand off from the shore, so as not to have any intercourse with the Spanish indians, who soon ran to the bank to stare at the visitors. They repaired to the missionary, whom they found in the refectory with the Franciscan, and as the house was built of mud, with an open door, the inmates could observe everybody that ̍passed. As soon as the Jesuit saw the Lemes, he called them in and made room for them at table, which they accepted courteously, saying at the same time that they would go outside and wait awhile, and after dinner they would converse more at leisure.

The missionary placed them at table, and some roast mutton and custard composed the meal, whereof the Lemes placed a portion on a plate for their new servant, who withdrew from the table and ate the food standing up, until the time arrived for the missionary to receive his guests, he having already dined. The Jesuit again began complimenting his friends, but he soon began to complain about their not having stopped when they passed, nor at least permitted a religious priest from having the consolation of performing those acts of piety which the prescriptions of the faith he professed imposed upon him, and that in this respect they had exercised great tyranny with that reverend gentleman, to whom, on account of his character, all respect was due.

To this complaint José Leme replied that, as regards having not stopped at the village, two reasons prevented their having the pleasure of paying their respects to his reverence: the first was the fear that some of their slaves should escape from them, as had occurred when they were performing the journey down-stream, as his reverence could well remember, because, when a carijó escaped to his port, he was unable to recover him; the second was that they feared to stop, lest, if they left that village without the carijó they might encounter near the river Mamoré the reverend visiting father who was journeying up-stream, who would endeavour to stop them and force them to turn back, saying that it was not permitted to the Portuguese to navigate that river; and the more so if the priest should attempt to use any violence, seizing upon the canoe so that it should not advance,

—that he supposed the said visiting father might leave orders at the village that such a stoppage should be imposed upon them. This was the reason that induced himself and his companions to hold aloof; and as for the priest, they had for the same reasons forbidden him to go to the village, adding, moreover, that he had come in the canoe of a very impertinent man, a resident of Pará, who was proceeding on business to Mato Grosso, who, from the above-mentioned things that he had heard, did not wish under any consideration to stop at the village, and that he, the said José Leme, seeing what had happened, and feeling for the reverend gentleman, had placed his little canoe at the service of the priest, in order that he might proceed in it to effect his reconciliation, and the father (who was then present) could prove whether that was the truth or not.

To this the Franciscan said that he perceived the idea very well, and for this reason he believed that what Leme had been saying was nothing short of the truth, and that his complaint was not against him, but concerning the gift of the canoe in which he had performed his journey. As for the excuse of not stopping at the village, he thought it a lame one, since the object in view was the service of God. To this Leme replied, that his visit there was with no other intention than to offer his reverence a passage in his canoe if he wished to continue his journey. Then followed the excuses of the Franciscan, and other words of little importance, from time to time.

Thereupon, the missionary, in turn, said to Leme that, as far as he understood, regarding the order restricting the passage of the Portuguese, none such existed, and the only instructions on the subject were those by the Superior, under pain of excommunication, that neither he nor any other missionary of the villages bordering on Mato Grosso should have any traffic with its residents, as regards purchases and sales; but there was no prohibition restricting them from bestowing as alms on passengers that which they might need for their sustenance; and thus, if he, José Leme, needed anything, he would attend to the matter with the greatest pleasure. Having thanked him for his offer, he added that, if his reverence would grant him permission to buy some live-stock, he would esteem it much. There was no doubt that, by this time, the missionary was restored to equanimity.

Quitting the field, so as to enter into argument with the Father Chaplain, he retired with the José Gonsalves referred to to a place suitable for the discussion as to whether or not to return to the escort, and after a slight resistance to our first solicitation and urgings, he agreed to continue his journey on condition that the indian fisherman who had accompanied him should not be punished. Our word having been given that the indian should not be molested, with the assurance that it was not he who deserved castigation, the priest inquired what had passed with the missionary regarding the escort.

He was assured that not a word was spoken on this subject, since the pretext he gave for undertaking this journey was that he had come in search of some alms for his hospice, hoping to be able to proceed with those wares to Minas. He easily believed this portion of the affair, seeing that the missionary had not given any intimation of being aquainted with the real state of things.

Accordingly, there only remained to make a careful examination as to what the village consisted of. To this we could obtain the consent of the missionary, if we could influence the minds of the Cacique* and Alcaide by bribes of trifling value, such as fish-hooks, needles, wild grapes, etc. Such gifts facilitated one's entry into all the houses which we thought it worth while to make a note of, and induced the greater number to sell us some birds and maize-meal.

Subsequently to the first voyage, in the year 1742, that the inhabitants of Mato Grosso, already mentioned, made, by way of the river Aporé down-stream, and after they visited the villages of Santa Maria Magdalena do Itonomas, and Exaltação do Mamoré, the Father Athanasio Theodosio, of the Italian nation, founded the village of Santa Rosa on the east margin of the Aporé, at no great distance from where the outlet called S. Miguel flows in on the west margin; but the locality not pleasing him, on account of the great number of ants which devoured the germinating plants, the said priest moved it down-stream to the place where it exists to-day, which is on the bank

* This is a name given to the indian the members of whose tribe consider themselves as his vassals.

of the river by the east margin, almost on the slope of the Cordilheira Geral, which in that place approaches the river and
causes the cachoeira that has been already alluded to in this
Diary. In the dry season it is rather complicated, and with the
river at half height is very dangerous, since the channels are
interspersed with rocks, but cannot be avoided, in consequence
of the furious nature of the current.

The plain on which this village is situated stands in 13 deg.
south latitude. It was cleared of vegetation, the rocks extending
on the banks of the river for the space of a quarter of a league
as far as the starting-point of the mountains which run along
the flank. Following the course of the river, and facing it, are
built the church, the house of the missionary, and of the indians.
It has a frontage of 600 fathoms, stretching along the river in
the direction of east-south-east. From the extremities on each
side of the village thick forest extends from the margin of the
river as far as the hills, forming almost a semi-circle. It is a
pleasant prospect when, from the top of the bank, the gaze
wanders over that plain, the woods, and mountains.

The church consisted only of a nave, without proportion between
the height and length. It is constructed of wood and mud, and
is completely destitute of ornament. The sole article of furniture was a silver lamp of rough construction and not much
workmanship, which shed its light on the host, which is
guarded in the principal chapel. Only in the sanctum itself,
but not on the exterior, was there anything to show that such a
sacred deposit was there. They say that there are valuable
chasubles, in which the missionary attires himself on the principal feasts of the year.

The sacristy runs at the same height as the rest of the church,
and so does the house of the missionary, which is composed of
three little rooms, consisting of a reception room, and there is a
smaller one as a sleeping compartment. Then comes a large
house, which is the only one that is tiled, and pillars to support
it. Here there was a carpenter's shop, where were some specimens of wood, skilfully worked, and the leaves of some doors
were already made, also windows. In the same building was

also a loom, in which cotton thread was being spun. The cloth
was not inferior in texture and closeness to the best production
of Guimarães. There were indians of that same village who
were skilled in both those manufactures. After this large house
followed—similar in height and construction—two houses,
the one a refectory and the other a dispensary. From here, as
far as the church, the roof was covered with tiles, repaired here
and there with a species of grass that resembles the reed of
Europe. This supplies the place of tiles.

The village is of a long shape, has two streets of houses
arranged in straight lines, and built in the portion opposite
the church. Between this and the former intervenes a tolerably
spacious piece of ground. The poverty of the indians is sufficiently manifested in the humble construction of the dwellings.
Few of them are built of mud and straw, and the greater part
of the latter material alone, which is woven in such a manner as
to form the walls and roof. To this outer poverty corresponds
the penury within, because, without any distinction of grade
among the indians, they huddle together in a miserable hammock to sleep, and they have a set of several earthern pots in
which they prepare the maize in many forms, all as insipid to
the taste as disagreeable to the sight. The females are occupied
in this service, and others in spinning cotton for the manufacture of tunics, in which the men as well as the women attire
themselves, with this difference, that the latter use the garment
as a kind of dressing-gown, without sleeves, which covers them
down to the feet, while, with the men, it reaches half-way down
the leg, without any opening in front, and is worn as an indication
of tribal fraternity.

These indians are very well framed, and have a commanding
appearance. Their arms are the ordinary ones, bow and arrow,
and we noted that they had not in their houses any other kind
of weapon offensive or defensive. The first tribe that was
catechised for this village is called Aricoroni, another being
added soon afterwards. Both were dwellers of the neighbourhood, and their numbers now united in the village amount to
500 persons all told, among whom 150 are capable of bearing

arms. The missionary who officiates and instructs them was the one who formed them into a community, and who was there on the occasion of our visit. His name and nationality have already been referred to.

All these indians are ill-satisfied with the administration of the Spaniards, because the Cacique who was best instructed of them all in the Spanish language complained of the excessive paucity in the supply of agricultural implements and of hooks for fishing, so that they have not even a knife for common use, and were often forced to use their stone hatchets to cut wood in the same way they had to do before they were civilised. Their possessions consist of cotton, some forest-wax, and cattle, which they graze on the small patch of land that intervenes between the village and the hills.

Asking the Cacique what was the object of the wrought woods and the cloth which was being woven near the house of the missionary, he replied that the priest intended to move the village to a central spot near the hill-range, and that those woods were for the new church and the houses of the priest, and the cloth was that which they had been accustomed to weave every year to take to Santa Cruz in order to obtain what was necessary for the church and missionary, and some articles for the indians.

The configuration of the perspective which this village forms is offered in the following sketch in which it is depicted.

The hours of meditative retirement of the Father Missionary being over by three o'clock in the afternoon, the chaplain prepared to embark. He carried with him, besides the breviary which only he took to the village, a well-stocked wallet, through the beneficence of the Italian religious order which that missionary represented, for which the Franciscan could not thank him too much. With affability and courtesy the Mineiros said farewell, recommending particular attention which they said was due to the religious gentlemen; and, after some more general words of leave-taking, the bank was reached, whither the guests were accompanied by their hosts. The canoe was ordered up, and the visitors embarked, continuing their journey until four o'clock in the afternoon, and joined the escort at

eleven o'clock at night, the priest entering his canoe as silently
as he had left it.

On the 20th day, at daybreak, we continued our journey with
the course of south-east for two hours, seeing the Cordilheira
Geral to the left, and, passing to the course of south-south-east,
the hills were lost to view. Having proceeded by this course
for little more than an hour, we turned to the south-east, and
soon to the east, and, passing to the north-east, we found that
the river opens out in two large bays, the waters of which flow
round an island almost triangular in shape. At the bank which
presently followed on the left, the river expanded into a lake,
after which comes a large marsh, the river now taking another
bend to the east. With this course we saw on the right the
first plain, along which we coasted for the space of two hours,
and halted the canoes after ten hours' journey, during which we
had gone seven leagues, in consequence of some currents which
we encountered, because the height of the river had already
increased considerably. For this reason the scarcity of fish
became continually greater, so that already at this height we
had to abandon the use of the line, nor could we hope to sustain ourselves by our guns.

This plain stretched inland as far as the eye could reach, and
the pilots said that it extended in the same manner as we saw
here for many leagues, and they were certain that in it, as well
as in most of the others that follow, is seen a portion of the low
land which stretches eastwards from the mountains of Perú.
This track is traversed by the rivers whereof mention has
already been made in this Diary.

The river is accustomed to overflow these plains at the point
of its highest flood, forming, every year, lagoons and very wide
swamps. This prevents the deer-hunting, which is usually
abundant. Nevertheless, these flooded plains are frequently
navigated in small *ubás* (canoes), and on the heights of land
which escape inundation they say that there is an abundant
supply of game. The lagoons also supply those fish that,
wandering with the flood from the channel of the river, find
this new habitation. Here, when the river goes down, and the
lagoons get dried up, those fish that remain in them serve as

food for the birds which gather there in troops, and in the course of time add to the fertility.

21st day. At three o'clock in the morning we began our journey with the course of east, and soon passed to the south and south-west, the plain continuing. On the west and and on the east are some lakes that expand themselves until they mix their waters with those of some pools that the river makes in penetrating the land by various mouths. In little canoes they enter by the lagoons, and, issuing to the pools, large bends in the river are avoided and the journey is made shorter. This is not possible in canoes of forty palms and upwards, because it is necessary to lop away the thick vegetation which, interlaced together, impedes the channels by which the water of the river is conveyed to the centre.

At this height the navigation of the river does not tend considerably to the south, since it inclines so much to the east, while at many of the bends it changes to the north and north-east, and, abandoning these directions, it passes again to the east and south-east, continuing to the east-south-east, which it preserves for some distance, and is the true course. With these points here indicated we passed the lagoons and pools referred to, and soon, after expanding for no great distance, the river began to form other marshes at its entrance on the left margin. These communicated with the preceding ones. Two little islands followed soon afterwards, of a long shape, extending in the direction of the river, which formed between them and the margins three channels, where there was some current. Having passed the islands and the bay in which they were located, we steered to the east, and now on the right plains no longer were to be seen, only a large river-mouth of equal size with the channel of the river. This we navigated in error for a quarter of an hour, and we found that that great volume of water divided itself into three branches: two along the margins of the river, and the other which dilates towards the centre. Having discovered our mistake by the absence of current, we turned the canoes, and, issuing again to the main stream, we continued to the left with the accustomed points above indicated. Presently, passing upwards to the mouth referred to, we came

across another also to the right, which we let alone, and proceeded on our journey to the left, and, navigating with the
course of east, we met with the canoe of the vanguard (which
was that of the chief of the escort), with a large *ubá* (canoe)
measuring sixty palms, with a cabin at the stern ornamented
with leather, having doorways cut in it, which presented the
appearance of a hired sedan chair. In it was travelling a priest
of Santo Ignacio from the village of S. Miguel to Santa Rosa.

The Spanish indians rested on their oars, so as to draw up
the *ubá* alongside our canoe. The loud salutations which the
indians exchanged with one another, such as " Friends, friends !"
etc., fetched the Jesuit out of his cabin, and, seeing the novelty,
he saluted the sergeant-major, to which the latter responded,
and soon the priest asked him whither we were going. He
answered, that we were going up-stream. " That is the truth,"
replied the priest, " but is your worship proceeding to Mato
Grosso ? " He satisfied that question with another in the
following words, " Is that district far from here ? " The priest
contented himself with replying that from there to the long
island would be eight days, and from this to Mato Grosso, he
had heard it said, would be twenty. The sergeant thanked him
for his information, and, presently putting an end to the interview by bidding farewell, he ordered that his canoe should be
rowed on. Those in the *ubá* did the same, and, on passing the
main body of canoes, which we were just coming up with, the
priest ordered three musket-shots to be fired as a salute, which
was received with general silence.

We continued our journey to the east, the whole of the bay
on the right being swampy, and, passing to the points of north
and north-east, we found an island which followed the bend of
the same river stretching a good distance lengthways close by
the left bank. The width is narrow here, and the whole of the
land is subject to be flooded. Soon afterwards, passing to the
south-west, we found a lake on the left, and, after proceeding to
the east, a very long reach. Nearly the whole of the land
along the right margin was level. Having passed two large
islands, we halted the canoes, after ten hours' journey, during
which we traversed eight leagues.

On the 22nd, we began our journey with a north-easterly course, which we soon changed for the east-south-east, and with this we saw the first plains on the left, which, near the margin, were already to a great extent inundated. These plains are densely overgrown with forest, descending from the slope of the Cordilheira Geral, which runs parallel to the river; and within view of these same plains and hills we continued our journey with the same course of east-south-east and at the same plain, to the left, we halted the canoes, after thirteen hours' journey, during which we had gone ten leagues.

At the beginning of these plains was the first site of the village of Santa Rosa, and because of the immense number of ants it was moved down-stream, as has already been mentioned in this Diary.

23rd day. We began our journey to the north-east, and presently to the east and south-east, with which course the plain ended in a very large swamp where the waters congregated, so they said, from the hills as well as the plain. We issued into the river by a mouth of equal extent with the river itself, and a little further on, having passed a small island, the river widens to a broad bay, in which an island that follows the course of the river formed two channels, either of which is of considerable width, and in that to the east of the preceding swamp it discharges itself by another mouth still larger than the one already mentioned; and into that on the west flows a wide-mouthed stream having a tolerably large volume of water. The large bay finished to the south, and about nine o'clock in the morning we saw a small *ubá* containing some domesticated indians from the village of S. Miguel, with whom we did not have any intercourse whatever, and only on passing them did we hear their customary salutations, increasing in loudness, " S. Miguel, S. Miguel," etc.

We went but a little way with the course of south, because we soon navigated to the south-east and east, and at four o'clock in the afternoon we found the bank on the left to be a high cliff, where we found some remains of there having been indian plantations in that place; and, in fact, one of our indians having landed saw from the top of the bank a wide extent of land planted with maize. After half an hour's journey from that high bank, the latter ran towards the hill range, whence issued a stream almost equal in size to the main channel of the river, and a little further on another similar one flows in, both combining to form the greatest unison of waters that there is on that margin. Continuing our journey with the same course of east and east-south-east, having passed the last stream at one hour's journey, we found another flowing in on the right margin far greater in size than the preceding, and that mouth is the stream accepted to be that of the river called S. Miguel, while really it is the stream Aporé, since from the water of the same river, without mixture with any other, it has its entrance half a day's journey from that mouth along the main channel of the river, and inside we expended more than a day in passing a very extensive island, as great in length as in width, which intervened between two large channels which the river formed here, the principal being the one we were navigating to the left. At seven o'clock at night we halted the canoes, after eleven hours' journey, in which we had gone nine leagues.

The village of S. Miguel was very near the spot where we halted the canoes, and as the villagers ran together from the plantations in order to hear mass on the following day, being the 24th of the month, dedicated to the Apostle S. Mathias, some *ubás* of indians passed. The natives were struck with the novelty of the event, and wanted to enter into conversation ; and, in order to avoid them, we passed the canoes over to the other margin of the river, opposite the village, whereupon we received no further visits.

On the 24th day, we celebrated mass very early in the morning, so as to proceed forthwith on our journey ; but we were obliged to delay, as the adjutant of the escort, Aniceto Francisco de Tavora, who carried in his charge the provisions to be served out to the soldiers and indians, represented to the sergeant-major in command that there was not sufficient flour in the commissariat for eight days' journey, even allowing each person merely the rations for one meal a day as we had been giving ; that the river did not supply fish, which had been carried into the forest by the floods, and that the latter did not furnish any species of game, for whatever they might have obtained had retired to the mainland. In that extreme necessity he required, in the name of the soldiers and indians, that, before passing forward from that village, some expedient for help should be taken so as to supply at least two meals a day, seeing that there was no other means of sustenance for the rowers, while on their healthy condition depended the completion of the voyage, and more particularly as some of the indians were already beginning to fall ill, and these could not improve or recover their strength without some chickens, so as to aid by nourishment the good effect of the remedies employed, which, without this assistance, were not only useless but became prejudicial.

The Mineiros, who on their journey were rowed by their slaves and some indians, stated that they also found themselves in need, and their staff had frequently to be supported solely from the forest-palm, and even this they could not always find, so that they needed some support during the voyage of a month and a-half that still remained. Otherwise they would be obliged to turn the canoes to land and form a settlement until they could meet with some luck or some forest tribe that would furnish them with some of the supplies they then needed.

These requisitions were substantiated by evidence. It was impossible that the indians should patiently endure the labour of rowing on one ration of flour a day. As a beginning of our troubles we found a difficulty in adopting some expedient which, without prejudice to our operations, should remove the present need and provide for the future during the considerable portion of the voyage still before us. Having occupied some time in deliberating upon this embarrassment, and it being seven o'clock in the morning without our having decided upon any course of action, there crossed over some *ubás* of indians, which issued from near the village that lay in sight, and they entered the canoe of the commandant, asseverating their peaceable and friendly intentions, and asking if we were going to Mato Grosso. To this he did not definitely reply, and merely asked them if they had any maize-flour and fowls to barter for fish-hooks, needles, cured grapes, etc. They readily acquiesced to the proposal and set off to fetch the supplies mentioned.

Meanwhile, more canoes gathered together from the same village, and in one of them was an indian servant of the mission-ary, a man well practised in the Spanish language. He soon recognised the Minreos José Leme and his brother, from having seen them on the occasion when they stopped at that village, in coming from Mato Grosso to Pará. This indian stated that without permission from the priest they could not sell anything of importance. In view of this, and the need in which we all were, the commandant resolved to send José, Leme and his brother to the village, to arrange for supplies for all; and as the commandant could not leave the canoes of the escort, he invited Gonsalves da Fonseca to make a second journey, disguised, ordering the said Lemes to follow his instructions, and that they should not hold communication with any person unless all three were present. Having arranged this expedient they got into a small fishing-canoe, accompanied only by three confidential slaves, and started for the village, when the canoes had withdrawn a good distance from it, in order to avoid the crowd of villagers, which was already great, but all to no purpose, as none of them had brought anything that would satisfy the requirements of the escort.

The skiff drew up at the port of the village, which is liable to floods, and there is in it such a strong current, that at any moment it would have swamped the *ubás* of the villagers, if they had not taken particular care in landing. From this low beach, which extends about thirty fathoms inwards, the land began to rise for a distance of four hundred (fathoms) a little more or less, until reaching a plateau on which the village is situated. On arriving there, information was given that the missionary was going round visiting the sick, and in this exercise he was met in one of the streets of that village.

This missionary, a member of the religious order of Santo Ignacio, was called Gaspar do Prado, a native of Germany. He was apparently eighty years of age ; of a penitent aspect, but very pleasant-featured and lively. With demonstrations of genuine politeness and pleasure, he greeted the two Minciros, assuring them that he was delighted at their return after such a long absence from their houses ; that he wished to speak to them more at leisure, but this could not be until after he had completed his round of visits to his patients. The Lemes offered to wait, and he sent word by one of the four indians that accompanied him, that the church should be opened for them to go and pray, and that afterwards he should take them to his room where he was accustomed to receive guests, and thus bidding farewell, the priest proceeded on the merciful task in which he employed himself every day after mass.

On the west bank of the river Aporé, there is a stream whereof mention has been made above. Here was located the village of S. Miguel, formed with the indians of the tribe called Moré, and it still existed there at the time when the inhabitants of Mato Grosso made the first journey by the Aporé, but in the year 1744 (they say), having suffered more ailments than usual, Father Gaspar determined to make a move from that place to the new site on the east margin facing the same island on the principal branch of the river, where the tribe referred to is at present located. Other tribes also of the neighbourhood, on the same right side, have been catechised, and live in the same village, making altogether the number of four thousand persons, all told.

The village is built in an oblong shape, with the streets laid out in straight lines, with such regularity that the site being very level they form a square inclosure of large extent, a view of the church and neighbouring dwellings forming one of the sides, while on the others were the houses of the indians, all equal in height, and made of wood and mud, with thatched roofs. In the middle of the square rises a pillar formed of a trunk of more than fifty palms in height, surmounted by a cross, which, though all of unpretending construction, nevertheless offers a pleasing appearance.

The church, although proportioned in size to the number of persons required to take part in the service, yet so roughly was everything constructed that two naves were formed by means of eighteen upright wooden props which supported the main beam at the apex of the roof. The principal chapel is divided from the body of the church by an arch which renders the interior of the chapel extremely sombre, while the whole church is

greatly in want of light. The main chapel contains an altar, at
the entrance to which hung an image of S. Miguel, very roughly
painted and devoid of expression, besides which the workmanship is grotesque. There were no pictures, and the altar lacked
that order and decency necessary to perform the sacred oblation
of holy service in it. In two collateral altars that are placed in
the interior spaces formed by the arch in the direction of the
epistle and evangel, there is equal disorder in the vestments,
and going further, there is on the side of the evangel an image
of Christ crucified, which is enormous in size, but so rudely
made that it seemed a grave impropriety that it should be
exposed to the adoration of the faithful. Another objectionable
feature noticeable in the main chapel was, that, instead of a
lamp and oil, a rough earthenware pan was placed in a corner,
in which burned a material unfit for a private house, so much
the more for that place in which, in the holy tabernacle, presided
the sacred Majesty of Christ. Finally, the whole of this church,
without examining into details, was of such a gloomy appearance that it needed a strong effort of the mind in remembering
that it was dedicated to God, so as not to lose reverence or put
a damper on one's devotion.

After the chuch came the dwelling-house of the missionary
divided into various apartments on the ground floor. These
covered a large extent of ground. From the roof in front descended a portico supported by pillars covering a space of
thirteen palms in width, roughly ornamented between the pillars,
and in this sort of verandah the priest used to receive his guests.
The remaining face of the square already described was occupied
by the engine-house for making sugar. This edifice was built
of very handsome kinds of wood, and all the other offices were
devoted to the service of that manufacture. The machinery
was not then working, but they say that the indians themselves
are the operatives.

The houses of these persons were far better built than those
of the village of Santa Rosa, because, besides being larger, all
were of wood and mud. The condition of the interior was of
equal poverty to that of the houses of the said village. The
only thing which there was in excess, was the multitude of

individuals; for each house was a species of slave-barracks, in which lived three or four families of different tribes without mixing one with another.

There is no general language for these indians as with those of Brazil: the members of each tribe act as if they were in their own forest, and the missionary has with unwearying labour to learn the language of whatever tribe he desires to catechise, in order, by this means, to instruct them in a Christian mode of life and political administration, to which they are admitted.

In the course of time, some indians manage to understand the Spanish language, but there are very few that make the adjectives agree with the substantives; and those who cannot express their meaning fully in words supply the deficiency by gestures, by which, more than any other way, they make themselves understood, and this is the more usual way whereby in the bush, for want of interpreters, they are accustomed to be catechised.

The commerce of this village is in the same articles of which mention has already been made in speaking of previous ones.

This village has communication with that of S. Simão (also newly founded, which will be noticed in its place), in three hours' journey by land to the first cattle-farm, and a day and a half to the town. Between the two, wide plains invervene, in which numerous horned cattle and horses are pastured, and, besides this fertility, plantations are attended to which produce maize and rice with abundance. There is also a vast number of domestic birds that multiply so plentifully that there is a great quantity of all in the same village; so that in several ways the people are better supplied and located than those of Santa Rosa.

The people of this village are, for the most part, well proportioned in height. There are tribes that, besides a good figure which nature has given them, are distinguished in colour from the rest of the Tapuias, because it inclines more to white than brown, or any gradation of shade down to black, which is the usual one in all these tracts. The dress is the same as that already described in this Diary, and only because it was a church-day did we notice on the women their species of finery, which con-

sists in binding the *tipoia* (the name which they give to the
tunics already mentioned) with a belt, so that, a part of the
clothing being gathered up, a small portion of the leg is left
bare, on which there were interlaced several threads of white
dried grapes, of the same sort that they wear on the bare parts
of the arms; and this uncouth arrangement constitutes the
whole of their adornment.

The arms which they use are bow and arrow, and there were
in this village about 800 indians of those baptised who were
capable of using them.

As this population was composed of the Moré tribes referred
to, which inhabited the lands that stretch from the western
margin towards the east, and after the new foundation other
tribes were added from among those living on the east margin
as far as the plateau of the Cordilheira Geral, many families
were accustomed to pine after the liberty of their forests, whither
they returned, abandoning civil life, and plunged into the
neighbouring forests in which they had left their relations. In
order to get them back again, the priest organised some expeditions of veteran converts, who went in search of the
deserters, and on that occasion they were going on a similar
mission, being recommended to two coloured Portuguese from
Mato Grosso who were obliged to take refuge in that village,
where they were admitted to the protection of the priest and
the benefit of this ministry on which they were engaged.

The missionary finished his daily round to the sick, and having
returned to his house he complimented anew his two guests,
asking them to tell him what progress they had made on their
voyage, and, to render more solemn the reply which he expected,
he ordered an indian to play upon the harp for the amusement
of those gentlemen. The Tapuio began by playing an overture,
and without a pause or alteration in the piece the entertain-
ment was changed to a performance on the bones. Meanwhile
the two Lemes were satisfying the question without speaking
about the escort, and passing to trivial matters which took up
time, and already the guests were on their feet to say good-bye,
when with an uncontrollable impulse the Tapuio maintained the fury
of the original harmony which greatly disturbed the conversation.

On saying good-bye, they asked the priest for permission to negotiate with the indians of the village for some provisions, of which there was great need, in order to complete the voyage. To this request he did not offer the least objection ; and soon the priest sent to call the Cacique and the Alcaide, and dispatched them to inform the villagers that they could sell to those two gentlemen all the provisions that they required ; and his benevolence went further, because he gave orders that they should quarter an ox, one portion of which was offered to the travellers, who, having thanked him for his attention, proceeded to the port of the village to await the negotiation for the provisions they had requested.

In a short time, the villagers ran together in large numbers to sell poultry, maize-flour, and fruit ; and, as money does not circulate among that people, bartering is adopted. This is based as follows : two needles for a hen, or, in the same way, two threads of white dried grapes, and so much more in proportion for this unheard-of commodity.

One boat-load was sent off to the escort while another was being negotiated for, and thus in a few hours a sufficient supply was laid in which with prudence would last until we arrived at the large island, where there were inhabitants of Mato Grosso, from whose plantations we hoped to obtain supplies of indian corn. As regards the flour belonging to these indians, it was not well prepared, and would not last all the rest of the voyage, and in order to buy the maize there in quantity it was necessary that the indians should bring it from their plantations, the transport of which occasioned loss of time, which in no case was suitable.

The purchase, or barter, being concluded about four o'clock in the afternoon, the two Mineiros went to say good-bye to the missionary, and they thanked him for the permission he had granted to his indians from which resulted the aid which they had acquired, and after some general words of farewell, without the priest remarking or paying attention to the silence preserved by the poor creature who had come with the Lemes, they withdrew and made for the escort. They reached it after dark, at an island distant from the village a little more than half a league, in an easterly direction.

On the twenty-fifth day, very early in the morning, we began our journey with the course of east, in which, with that of south-east—with a slight change of twice to the south—we navigated during the whole day. As soon as we started from the island mentioned, we went along an almost straight reach. This took two hours, and soon there followed a bend, at the beginning of which, on the right, a stream of pretty considerable volume discharges itself. Presently, after this, came a plain, which extended along the same margin of the river, and, having passed a small island, we also found a plain to the right, from which flows in a stream proceeding from a swamp in the neighbourhood. This plain lies between the village S. Miguel and that of S. Simão, which was still further up-stream. In the same plain, that extends inwards as far as the cordilheira, are pastured the cattle which supply these two villages; these cattle having been moved to this district from the large stock there was at the first site of S. Miguel, where they are still preserved in large numbers.

With the plains referred to in sight on both sides of the river, we continued our journey that day, and, after passing the mouth of a very large swamp that there is on the left, there followed another small island on the right, opposite which is the mouth of the stream commonly called the Rio de S. Miguel, of which mention was made on the twenty-third day. In the dry season this mouth can only be navigated in a very small canoe, because there are sandbanks that embarrass and totally impede the navigation of larger canoes. Opposite this bank, on the left, a very wide lake empties itself, which was formed in the plain and affords navigation to any canoe. On the right there are plantations belonging to the villagers of S. Miguel. There followed presently a pretty extensive reach, and, almost at the end of the bend which succeeded, we halted the canoes, after thirteen hours' journey, in which we went ten leagues.

On the 26th we began our journey with the course of north-east, to complete the bend in the river where we had halted on the preceding night; and, afterwards, with the course of east and south-east, which were the most frequent, we continued our journey for the space of three hours. The margins of the river were clothed with vegetation, and afterwards plains on both sides disclosed themselves, continuing as far as those already mentioned on the previous days; and the reason they were not always seen arises from the impediment of the trees that lined the greater part of the bank. Four mouths of lakes were seen to the left, the waters of which communicated with those of the river, and the lakes themselves communicate with one another, so that it seems as if there were a single bay capable of navigation to any canoe whatever; and so widely does the river expand towards both margins, that now it was not easy to find a small portion of land affording space enough to cook or eat, or even to sleep at night, and the members of our party were put to very great inconvenience by having to use the same canoe for both services.

Already, at this place, sickness broke out vigorously among the indians, because the change of water, as well as of air and provisions, all combined, produced such disorder in the weak constitutions of the indians, that, at the same time, nearly all of them were attacked with a malady which, besides being unpleasant and reducing them in a short time to a complete prostration of vital energy, is highly contagious among that unfortunate people.

On this first alarm, we assisted with such remedies suitable for the attack as could be obtained from among our medical stores and the forests, which latter were more effective. With these and more nourishing food the sick began to recover, and as they grew stronger the misfortune did not so greatly impress them, though the first outbreak had killed two, who, through their extreme weakness, were not able to resist it.

On this same day, it being already dark, we halted the canoes at the left margin, where a swamp flowed in by a mouth of equal width and current to the main channel of the river; and on this day, with eleven hours' travelling, we went nine leagues.

On the 27th day we began our journey very early in the morning, following the part to the left; and, after having navigated for an hour and a half, on our way, it being already broad day, we found that we had mistaken our course, and that was not the river, because the water and current were less, and it divided itself into several branches along the plain. We retraced the distance we had traversed, and proceeded to the right with the course of south, and afterwards to the east and south-east, which were the principal ones.

On one of the branches, in which terminated the river that we navigated by mistake, is situated the village of S Simão, distant from that of S. Miguel a day-and-a-half's journey by land, and one down-stream in a canoe. It was founded in the year 1746 by Father Filippe, a Spaniard, who catechised the indians called Causinos, and afterwards some families of the Cagecerés and Morés, all of whom are inhabitants of the east bank of the Aporé as far as the Serras Geraes.

At that time, the missionary of this village was Father Raymundo Laines, a native of Navarre, and there were in it some three hundred able-bodied indians. From this village to the locality called Corumbiara, the discovery of which will be mentioned further on, are twenty days' journey by land along the Aporé, and crossing the habitations of the tribes called Jaguarotás, Mequens, Guatarós, and those already named, the Cagecerés and Morés, and for the conversion of some of these, the said missionary, Raymundo Laines, had gone and was then personally labouring.

In the directions referred to, we continued our journey, sighting level land to the right, where flow in the mouths of large lakes; and, after passing a small island, the river forms three branches: one, which was our right course, turned to the east, another to the south, which we could not examine to ascertain for certain whether it was the result of a large lake, and the third, which is smaller, inclines to the south. Navigating the last named, towards the left, for a quarter of an hour, it was found that the river, on the same right margin, has two channels opening towards the centre, and it was evident that after receiving the waters of the plain it found an exit by the two preceding mouths, forming a large island in the middle by the blending of these waters.

At a little more than half a league's distance from the place mentioned was found, on the right, the best situation for a town

that had been met with up to this place from Santa Rosa, because it was a plain well adapted for cattle-farms. From it we
passed to high firm land covered with leafy trees, affording not
only a good retreat for the cattle on occasions of great floods,
but also on it could be performed the labours necessary for the
sustenance of the inhabitants, and to this advantage of soil must
be added a remarkable freshness of the breezes, all rendering
that district very pleasant.

Opposite this plain, where it was already covered with forest,
we halted the canoes after six hours' journey, during which we
went four leagues.

On the 28th, and last day of February, we continued our
journey with the course of east-south-east, between a small
island and the mainland on the right, which was a high bank,
with good land for a town and for cultivation, and here the
river narrows more than ordinarily, and soon, in the directions
of east and south-east, the river stretches into a small reach, in
which there is an island of tolerable length lying close to the
left bank. We continued by the right along the same high
land, which terminated at the distance of more than half a
league.

There soon followed an island to the right, after which another was found to the left, where a stream flows in, the mouth
of which extends from there for a quarter of a league, and higher
up on the right is a large lake which flows in with the course of
south-west. Continuing our journey towards the east, at bends,
which did not exceed a southerly direction, plains were discovered to the right, in front of which, at an island near the
left margin, we halted the canoes, after six hours' journey,
during which we went four leagues.

These marches were made shorter on account of the indians
being very feeble through ill-health, and as the greater part of
them gained strength by working at the oar, while, at the same
time, others fell sick, it was necessary to moderate the work so
as not to impede the navigation altogether, since we had to
depend upon manual labour alone for its successful conclusion.

At the beginning of March we continued our journey, with
the course of east, east-south-east, and south, and soon there

followed a turn to the east, which inclined a quarter to the
north ; and thus there are bends one after another, in which the
course varies but little—that is, east-south-east, being the direction which the river takes.

At the end of four bends, which we navigated with the courses
referred to, we began to journey to the south-east along a very
extensive reach, towards the end of which was found the locality
called Das Pedras, because there terminates in that place a rocky
headland which stretches from the interior to the river at its
right bank. There it forms two bays, the promontories of which
terminate in huge boulders, among which is one having three
faces, resting on a fourth as a base, thus forming a figure almost
pyramidical, which, although a work executed by Nature itself,
appeared, from the symmetry of its construction, to be a production of art. It was rendered a more picturesque object by the
extensive foliage that grew at its apex, and, spreading all round,
hung pendent from the sides.

Having passed these rocks, the whole of the succeeding portion
of the margin was of high land, which sloped down to the river.
Here there was a reach that extended for half-an-hour. On the
left bank an immense volume of water flowed in from a stream,
the mouth of which was at a distance of two hours' journey from
the rocky locality.

On the west margin, almost opposite where the said stream
flows in, we halted the canoes, at a place off which the inhabitants of Mato Grosso take up position when they come out
fishing. After eight hours' travelling, we proceeded seven
leagues.

On the 2nd we did not travel in the morning, it being necessary to give the indians some rest, and as, in the neighbourhood
indicated, an attempt was being made to obtain a supply of
game, that place consisting of firm land continuing inwards,
while, as for fishing, we had already given up all hope of such
aid, because the river was completely full, and all the fish were
removed to the swamps and lagoons.

Accordingly, we set out on our journey about two o'clock in
the afternoon, with but a small result from the hunting excursion which was made ; and, proceeding to the east, south-east,

and east-south-east, we passed with this course the mouth of the stream which on the preceding day we had noticed on the left margin. There soon followed on the west a bank of high land well disposed for cultivation, and it finished where the river divides itself into two channels of equal size and volume of water; one of them flows to the south-west, from the mouth of which were seen plains and some hills at a distance. We did not navigate this branch, because the current was very feeble compared to that which there was in the river running to the east. With this course we journeyed to the left. In the middle of the mouth there is a small island, after which followed a bend terminating with an island, at which we halted the canoes when it was already night, and in five hours' journey we went four leagues.

On the 3rd day we began our journey at five o'clock in the morning round a bend in the river, which terminated towards the south, and from this we passed to the east, south-east, and east without there being anything on the margins worthy of noting, except we met with four small islands, and at the last we drew up the canoes after eight hours' journey, in which we went six leagues and a half.

On the 4th we continued the navigation with the course of east, and soon the river took the same direction, and thence round to the north. Without delay we returned towards the east, and passed to the south-east. Here the river widens, and a large swampy lake flows in on the left margin, where an island began and another of considerable length succeeded, which followed the windings of the river on the west margin. Nothing further of any note occurred. We halted the canoes after six hours' journey, during which we traversed five leagues.

On the 5th, we started on our way with the course of south-east and east, with which we passed two bends of the river, and only noticed that a lake flowed in on the west bank. After this came an island, and then a small reach, at the end of which a stream discharges itself, the waters of which unite with those of the preceding lake. The river presently took a bend eastward and thence to the north, and worked round in a semi-circle to the south.

There extended to the left margin a very wide plain, terminating inland with the Cordilheira Geral, which was seen from this place running like the preceding ones with the course of the river. Here, near the level bank, were two islands, one after the other, following the directions of two bends in the river, which reached round from east to north and terminate towards the south, but the direction of the river was steadily east-south-east. In the same manner we navigated another bend, and stopped the canoes after seven hours' travelling, in which we went five leagues and a half.

On the 6th day we began to double a bend with the course of north which terminates to the south. On the left margin was a plain extending to the Serras Geraes, which were visible from the river. The plain continued, while we proceeded to the south-east and east along a very extensive reach, which ended where the river appeared to be divided into two channels of equal size and volume of water. That on the west was formed by a large quantity of water which issues from the swamps and lagoons to form that estuary. That on the east was the one we had to adopt, being the main channel of the river. The margin consisted of land raised into several hills of red earth clothed with leafy vegetation, which rendered the place cheerful and pleasant.

On losing sight of these first hillocks, we met with another small reach, and on the west bank was the mouth of the stream which, receiving the waters flowing from inland, forms the great channel mentioned, and after half-an-hour's journey, having passed the mouth, the squadron drew up on the left bank, after nine hours' journey, during which we had gone seven leagues.

On the 7th of the month we continued our journey, with the course of south for one bend, which reached round to the north-east and finished with south. On the east margin of this bend there was a very beautiful plain, which ended inland at a little more than half-a-league's distance from the high and well-timbered land, behind the shoulder of which could be seen the Cordilheira Geral running with its accustomed course. This plain would be suitable for raising cattle, and the neighbouring lands towards the centre appear to be well adapted for cultivation.

We continued our journey, with other bends of the river similar to the preceding, and, after having passed three, we saw the large island, where the squadron halted about five o'clock in the afternoon, having gone five leagues in six hours' travel. This island is called "large" or "long" island, being the most lengthy one of the many in this river. Almost the whole of it is flooded at high water, and only at its beginning, which is where the squadron halted, is there a small sandy height, a slight portion of which is free from inundation. Here were nine thatched huts in which lived a dozen persons—six whites, including three Portuguese, and six mestizoes, one of them being named João de Sousa, a mulatto, whom we found in the village of S. Miguel, with another companion of the same colour, as has been already noticed in this Diary, collecting together the dispersed members of that community.

The inhabitants of this island are deserters from Mato Grosso, who, being guilty of complicity in some crime or other, sought that retreat to pass there an estranged life, as of beings abhorred by human society. Their practice is to enter into the neighbouring woods and proceed until they discover some trace of human habitation, and thereupon they endeavour to ascertain particulars as to the force and situation of the village; or they await the inmates of some malocca (dwelling), who sally forth into the country for provisions. Rushing out suddenly upon them, and scaring them with shots, they make them prisoners. Or else they single out some village or settlement the people of which are few in number; they make an attack in the dead of night, and, after killing some, they terrify those who cannot flee, and retire with the plunder so iniquitously acquired; reserving the better part of which for their own use, they sell the remainder to the villagers by means of travellers who stop at the swamps (a place which will be mentioned further on); so that from this detestable transaction they obtain what they need in the way of clothing, powder, and lead for the repetition of further raids.

Thus this lost people established themselves in that most vile habitation, where, mingling with the familiar treatment and social customs of barbarians, those miserable persons who receive them taught them to become so fond of the style of life led by these freebooters that to-day they serve them as interpreters, so as, in a friendly way, to obtain great plunder at small cost. It seems even more infamous to practise such things among people of peaceable and friendly disposition, and that persons trained under the conditions of honour and Christianity should be base enough to adopt the same course practised by those who are banded together by force of circumstances. They are insolent to those tribes for which they have no interpreters, and these people in the neighbourhood, on the west side of the island, being harassed, take refuge, numbers of them, in the village of S. Nicolao, situated at the headwaters of the river Baures, which is distant three days' journey by land from the said island, going to the west.

The tribes that inhabit the tract on the eastern side of the river, and among which the inhabitants of the island caused the havoc alluded to, are, in the first place, the Mequens, a warlike people, among whom the missionary of the village of S. Simão had actually gone, having started, as has been already noticed on the.........day, in order to bring them within the pale of the Church, the Ababás, Paivajaes, and Urupunás, a peaceable people dwelling on the plains, like the Mequens. Travessões and Pataquis, nations distinct, but neighbouring and well-disposed for evangelization, populate the plains and the Serras Geraes looking towards the west.

All these tribes dwell among those wild woods and rugged rocks, without any kind of civilization. Natural law is trampled under-foot by barbarous customs with only the exception of eating human flesh. They have no idols, nor do they attribute divinity to anything visible or invisible; and, therefore, all this multitude of pagans lived in the lamentable condition of atheism, from the dark blindness of which the piety of the evangelical ministers who were catechising and instructing them will bring them to a knowledge of the true light, because all these nations are easily persuaded. This was verified with the tribes Ameões and Guaiorotás, almost bordering on the Pataquis, who, with a very little urging, have permitted traffic with the Portuguese more readily than with the Spanish priests, who began catechising them without making them presents of cutlery, which they more especially value; such things, and knick-knacks of still smaller value, usually prove most attractive for conciliating them to the new mode of life.

On this island at which the escort drew up it was necessary to make a halt of six days, in order to replenish our stock of maize in grain so far as possible, so as to last until we arrived at the first settlements near the swamps; but, as the inhabitants of the island had planted only a very limited quantity of that commodity for their own sustenance, the assistance we could get from them was but small, and this had to last us so many days that, apart from the slight utility, this delay produced terrible consequences; because, presently, on the first night, very suddenly a sergeant of the escort died, having suffered two days previously with a slight fever. On the second day some slaves of the officers of the escort going (the indians of the party not being able to go) to the mainland on the west for the purpose of hunting, they found the whole of the country so bare of game that one of them, a mestizo of José Gonsalves da Fouceca, venturing too far inland in search of sport, met with his last misfortune under the paws of a panther that devoured the miserable slave.

On the fifth day, fifteen indians, who were already convalescent from their past misfortune, joined together and fled by night in a *ubá* belonging to the inhabitants, committing themselves to the current of the river, and had the good luck to reach their village on the river Xingú, missioned by the priests of the company. Through these and other calamities of less importance that happened at the port of that unhappy spot, the escort was greatly alarmed, it now being more apparent that we could not replenish our stock of provisions sufficiently without depending on the river and forests, which were already very sterile.

It was not possible to receive the advantage of hospitality in that place, in which, besides the inhabitants being guilty of the atrocities, it appeared that the same malignity was diffused among the inhabitants of the neighbourhood, so that poor travellers indifferent as to their injustice, paid the penalty of repeated misfortunes for holding communication (which they are obliged to

do) with men against whom the priest of Mato Grosso keeps on
hurling his censures, but leaves them intractable as to human
intercourse.

Notwithstanding this unfortunate position of public excom-
municants, the Spanish priests admit them to the sacraments,
when some one of these freebooters seeks the neighbouring missionaries for the annual confession and communion. They
accept him under some pretext that, perhaps, moral theology
may permit to them, which is none of my business, nor is it
proper to ventilate the same here. However, despite all the
good reception they meet with from the said priests, it happened
in the year 1749 that Father Raymundo Laines, a native
of Navarre, navigating from the village of S. Simão, already
mentioned, as far as that island, and attempting to raise
the cross and portable altar, in order to celebrate mass,
the said inhabitants did not wish to allow it; believing
that such an act (although they were religious and of the
Catholic persuasion) performed in that place by a stranger,
and tolerated by the Portuguese inhabitants, would prejudice
the right of dominion of the Crown of Portugal; and only by
force of the persuasion of one João de Sousa de Azevedo, who
was found on the voyage up here, they permitted the said celebration, and, having concluded the ceremony, even in the
presence of the priest they took down the cross, requesting him
that he should not return to that place. The truth is, that but
for those lawless men who dwell there, the Spaniards would
have advanced further up-stream. This opinion was substantiated
by our hearing it remarked to the said fathers that those men
were supremely prejudicial to them, and that from Santa Cruz
de la Sierra would come a punishment on them that would
dislodge them. José Leme do Prado, the surgeon Francisco
Rodrigues da Costa, and Tristão da Cunha Gago, put down this
Castillian bravado, which they had heard them give vent to in
the course of conversation. From this large island up-stream the
river is so cut up by the repeated channels into which it expands on
both margins, so great in volume and of equal current to the
main body of the river, that the Mineiros would not undertake
to guide the escort without the risk of straying from the right

course several times ; and, in order to avoid this disadvantage, which would be attended with fatal consequences in the loss of time which would have to be consumed in useless navigations, the want of provisions being such that they were not even sufficient to last the ordinary time of the voyage without stinting, the sergeant-major took the expedient of engaging one of the inhabitants of the island called José Martins, a native of the town of Obidos, as guide for the voyage as far as the port of Sararé, at a salary of twenty-three *oitavas* of gold, with board, a clause being inserted in the agreement stipulating that he would not be obliged to go beyond the port of Sararé to the settlements. And on the faith of this understanding he embarked in the sergeant-major's own canoe, which led the rest of the squadron.

By the channel on the west, which tends to the south-east, we began our journey on the 13th day of March about nine o'clock in the morning, and, the river continuing towards the east-south-east in bends one after the other extending from north to south with a small reach to the south-east, we navigated until seven o'clock in the evening, and, after eight hours' journey, had gone seven leagues without there being anything noteworthy, further than a swamp on the bank of the mainland so extensive that it occupied three bends of the river, which in the channel we were navigating would have a width of three hundred fathoms, that on the opposite side being less than two hundred, while frequent banks of sand and impediments such as fallen trees rendered its navigation very difficult.

On the 14th day we began our journey early in the morning with the course of north, and soon we turned eastward, with which direction we finished passing the large island, after a journey of four hours, during which we accomplished three leagues, which, added to those of the preceding day, give a length of ten leagues. From that part, where the river divided itself into two branches that are formed by the island, we saw the land on the right through which the smaller channel of the river takes its way, and we observed that the locality was well adapted for a town and plantations, the land being free from the inundations of the river. We continued our journey with the course of east and south-east, and in six hours' journey we went four leagues and a half.

On the 15th we continued the navigation, with the course of east and south-east, of a very wide bend extending on the east to a very extensive plain, within sight of which, after passing two small islands where the river forms a wide bay, we noticed the Serras Geraes at a distance of four leagues, more or less, at the foot of which terminated the plain referred to, that continued along the east margin of the river with some interruption of trees in parts on the bank, but the vast plain that extends inland appears free from forest so far as the eye can reach.

With the same course of east, when already nearly night, we found that the river divided into two parts : one tended to the south with an island in the middle of the channel, and the other to the east, which the guide declared to be the way ; and that which expanded to the south was a very wide bay, and by this route, in light canoes, three bends of the river might be avoided.

At the margin opposite to this bay the canoes were halted, after nine hours' travelling, during which we went eight leagues.

On the 16th, at break of day, we continued our journey with the course of east, and the river turning to the left arrived at the north, and soon an extensive reach beginning towards the south and south-east, we finished at nine o'clock in the morning, after passing two small islands contiguous to each other, the spot being called Casa Redonda. This place on the west bank consists of high land free from ordinary and exceptional inundations. Twelve villages were found established in this district in the year 1743, when the inhabitants of Mato Grosso explored the Aporé for the first time. The principal of these habitations was one built in a round figure, and in the middle was built a house after the fashion of an amphitheatre, and for that reason they called that place Casa Redonda (round house).

Notwithstanding that these people caused no harm, nor did they receive any hurt from those first travellers, still, being afraid of some hostility, they withdrew inland, seeking the refuge of some neighbouring hills (whereof mention will be made further on), where they settled themselves afresh with such energy and celerity, that even on the second voyage that the said inhabitants made they did not find more than some confused ruins overgrown with foliage, to dispel the thought that all was an illusion of the fancy, as to what they had seen in that place on their first trip.

In this place the Aporé makes a spacious ostentation of its waters, swollen with the tribute which the river Cavalleiro pays it. The latter is two hundred yards wide at its mouth on the east margin, facing the site of Casa Redonda. Its direction is from south-east to north-west, and it expands into various courses along the plain which intervenes between the hill-range and the margin of the Aporé, until it loses in it its individuality and name, which was given to it by the people called Cavalleiro who inhabited its margins with the tribe Guaraiutá, and that name was extended to the Cordilheira Geral, in which it rises in the higher part towards the west. By this river issued Antonio de Almeida and Tristão da Cunha Gago, the first discoverers of the lands belonging to the tribe called Corumbijara, where gold was found, of whose exploration by the stream bearing the name of this people an account will be given in its proper place further on, with all possible exactitude, in accordance with the enquiries that were made on this subject in order to verify the truth, which was the principal object of all this narrative.

Leaving, on the right, the site of the Casa Redonda, and on the left the river Cavalleiro, we continued our journey with the courses of south and south-east, in which the reach finished, after which there soon followed another shorter one in the direction of east, and at the end of it flows in a large quantity of water from the stream that traverses the plain on the west which was seen; a little further on, on the left, another mouth, equal in width to the channel of the river, results from a large lake, the waters of which surround an island of lengthy shape. We did not proceed up it to ascertain whether the banks were subject to floods, and after two more bends in the river we halted the canoes, after eleven hours' journey, in which we went seven leagues.

On the 17th we continued our journey with the course of north, in which terminated a bend which we doubled, as well as another until arriving at the south, and from there we journeyed to the east, where began a reach on the right margin of which flowed in a spacious channel equal in width to the main river, being a stream with a very wide estuary, which permitted navigation to every kind of *ubá* many leagues up-stream, and these are the marks by which travellers who have not had great experience are deceived when proceeding upwards, so that when they discover the mistake it is after many hours of lost navigation, so vast is the extension of the waters over those plains when the flood succeeds in overflowing the bank of the river.

After the plain discovered on the right margin, we proceeded along another small reach, terminating in a bend to the left which turned to the north and veered round by the right to the south. Presently, with the course of east, we found that the river was divided into two channels by reason of an island which stands opposed to the current, so that three hours' continued navigation was necessary to pass it. This island follows the end of the river close along the mainland on the left. There soon followed two more channels of the river, without anything noteworthy on the margins, and we halted the canoes on the right after ten hours' journey, in which eight leagues were traversed.

On the 18th we continued our journey, without a change of course or anything notable on either margin, until arriving at a very extensive reach in the direction of south-east and south, in the middle of which, on the right, we halted the canoes, after eleven hours' journey, in which we advanced eight leagues.

On the 19th we began our journey with the course of south-east and east, until passing the reach and two islands, one after the other, the first lying along the left margin and the second to the right, on the low bank of which flows in the river called Paragaú, which, they say, has its sources in the province of Chiquitos—domesticated indians missioned by the Spaniards, religious members of Santa Ignacio, whereof individual mention will be made in the proper place. The river Paragaú appears with a mouth seemingly three hundred fathoms wide. It has its direction to the south through level ground, and there is nothing else worthy of notice in connexion with this river. A small island lies at its mouth, having passed which, the Apor already making its accustomed deflections, we found, on the more concave part of the first, that a lake of considerable size flows in on the right, and, on the same side, at the end of the bend, a stream divided into two channels of greater width than the river itself. Continuing with the ordinary courses, without having anything more noteworthy to record than that we discovered on the right the Campina Geral (general plain), we halted the canoes, after eleven hours' journey, in which we went eight leagues.

It was not possible yet, at this height, to use great force·in rowing, because the indians destined for this work still suffered from the previous complaint, with which became complicated another that came upon them, and in general upon all, namely, intermittent fever—simple, double, and quartain—so difficult of expulsion in the debilitated state of that people, that, although with some only emetic remedies would avail, while with others it was necessary to adopt bark (quina), none of them gained their health permanently; because after a short rest they would again relapse into a feeble condition, resulting more from the disordered constitution, produced by the patients who would not abstain from what was injurious to them, than by the malignity of the attack, which easily yielded to the repetition of the medicines referred to even in cases of relapse.

The greatest consternation which resulted from this fresh attack of diseases was from the want of provisions, and most lamentably that of bread, on sharing out which for the term of the voyage still remaining it was necessary to give half a ration even to those who needed it, although the want of this article was general, all being reduced to a state of debility. If, by chance, we managed to secure, by hunting, a tapir or a bird called mutum, it was necessary to take great care in the distribution of these pleasant viands, in order to keep the sick and convalescent from them, because they are infallible, in however small a portion, to bring on a relapse. In addition to the calamity of illness must be added, on the 18th day, an unfortunate accident which was nearly fatal to an indian sergeant-major of the village of Parijo de Cametá, the most esteemed and bravest man that accompanied the escort. He was following up, in the forest, a small animal, a little larger than a rabbit, called " paca." It had run to earth in a part where he found it necessary to insert his arm into the hole to secure the prey This he did at the cost of a bite on the right thumb from a most venomous snake called " surucucú." This species is frequently found in company with the " paca," so that from the frequent companionship of those two animals arises a fable that exists among the Tapuias, that the " pacas " proceed from those snakes. The bite is fatal to every kind of animal, and it was very affecting to see that indian bear it with so much patience and fortitude. Though the bitten part was quickly cut out and cauterised with fire, so as not to allow the poison to communicate itself through the blood, these prompt precautions were not sufficient —nor were some antidotes that were applied—to revive the patient, who, after three hours of anxious suffering, lost his strength and power of speech, and entered into the agonies of death.

At such a terrible symptom, for want of Venetian theriac we tried the beak of the *Acauan*, and *Unicornio de Inhuma*, reducing them to powder, which were mixed with water, and got him to swallow with great difficulty; but immediately this cordial began to circulate, the lethargy disappeared and the agony was lulled. It became apparent to the indian that he should relieve himself of the malignant oppression which weighed on his spirits and lie on his stomach. No new accident resulted from repeating the wonderful remedy which checked the first attack of the poison. The wound being healed, by taking every care, in a little more than ten days he was completely restored. What greatly aided in his cure was the temperament of this indian, who resigned himself obediently to a very austere diet to which he was subjected during the first five days of his course.

It appeared suitable to give here an account of this success for the credit of the marvellous antidote, which Providence conferred by Nature on those two birds, which are numerous (principally *Inhumas*) among the lakes on our present route. The salutary effect produced on this occasion pervades the whole frame of these two animals, so that any bone, or part of the flesh impregnated with myrrh, belonging to them has equal efficacy in destroying the malignity of a body infected with poison, whether proceeding from the bite of a cobra, as in the case related, or apparent "propinação," as often happened.

On the 20th day, we continued our journey with the course of south and south-east, the river making a swerve to the north, and soon doubling round to the right we found, towards the south-east, after leaving the east side, a very wide plain terminating in the Serrania Geral, which was here of very great elevation. With the same preceding courses we doubled three wide bends which terminated in a small reach, and the latter in a very spacious bay, where the river divided itself into three channels by reason of two islands which stood opposed to the current. Having passed these, we were navigating round a bend, in which, at its termination to the south-east, we saw a plain on the right traversed by a stream of considerable volume, and there soon followed a reach to the east, in the middle of which we halted the canoes, after eight hours' journey, in which we went six leagues.

In this place it was necessary to give a rest to the party on the 21st day; and, although their debility required a long delay, it was not possible to grant that benefit, because it was necessary to push forward, in consequence of the provisions, which diminished every day; and, in that alarmed state, any detention would have been prejudicial, because it would not stop the great amount of sickness, and, as some were free from these ailments but could not resist the horrors of famine, all would fall victims to extreme misery, in a country destitute of any human aid. Accordingly,

On the 22nd we continued our journey without a change of course, because in the direction of east-south-east the river made its customary turns from north to south, and, without having anything memorable to notice on either margin, we halted the canoes after seven hours' journey, in which we went four leagues.

On the 23rd we doubled two bends in the river with the usual courses, and entered a very extensive reach to the east, which terminates where a lagoon discharges itself on the right margin. Without having observed anything noteworthy, we halted the canoes after ten hours' journey, in which we went eight leagues

On the 24th, continuing our journey to the east and south-east, passed some turns, we found, on the right, a high bank of red earth, covered with a virgin growth of lofty trees of thick foliage; and, having passed this locality, near the end of a bend, having a southerly direction, a stream of considerable volume flows in on the left margin, and soon there followed an island contiguous to the bank, sweeping round with the river itself to the east. With this course we saw a spacious estuary formed by an extensive inundation of the low land on the left margin. After passing this locality, and two more turns of the river, without anything worthy of recollection, we halted the canoes after eleven hours' navigation, in which we went eight leagues and a half.

On the 25th, we journeyed without a change of course; and, the river continuing with its accustomed bends, there was nothing further to notice than a continuation of inundated land on both margins, without the slightest height of land appearing at which to halt the canoes, it being necessary, on account of the luggage to accommodate ourselves in a narrow space in the canoes, the healthy mixed up with the sick. This, of all the calamities that they endured, was not the least, for minds oppressed like theirs; because the navigation was on a river which the season had converted into a sea—without fish—which expanded, as if by a similar freak of nature, among immense trees without fruit of any kind, so that the three procreative elements combined towards the sterility of that region, for nothing winged through the air, unless occasionally a number of macaws, emitting hoarse cries, as if, perhaps, they were complaining of not finding the slightest morsel wherewith to appease their hunger. This fact seriously affected some of the patients who were convalescent, because the stock of provisions was very low, and we only looked to obtaining an occasional macaw or parrot,* to extricate us from that most lamentable position.

On the 26th, we proceeded with the courses already mentioned, without anything further worthy of note, except that after eight hours' journey we found level land on the right, off which we halted, having traversed five leagues that day.

On the 27th, we began, with some effort, to proceed with the courses of east and south-east, in order to arrive at the celebrated locality of Corumbijara. Accordingly, at nine o'clock in the morning, we had the fortuitous pleasure of seeing a hill in which terminates the range that to the west has its origin parallel to the plains of Mato Grosso, and accompanies the Aporé as far as this place, when it inclines westward. Hence it continues for a distance of three days' journey, where it attains its greatest elevation.

The canoes having arrived at the place where the Aporé laves with its waters the foot of the lofty mountain that stands at an angle in the direction of that range, we saw, in the part opposite, the stream whereof mention has been made on the 16th day in this Diary, when we treated of the river Cavalleiro; and as this place was remarkable for the opportunities it afforded for the discovery of gold from that spot as far as the headwaters of the Cavalleiro, it will be necessary to give here an individual notice of this success, which was not without important consequences.

In the year 1741, the number of workmen in the mines of Mato Grosso being greatly reduced, these mines having fallen off from the great wealth they showed during the first few years after their discovery, several of these workmen from the mines, who were experienced miners, resolving among themselves to make further explorations in those extensive regions, undertook the task of subduing the wild inhabitants in order to obtain from them some information as to the discovery of gold, so that these adventurers might improve their fortunes, which had diminished through the decadence in the mines of Mato Grosso.

Having taken the measures adapted to the execution of this idea, fifty men banded themselves together, among whom were fifteen whites, Portuguese, and Paulistas, appointing as their director or chief, Antonio de Almeida e Moraes, as he had had great experience in similar exploits. In the year referred to they pushed forward down the Aporé; and these were the first persons who ventured to unfold to the inhabitants the secret in which the navigation was enveloped.

* These two birds, as well as one called *cujubi*, supply, in those desert tracts, the want of fowls for the sick.

After passing the hill-range mentioned, leaving the canoes in a convenient place, they plunged into the forest which extends westward, so impenetrable, by reason of the terrible swamps, that it was with great difficulty they succeeded in wading them breast-deep in water, and finding their canoes, in which, crossing the river towards the east, they entered the stream above mentioned, with the course of south-east, and after a few days' journey they met with the opposition of the tribe called Guaraiutá, who were of such a warlike disposition, that instead of waiting to be surprised by the miners, they took the initiative by attacking the baggage. In this action, the allies exerting all their force, put the indians to flight, taking prisoners those who were not killed in the conflict.

Elated by this victory, they penetrated further inland, without any impediment, where, in various streams or affluents which flow into the river Cavalleiro, they found veins of gold, the trace of which they were not able to follow up in consequence of the immense multitude of barbarians who, with the object of disputing the ground with them, placed themselves in a position to resist the adventurers, so the latter returned to Mato Grosso to give an account of what had happened to the *guarda-mor* (superintendent), who was at that time in charge of the mines of Salvador de Espinha Sylva.

The samples of gold having been received and the information divulged of the fresh discovery in those districts, already, in the year 1743, some inhabitants assembled at the spot, including the *guarda-mor* himself, who determined to go personally to verify the data, in compliance with the duties of his office. They eagerly sought the large brook or stream of Corumbijara, and penetrated as far as the affluents alluded to. Among these they formed their settlement, where, having worked for three months at excavating, there only appeared veins of gold similar to the first specimens. For this task Tristão da Cunha Gago was employed as miner, with other experienced persons, who, after having carefully examined the land, and not finding any further trace than that indicated (which was not reckoned), they concluded that the hollow portion of earth there did not produce that metal, but they were unanimously of opinion that the gold had been washed down by the rains from the Cordilheira Geral where it had its origin to the westward. Moreover, in order to follow up the trace as far as they desired, it was necessary either to subdue or pacify the tribes of wild indians that intervened between them and the hills, for which they were not properly prepared, because the tribes were various, very numerous and warlike. The *guarda-mor*, therefore, adopted the expedient to withdraw the party, together with other private individuals who accompanied it.

However, Antonio de Almeida and Tristão da Cunha did not retire. They were the first discoverers who refrained from attending to their plantations, and various other persons followed their example; so that after more than eight months from the appropriation of those districts they could neither arrive at what they supposed to be the source of the gold, nor had they any greater result for their labour than a loss of time and of many slaves who died or fled.

After this last proof or undeception all resolved to abandon Corumbijara, some returning to Mato Grosso, and others, with the two discoverers, seeking the land to the north, crossing the headwaters of the river Cavalleiro, and always finding good ormations of gold, even near the hill ranges which shelter the villages of S. Simão and S. Miguel. After subduing the tribe of the Amiós, which they entirely routed, their progress was opposed by indians called Guazaités, that in great numbers inhabit the plains which extend as far as the hills. Being so numerous, they forced the adventurers to recede by the way they had taken. They issued by the said Cavalleiro in new *ubás* which they made, and in them they returned to Mato Grosso only having gained some prizes captured from the tribe Amiós.

Up to that time no entry whatever was made by this part of the Corumbijara, because it was necessary to have a troop of great force to conquer the multitude of tribes inhabiting that tract.

The nations best known and most formidable are the Membarés and the Guiniás, their neighbours. They are not wild, of a docile temperament, and well disposed to be civilised. Next follow the Guazaités, who, although they are more daring, nevertheless have a capacity for being catechised. The Ababás and Vrupunás, devourers of human flesh, are somewhat rebellious to negotiate peaceably with. The Guatarós, Maurés, Taquaras, and Causinos are of the most facile disposition to be converted. Many of them have been brought into the villages of S. Miguel and S. Simão. From the last-named to the site of the Corumbijara there are, it is said, eleven days' journey by land.

That which is written regarding these discoveries was heard from the two discoverers, Antonio de Almeida, Tristão da Cunha, and also Rodrigo Francisco, a man of thorough reliability, who was an excavator, with his slaves, up to the last desertion.

The escort continued on its journey, with the courses of east and south-east, within sight of the hill-range on the right, so close to the river that there was an opportunity of examining with the eye their rugged bulk. Its composition is for the most part of bare rock. In some parts, where there are fissures, very crystalline water is precipitated from the summit of those heights. Through the continuous moisture some verdure is produced, which renders the site of those clear-descending volumes a very agreeable one. At the foot of one of these mountains we drew up the canoes, having, in eleven hours, gone six leagues.

By this time the commissariat of the escort already became so low that, even observing the strictest economy, it was not possible to reach the first settlements of the swamps; and as it was practicable to go to these in a light canoe in order to negotiate for some aid in six days going and coming, we took the expedient of despatching, that same night, a commissary to treat with regard to this matter, with orders to return to the squadron as soon as possible, which was intended to continue on its way, though somewhat, slowly on account of the convalescents being extremely weak.

On the 28th day we proceeded with our journey, without a change of courses, in an almost continuous reach, hills being visible towards the west. One of these we noticed to be less in height than the others, having on its summit, as if a termination of its crags, a rock of such union and very fine symmetry that they formed a perspective as of little towers, like those which artificers erect as belfries; and from this circumstance of nature it resulted that the first travellers on this river called the whole range the "Hills of the Towers." Having passed this place, the river makes a little turn, and in it we halted the canoes, after nine hours' journey in which we went five leagues.

On the 29th, the day of the glorious resurrection of Christ, our Lord, after celebrating the bloodless sacrifice of mass, we proceeded on our journey in a continuous reach, having the same hills on the right, until eight o'clock in the morning, when they were lost to view by turning further inland, and soon the river took its direction to the customary course of east-south-east in bends. like those already often mentioned, from north to south, the land on both margins being inundated at high water, without discovering on the bank the slightest portion of it capable of halting at, although we searched until eight o'clock at night, so that we had to rest with the great inconvenience of no other place than the canoes. In twelve hours' journey we went seven leagues.

On the 30th, we journeyed round the usual turns of the river, without a change of course or of margin, for the great space of five hours' journey, until on the left side we found that a large lake discharged itself, after which presently succeeded another estuary of water of the same kind, which the pilots said was of the preceding lake itself. On the right margin we discovered the mouth of a swamp, after which followed a plain on the right, and soon on the left a stream, past which we navigated, and found that the plain continued on the right, high and free from inundations, at the end of which, on the opposite margin, we found the mouth of the stream mentioned, in front of which we stopped, having in eleven hours' journey travelled five leagues.

On the 31st, we continued on our way, with the course of east and south-east, with the same bends of the river, in which intervened some small stretches of swamp on both margins, finishing in two large lakes, the mouths of which opened parallel to both sides. Having passed this place, we saw on the right the Hills of the Towers (Serras das Torres) at no great distance inland; and without there being anything further worthy of recollection, we halted the escort after ten hours' journey, in which we went four leagues and a-half.

On the 1st of the month of April we continued our journey without alteration of the preceding courses, with the customary turns, and in one which reached towards the north and ended to the south, we found a plain on the left which terminated inland with the Cordilheira Geral, which was seen following the usual direction. Before finishing the turn mentioned there flows in on the left margin a large body of water, part of the vast quantity that collects over the plain and descended to that spot. Thus this stream originates from the flood of the rains, which in various channels descend from the mountains to disperse themselves over those wide plains. After this locality followed two reaches, the first to the south, the second to the south-east, and after these the accustomary bends. Without there being anything to note, we halted on that day in front of a bank of high land that there was on the right margin. With twelve hours' journey we went seven leagues.

On the 2nd, navigating without a change of the ordinary courses, we met an extensive reach to the east, where the river being divided into two channels forms a long island, the Serras das Torres being seen on the right, from which descends a river called the Rio Verde, the direction of which is from west to east. It is not large and has several cachoeiras. It discharges itself on the right margin into the channel the Aporé forms between the hills and the island referred to. There is no other note-worthy point about this Rio Verde. At a slight distance from this, the island terminates the reach, and we doubled round the customary bends. Having passed six, we found on the right a mouth equal in width to the main river, and presently, further on, another almost similar, both communicating inland with a very extensive swamp. We proceeded along a small reach, at the end of which, on the left, there was an outlet resulting from the preceding swamps, and there followed some high land on the bank which was already cultivated by the fishermen of Mato Grosso. From this place we saw the Serras das Torres, and continued our journey without further novelty, halting the squadron after ten hours' journey, in which we accomplished six leagues.

On the 3rd, we proceeded with the customary courses along

two reaches, after which followed five turns in the bank, which was inundated on both margins, with so many inlets, that it was with great difficulty we could adhere to the main channel of the river, added to which there were continuous plantations of rice and other grains, which, interwoven one with the other, on the surface of the water, leave the passage very narrow for navigation. The rice of which mention is here made abounds not only in the main channel of the river, but among the lagoons and swamps. It is a spontaneous production of nature, and, after the season, supplies abundant provision for various birds, because there is no inhabitant to take the advantage of gathering it, for which no more labour is necessary than to go in little canoes among the plains, when the slightest movement shakes the grain out of the ears, so that in a short time this species of cereal might be enjoyed long and frequently, but this permission was entirely useless to the need of the escort, as it poured forth at a season unsuitable for taking advantage of it.

We continued our journey by various bends and reaches, having on both margins many estuaries resulting from the immense volumes of water which, at the flood, reduce the low lands to wide and tranquil oceans. At the margin of this kind of sea, as we could not discover land, we made a halt after twelve hours' journey, in which we traversed seven leagues.

On the 4th of the month we continued our journey along a reach to the east which terminated in a bend to the right, finishing to the south-south-west. Presently, there followed the usual courses, without the inundation permitting either margin to be seen by reason of the water, and so extensive and continuous are the estuaries, that it is difficult to perceive the proper route unless care be used ; and in this, and other similar places which have been indicated, the escort would have lost its way but for the benefit of the experienced guide whose skill did not fail him in that labyrinth of waters. From this difficulty we were extricated by a miracle of the most beneficent Providence, for we might be able to get through without any fatal catastrophe, notwithstanding the want of provisions experienced in those most sterile regions, abundant only in water and wild forest.

On this day, in eight hours' journey, we went four leagues.

On the 5th day we travelled rather slowly, without change of course until eleven o'clock in the day, when we halted, having found firm land on the right. Here we found it necessary to cut down some wild palms, so that those who were not sick might derive from the heart—which is of a superior quality—of these trees sustenance for that day, and lay by for another, because there was not a quart of maize-grain left, without any other food for some of the needy sick. It was on this day that the escort was in the greatest consternation of want, not having any other resource than the palm referred to, because the expected aid had not arrived.

On this day, of five hours' journey, we went only two leagues.

On the 6th day, on which the church celebrates the services of Our Lady the Virgin, we began our journey while it was still very early, when, at the break of day, we met with the little canoe which had gone to negotiate for aid, the sight of which resulted in general rejoicing. In fact, what it brought consisted of maize, rice, beans, and some fruit. A distribution was made so as to compensate for the past privation, and we halted for an hour, so that the debilitated might take some refreshment. Presently, we continued on our way without change of course, the greater part of the banks being swampy land, and halted when it was already night, after ten hours' journey, in which we went eight leagues.

On the 7th, with more vigorous effort, we continued our journey, without any change in the courses or bends of the river, in which we found a stream called Capirari flowing in on the right, the entrance of which we ascertained to be to the west, and soon it turns to the south-west. It is about forty fathoms wide at the mouth, but it soon assumes a greater width. Its source is in the Terras das Torres. Its waters are crystal-line, and it is frequented by fishermen from Mato Grosse, who state that there are some cachoeiras in it, but they do not give any further information that could render it memorable. Having passed the mouth of this stream, we continued our journey, following the bends of the river, which were more frequently of smaller size than the preceding, but always with the same direction and courses. The rice-fields continued along the margins, and all the part to the left was swampy. The river, at these parts, has some three hundred fathoms of width at most. After twelve hours' journey, we accomplished, on this day, nine leagues.

On the 8th, we began our journey very early in the morning, with the object of arriving while it was still morning at the first settlements called Pantanaes (swamps); and, in fact, at eight o'clock in the day we found the mouth of a swamp on the left, entering which is a port called Bello, where a transport might be effected in the dry weather to the plain of Mato Grosso, but at that time it was impracticable, as the waters had inundated the whole plain. Having passed this place, there followed three bends of the river, and soon a reach, in which on the right, on high land, are the first settlements. We passed within sight of the first, which is inhabited by a mulatto, and in the following one, at a quarter of a league's distance, is the dwelling of one Joaquim Ferreira Chaves, who fraternised with the first navigators who proceeded to Pará, where being engaged as a soldier he deserted from Mato Grosso, making his journey by Goiazes, whence he passed to Cuiabá, and from this town to his dwelling.

In this situation from which was dispatched to the escort the aid mentioned, it was necessary to halt the canoes on this day, after four hours' journey, in which three leagues were traversed. Here we rested the remainder of this day and also the whole of the following. The land on which these settlements are situated is high, free from inundation even in the very high floods. It is level, and produces very fine groves, which continue as far as the Serras das Torres that lie to the westward. These settlements produce the vegetables of the country in abundance, such as maize, and also rice of good quality, which in size of grain and flavour is not inferior to that of Venice; but that which is gathered from the swamps produced by Nature, without cultivation, is not good enough to be desirable, and it is only through necessity that its use is adopted.

All these habitations among the swamps are favoured with a more temperate climate than the plain, in which there are variations, whereof mention will be made in its place, and the inhabitants of those settlements are less afflicted with illness, and pass their lives in greater plenty than those of the plain, who are often helped from those swamps.

On the 10th day, about one o'clock p.m., we proceeded on our journey, with the course of east-south-east, and sometimes to the south, round small bends, one after the other. There followed two more settlements on the same continent as the first, but the whole of the opposite bank was swampy, from which proceeded some outlets communicating with the river, passing along which we entered, when it was already close on nightfall, a branch of the swamp in search of a settlement on which resided Tristão da Cunha Gago, whence there is also a road to Mato Grosso, but those who know the neighbourhood affirm that it is all inundated in such a manner that not even the inhabitants of the locality could issue into the country to hunt. The escort passed the night in the same place, having in five hours' journey accomplished three leagues.

All these settlements have their labour supplied by some Tapuois who are attached to these districts, or who work in the manner stated in this Diary, and only the plantation of Joaquim Ferreira Chaves was cultivated by slaves from Guinea.

On the 11th we determined to continue our voyage, and to enter the river Sararé, and to seek the halting-place in it called Pescaria (fish-market), that being the place of embarking and landing for persons proceeding to or from Mato Grosso, and even from the swamp, which was very extensive, and deep enough for a permanent lake. We continued our journey by the Aporé, with the customary courses, in the same direction of east-south-east, noting always that on both margins there were several estuaries of water, that left it doubtful which of those on the left was that of the Sararé, and as its mouth is on the swampy margin, we passed the canoes to the part opposite, in the main channel of the river, where we stopped at a margin with good elevation.

About two o'clock in the afternoon we entered the Sararé, with the course of east-south-east, and soon passed to the east, and continued thus until five o'clock in the afternoon, when we made a halt. The mouth of the Sararé was two hundred fathoms wide, and inside it floods more, especially where there are islands, with which it is studded. The margins are almost all of swampy lands. At times of high water there grows along the margins, on the surface of the water, a species of plant called the *auapi*, with large thick leaves, which ramify and form a mass so closely woven that in parts, when crossing the river, it is necessary, for large distances, to clear away these impediments by force of large knives and *machetes* (a kind of cutlass) so as to open up a road for canoes of greater draught than those used for fishing. Sometimes the growth of this impediment assumes such proportions that the only passengers frequenting that river were fishermen, who navigate without the necessity of bursting through that network of such tenacity that it costs great labour to break through. To this must be added many logs that fall from the margin and drift across the river. It is sometimes necessary to chop these through in order to continue one's journey.

Overcoming these difficulties, we navigated, on the 12th and 13th days, until, on the 14th, at three o'clock in the afternoon, the squadron rested at the port Pescaria, having completed nine months' voyage from Pará, where we had begun our journey on the 14th of July of the previous year.

On the same day the sergeant-major of the escort intimated, by letter, to the prefect (*Juiz Ordinario*) of Mato Grosso, Antonio da Sylveira Fagundes Borges, that out of service to his Majesty, he should provide quarters at the settlement of Matriz for the officials of the said escort, and that he was also awaiting in that port some saddle-horses and commissariat, as soon as possible, for the transport of those officials.

With thorough punctuality, the horses arrived on the 15th day, and, on the 16th, at eight o'clock in the morning, we began our journey by wading through a forest, and soon a wide swamp, formed by a flooded plain, with some mounds, at the end of which the said prefect (*Juiz*), accompanied by the Vicar of Matriz and other persons of distinction, were waiting for their guests, whom they saluted with the greatest courtesy, and accompanied them along the two leagues' level travelling which remained, as far as the slope of the hill that has an elevation of half a league, and arriving at the settlement of S. Francisco Xavier, all sought the central church, which is under the tutelage of that saint, where, having rendered to God the thanks due for the benefit of delivering that escort from so great an amount of dangers, the officials withdrew, each one to the quarters which were destined for him.

REPORT TO THE DIRECTORS

OF THE

MADEIRA AND MAMORÉ RAILWAY COMPANY, LIMITED,

BY

EDWARD D. MATHEWS,

Resident Engineer,

UPON HIS RETURN FROM

BRAZIL AND BOLIVIA

IN

1875.

LONDON:
WATERLOW & SONS, PRINTERS, GREAT WINCHESTER STREET.

1875.

TO THE BOARD OF DIRECTORS OF THE MADEIRA AND MAMORÉ RAILWAY COMPANY, LIMITED.

GENTLEMEN,—You chairman has done me the honour of suggesting that I should present you a report upon the various matters that have come under my notice since I left the head-quarters of the railway at San Antonio, in April of last year, and I therefore beg to offer the following remarks :—

INTRODUCTION.

Upon the management of the Public Works Company of their contract I have already fully reported, and doubtless the board is in possession of my ideas upon the subject. In December of 1873 I found it necessary to request the Brazilian authorities to place an embargo upon the plant, tools, and other materials that had been placed at San Antonio by the Public Works Company, as the agents of that Company seemed desirous of taking away the articles that would be saleable in Pará. This embargo was granted me, and nothing was removed.

The engineers sent by Messrs. Dorsey and Caldwell arrived at San Antonio in January, and at once commenced their survey, being successful in finding a very fair line for the first section from San Antonio to Macacos. Their survey and profile for this section would, with some slight modifications, have been accepted by me on behalf of the Company. Thus the allegation made by the Public Works Company that this portion of the line could not be made for less than £20,000 more than their contract price is disproved.

Finding that Messrs. Dorsey and Caldwell's engineers were not prepared to carry on their surveying operations over the whole length of the line, I decided that it would not be my duty to hand over to them the material and plant; but a correct inventory thereof was drawn up and agreed to by Mr. Steele, the contractor's chief agent, and myself. A copy of this inventory is extant in this office.

The advices that I received from London in the early part of 1874 led me to believe that the steps taken by the Bolivian Commissioners in London would lead to a long delay in the prosecution of the works, and I therefore decided to make a journey up the rapids and into Bolivia in order to obtain, by personal inspection, a knowledge of the ground over which the railway had to pass, and of the country it was to be built for.

I arranged to leave Señor Ignacio Arauz in charge as agent of all the Company's property, and left him a proper power of attorney and full instructions as to his duties. From the letters that have been received from him in this office during my absence, it is clear that he has faithfully kept the trust reposed in him by me.

On the 24th of April I left San Antonio and commenced the passage of the rapids, and thus having brought my report to this point, it will be well to make a sub-division of the subject matter of the remainder. This I will do as follows —

1st. The right bank of the river, and the passage of the rapids.
2nd. The river from Guajará Merim to Exaltacion.
3rd. Exaltacion to Trinidad.
4th. Trinidad to Coni.
5th. Bolivia. Coni to Cochabamba, Cochabamba to Sucre.
6th. Sucre.
7th. Sucre to London.
8th. Results to be expected from the enterprise.

I.—THE RIGHT BANK OF THE RIVER AND THE PASSAGE OF THE RAPIDS.

ORGANISATION OF THE EXPEDITION TO BOLIVIA.

A party of Bolivian merchants had organised their return to Bolivia in the early part of 1874 ; having arranged to commence the ascent of the rapids in the month of April. I requested permission to accompany them, it being unwise to attempt the passage of the rapids with single canoes, for one crew is not able to pull the canoe through some of the stronger currents, or drag it over the dry land portages of Theotonio, Girão, or Riberão.

Cargoes taken up to Bolivia.

The expedition consisted of seven large embarcations called "gariteas," two small canoes or "montarias," and probably about 100 persons, including Bolivian Indians and whites. The merchants were taking to Bolivia general stores of merchandize, such as dry goods, guaranà, iron-pots and kettles, ordinary claret, bar iron, and a few other odds and ends suitable for sale in the Mojos and Beni districts. I should estimate the total weight of merchandize thus taken up the rapids at about thirty tons, and its value at about £10,000.

State of the River at the time of the Journey.

The state of the river was considered favourable for the ascent of the rapids, as the water had fallen about six feet from its highest flood level, and was falling daily. Consequently the currents were not so strong as at high water, and many side channels were available which could not be used either at high or at low water. At high water they would be useless, as the currents would carry the canoes into the overhanging branches of the trees bordering the bank ; and at low water they would be entirely dry. At many of the rapids the use of these side channels avoids the danger of hauling the canoes through the main rapids where the power of the water is greatest.

Start from San Antonio.

My own canoe had fourteen paddles, but as I had to take whatever men I could get, several sick Indians who were desirous of returning to Bolivia fell to my lot. This caused me generally to be far in the rear of the other canoes, and often obliged me to take a long spell at the paddle myself, for on many days during the latter portion of the journey I have paddled for eight hours at a stretch.

At starting from San Antonio the baggage had to be taken over to the left bank and there passed overland, while the empty canoes were sent back to the right bank and passed up the channel which separates Holy Island from the mainland. This work of transferring the cargo took up the whole day, and at dark we had not succeeded in re-embarking all the effects, so made the canoes fast to the bank, and passed the night of the 24th still in San Antonio, but above the fall on the left bank of the river opposite the settlement.

SAN ANTONIO TO THEOTONIO.

The next day the 25th, and part of the 26th, was passed in ascending from San Antonio to Theotonio, which was reached in the afternoon of the 26th. The currents on both sides of the river being very strong we could only paddle for short distances, and it may be said that the whole distance was done by "roping," that is, sending forward the small canoe, or montaria, with a light cable, the end of which to be made fast to any available stump or tree, then the crew of the larger canoe, or garitea, haul on the rope until the point to which the rope is made fast is reached, and then the montaria goes ahead again and the process is repeated. This slow mode of progress will account for the great time occupied in getting over the short distance of about nine miles by water from San Antonio to Theotonio.

THEOTONIO.

Arriving at the Theotonio Falls, one has to cross a large bay formed by the river at the foot of the fall, and here the waves are almost always high, especially when the wind blows up the fall. Here my garitea had a very narrow escape of swamping with all on board, for being low down in the water the waves dashed over the quarter, and many of my Indians being Cruzeños, and therefore unaccustomed to canoe navigation, took fright, and by rising up and ceasing to paddle endangered greatly the safety of the canoe. However, by encouraging them by words and prompt action in baling out the water that had entered the canoe, we happily got to the foot of the fall, on the right bank, where the Bolivian merchants that had preceded me from San Antonio were waiting to assist me in passing my garitea and baggage over the portage.

RAILWAY LINE BETWEEN SAN ANTONIO AND THEOTONIO SHOULD BE KEPT INLAND.

The country between San Antonio and Theotonio needs but little description, as the surveys have already been completed to Macacos, and the path has been cut inland of the fall. I observed that between Macacos and Theotonio, the right bank of the river is hilly, the banks rising in places with a straight wall of twenty or thirty feet elevation above the highest water level. The line, therefore, between these points must probably be kept away from the river, and a cut should be run inland from Arauz's Road at any favourable place on the first two miles from Macacos, in order to determine how great is the extent of the rough ground on those miles. The result of this cut would also determine how the drop to lower land at station No. 3 or Ross Town should be overcome, and indeed determine whether such descent be advisable or not.

INLAND LINE ADVOCATED.

An investigation of the surveys and sections made by the Public Works Company shows that the road had found an elevated plateau at about three miles from Macacos, gently rising from 100 to 180 feet above high water at San Antonio. The engineers of the Public Works Company left this plateau and turned down to the river, on nearing which the line encounters a drop of 150 feet in 27 chains, equal to about $8\frac{1}{3}$ per cent., or 1 in 12, but the necessity, if any, for leaving the plateau and seeking the river bank, does not appear.

From this plateau, I think, a strenuous effort should be made to keep the high ground gained, and run the line as direct as possible for Ribeirão. An inland line would probably secure good crossings over the rivers Yaci-Paraná, Tres-Irmãos, and Ribeirão, and avoid any low lying lands that may exist near the outlets of those streams. A glance at the levels given by the Public Works Company's sections would lead to a belief that a great inland plateau, about 150 to 200 feet above high water at San Antonio, exists a few miles from the course of the river through the rapids, and having gained this plateau it would certainly be advisable to endeavour to keep on it. My remarks upon the country through the remainder of the falls, should therefore be considered as remarks upon the river bank, and not necessarily upon the site of the railway.

THEOTONIO HEALTHIER THAN SAN ANTONIO.

To resume then the account of my journey up the rapids. The afternoon of the 26th of April was occupied in transferring my garitea and baggage to the upper side of the fall. The portage was about 600 yards in length and passes over a rocky hill about fifty feet high. Theotonio is reported to be much healthier than San Antonio, and it might be found advisable to make a sanitarium at this fall, or perhaps even make the Drawing and Accounting Offices there.

From Theotonio the canoes started in Indian File, with the small montarias ahead, the passing of the strong currents being much facilitated if this order is kept. The practice is that each garitea should be furnished with a strong but light cable of about fifty fathoms length, and on arriving at a current that cannot be passed by paddling, the montaria goes ahead with the rope of the foremost garitea as before described, and then each garitea embarks the point of the cable of the garitea next behind him, so that a continuous chain is formed over the current, which is surmounted by these means. If the current is a long one and requires more than one "roping" to ascend it, then no garitea is allowed to leave a point gained by hauling until the garitea next behind it shall have reached that point and given the end of its rope again to the foremost one.

THEOTONIO TO MORINHOS.

The 27th and 28th were occupied in ascending from Theotonio to Morinhos, which was reached on the morning of the 29th. Several strong currents are met with on this length of the river, and in one of them my garitea became fixed on a rock. A montaria that was sent to my help by my Bolivian friends was driven with great force against the bow of my garitea and sucked under water by the current, the four men who were in her with difficulty escaping by clinging to branches of trees, while the montaria did not appear, until some distance below us it came up to the surface of the water.

ARAUZ'S ROAD.

The river bank between Theotonio and Morinhos is very level, and Arauz's road, which runs close on the river edge, would serve very well for the site of the line should it prove desirable to leave the higher ground before referred to. I may here mention that Arauz's road was in most parts where I visited it much overgrown, but it was evident that the work of clearing had in the first instance been well done, all the large trees having been felled. The overgrowth was of quite recent character, and being entirely composed of soft and succulent plants, would be easily cleared away ahead of any exploring party or travellers by a very few Indians.

MORINHOS RAPID PASSED.

The Rapid of Morinhos was passed with gariteas half unloaded and in two rope's lengths through a channel on the right bank of the river, about four hours being occupied in passing all the seven gariteas through the currents. Above the real fall, and perhaps only 200 yards further up stream, very strong currents were met with and the whole of the rest of the day was occupied in overcoming them.

The river Morinhos passes close by the foot of a hill which rises to perhaps 100 feet, but Arauz's road runs at the base of it with plenty of room for the line without causing any considerable work.

ABOVE MORINHOS RAPID.

From Morinhos upwards we continued paddling and roping up the right bank which presented a uniformly level appearance, and on the afternoon of the 30th we crossed over to the left bank of the river, in order to avoid many strong currents which are met with near the mouth of the Yaci-Paraná.

YACI-PARANÁ.

From the section shown by the Public Works Company's Engineers of the river bank in the neighbourhood of Yaci-Paraná it would appear that some two miles and a quarter of the bank is liable to flooding at high water. The exigencies of the ascent forced me to keep in the company of the other gariteas on the left bank of the river, and from that side I could plainly see that although the land on the river bank upon the opposite or railway side was low lying, there was high land about a mile and a half to two miles inland. This also is another argument in favour of keeping, if possible, on the high plateau found by Arauz's road between Macacos and Theotonio.

CALDERÃO DO INFERNO.

On the 3rd of May we arrived at the foot of the Rapid of Calderão do Inferno, having once visited the railway side of the river finding the bank as usual very level and suitable for the site of the line if necessary. The fall was passed on the left bank or opposite side of the river, and it took us the whole of the 4th and part of the 5th to get the cargoes carried over the portage of nearly a mile in length, and the canoes hauled empty up the rapid. Of the right bank of the river at this fall I cannot speak from actual observation, but my assistant Mr. Dalton who was over this portion of Arauz's road in July and August of 1873, reports it to be "uneven and very rocky from the 'Rabo' or lower fall to the top fall a distance of three or four miles." This description would also do for the left bank that I was over, but the rocky ground there, was principally composed of large boulders and a section over it would not be very heavy. The section given by the Public Works Company doubtless indicates correctly the right bank of the river at the fall, and shows that a line along the river margin would not meet with any more serious obstacles than a short cutting of about ¼ mile in length and 10 feet average depth.

From Calderão do Inferno to Girão.

On the 6th of May the canoes were re-loaded, and the upward journey was recommenced, and by night fall the foot of the Girão falls was reached. When nearing the falls the hills above Girão show bold and high, but appear isolated and have not the appearance of forming part of any continuous range. It seems therefore probable that the inland line that I have suggested would avoid the broken ground at Calderão, Girão and Tres-Irmãos. There is nothing special to remark of the river bank between Calderão and Girão. Arauz's track was taken about two miles inland and ran over a mile or so of very uneven ground, the section showing cuttings and banks of 20 and 30 feet in depth. My assistant Mr. Dalton describes this portion of the line as "passing over very bad country," but there was no reason for taking the track away from the river bank where a much lighter section could have been obtained.

Deaths of two Indians at Calderão do Inferno, causes explained.

At this part of the journey I had the misfortune to lose two of my crew, a man and a boy, and as the fall of Calderão do Inferno bears nearly as bad a name as San Antonio for fever, I shall explain the causes that in my opinion account for these deaths. Firstly one should observe, that in ascending the rapids, the crews are composed of Indian sthat have been for some time engaged on the Madeira river in rubber gathering. Those of a weak constitution suffer from fever and are naturally anxious to embrace an opportunity of returning to their country, they get over San Antonio, Theotonio and Morinhos falls and it is at Calderão and Girão that their strength begins to fail and the weak ones of the crew die. I do not believe that there is any special danger to life at either of these falls, save only that it is thereabouts that the constant wettings and hard labour of the ascent commence to tell upon the Indians. The boy that died in my canoe shortly after leaving Calderão was an earth eater and having indulged too largely in that disgusting vice of certain of the tribes, I believe that the intense heat of the sun killed him, his stomach being in a thoroughly disorganised state. The man, had stolen cachaça (native rum) at Morinhos and while intoxicated had fallen in the river, this brought on a severe cold. The nights were wet and cold and he according to a custom that exists with many of the Indian tribes determined to die, and received his relatives' commissions to those members of his family that had preceded him in their last journey. I therefore found it impossible to save his life and during the night at Girão he died. I myself was very unwell at these falls, but it was merely a return of a bilious fever that I had suffered from in San Antonio and could not be anything special to the part of the river about Calderão and Girão.

Passage of Girão Rapid.

The passage of the Girão falls required that the unloaded craft should be hauled over the portage of about ½ a mile in length, the descent from the summit being down a steep rock. This work occupied the whole of the 7th of May and on the 8th the upward journey was continued.

Girão to Tres-Irmãos.

The 8th, 9th and 10th of May were occupied in passing from Girão to Tres-Irmãos, the railway bank of the river appearing to be almost level and with few streams or ravines. Of the many streams marked on the survey of the Public Works Company on this length, I only noticed two that were of any consequence and for these openings of 60 or 80 feet would probably suffice.

Passage of the Rapid of Tres-Irmãos.

The rapid of Tres-Irmãos was passed on the right bank on the 11th without any of the canoes being unloaded, and without greater trouble than that encountered in strong currents between many of the falls.

The river of Tres-Irmãos is a considerable stream, I ascended it for about a mile. The banks are of solid earth and suitable for receiving screw piles. Openings of an agregate waterway of !80 feet would, if the river is to be passed anywhere near its outlet into the Madeira, have to be provided.

PACAGUARA SAVAGES.

Here we met the first savage tribe that we had seen, consisting of men and a child who belonged to the " Pacaguaras " that have their " Malocales " or settlements on the river Tres-Irmãos. The men had on shirts and trousers received from the engineering parties that had been up the river, and were very well behaved and at our request brought from their plantations a supply of plantains and fresh maize for which they received the payment of an axe, machete, or a few fish-hooks and some line. These savages might if properly treated be made useful in fishing and hunting, and in scouting for the few tribes that are hostile.

TRES-IRMÃOS TO THE BIG BEND.

On the 13th we reached the point where Arauz's road leaves the river and crosses the big bend. This point has been named "San Louise." The river bank from Tres-Irmãos to this place needs no description being in every respect favourable for the railway if the line has to follow the river's course.

ACROSS THE BIG BEND.

The track across the big bend was described to me by several of the Indians who had accompanied Sr. Arauz, as perfectly level and dry, There has been some idea of its being swampy during the rainy season, but enquiry proves this to be incorrect, for the section given by the Public Works Company shows rock at the bottom of all the small streams crossed. I had in my canoe a very intelligent Bolivian of rather a superior class, who had been one of the mayor-domos or foremen under Senor Arauz, and this man assured me that from the track over the big bend, looking eastwards, no sign of hills could be seen, but only a vast undulating plain. Senor Arauz has also assured me that there are no hills to be seen, and all the evidence that I could collect leads me to the belief that a cut across the interior, entirely avoiding all the bends of the river, would not only shorten the length of the line, but would locate it where the smallest amount of earthwork and bridging would be met with.

PAREDÃO AND PEDERNEIRA RAPIDS.

Four days were occupied in passing round the big bend, and overcoming the rapids of Paredão, Pederneira, and another fall above the latter that is not named or mentioned in Messrs. Keller's maps.

FROM END OF TRACK OVER BIG BEND UP STREAM.

On the 17th we passed the huts called "La Cruz," at the point where Arauz's road struck the river bank after passing the big bend. The land hereabouts is fairly level, but there are several ravines, three of which would require sixty feet spans. Mr. Dalton reports the first mile up from "La Cruz" as "low-lying," but I did not notice this considering it fairly suitable for the line, and on this length I notice in my diary, that I landed twice.

AS PENHAS COLORADAS.

About eight miles above La Cruz, occur on the right bank, the hills called "As Penhas Coloradas" the "Red cliffs" which rise straight up from the water a height of perhaps a 100 feet. The Public Works Company's Engineers surveyed Arauz's road over these hills, but do not appear to have taken any levels, for in their sections about 25 miles from stations 17 to 19 is entirely missing, neither can I find in their level books any record of levels having been taken over this length. These hills continue for nearly three miles when the land returns to its natural level. The ascent and descent of these hills would, if the line has to go over them, be severe and give heavy work, but there summit is tolerably flat. To cut a side benching round on the river side of these hills would be an almost impossible work, for the current is very strong at their base, and therefore a bank would require to be protected from scour by expensive piling or revetment.

ARARAS RAPID.

The Rapid of Araras was passed on the right bank on the 18th, without unloading canoes, and with only one "roping" of about 80 feet. It is said that when the river is low the left bank is the best for ascending this rapid, but with water the right bank is preferred, as the " olada " or wave is high on the left bank when the river is at all full.

ARARAS TO THE RABO DO RIBEIRÃO.

The right bank from Araras, past Periquitos and to the " Rabo," or the " tail " of the Ribeirão falls requires no description. It is all fairly level, and the section given by the Public Works Company shows that there are no earthworks of any consequence. This portion of the journey was passed on the 19th, the rapid of Periquitos being ascended on the right bank without unloading canoes.

THE RABO DO RIBEIRÃO.

The passage of the " Rabo " and Fall of Ribeirão occupied the 20th, 21st, 22nd, and part of the 23rd. The Rabo extends for about five miles below the real Fall of Ribeirão, and is a succession of whirlpools and currents, extremely dangerous to canoes, either on the upward or downward journey. The downward journey is by far the most dangerous as the canoes have to be steered in full course through the boulders and rocks scattered over this length of the river, which here has an average fall of about $4\frac{1}{2}$ feet per mile.

NAVIGATION OF THE RABO DO RIBEIRÃO IMPOSSIBLE.

If any scheme of partial navigation between the rapids is ever carried out, a tramway would have to be laid down over this length, for no steamer could live in these broken waters. The bank does not offer any material obstruction to the building of the line up to the fall, at which place the levels taken by the Public Works Company's Engineers appear to have concluded, for no section is given above the river of Ribeirão.

THE RIBEIRÃO FALLS.

At the foot of the fall the river forms a bay somewhat similar to, but smaller than that below Theotonio. Round this bay the water runs in a return current which flows with great force up stream, and if the crew do not paddle with great strength, the canoe is in danger of being carried into the rough and broken water at the foot of the fall. To pass the upper fall it is always necessary to take the right bank of the river, and unload the canoes entirely, hauling them over the portage of about 500 yards in length, and re-embarking the canoes above the fall.

The River Ribeirão.

The River of Ribeirão, which is at the foot of the fall, is a consider-able stream at the outlet, but narrows a very short distance there-from—a sixty foot span would probably suffice.

From Ribeirão to Misericordia.

The land on the railway side of the river from Ribeirão to Miseri-cordia is rocky and rather uneven, having stretches of raised lands for distances of a quarter to half a mile intermixed with reaches of the ordinary level. The elevated portions are not, however, more than twenty feet above the level ones, and therefore the works would not be heavy as the elevations could be graded up and down.

The Misericordia Rapid.

The Rapid of Misericordia, passed on the morning of the 24th, is especially dangerous, although at first sight it appears a mere " corriente." The river swirls over a point of rock and forms a suc-cession of whirlpools from which no canoe, if once drawn in, can escape. In ascending, this rapid must be passed by a creek on the right bank. It is said that at low water this rapid does not offer much obstacle, but I did not receive any information from my Indians to that effect. The channel on the right bank offers no obstacles that cannot be overcome by hard work in unloading or hauling, and is there-fore preferable. The descent should be made almost in mid-channel, while if anything, steering to the left rather than to the right bank.

Captain Miller of the Steamer "Explorador" met with.

At this rapid three canoes descending from Bolivia came up to us and in one of these was Captain Miller, who had assisted Dr. Velarde to take the " Explorador " up the rapids. Captain Miller had left the Company's service, served the late contractors the Public Works Company for a time, and was then on his way to San Antonio or Manaos in search of employment.

Sr. Barros Cardozo, Brazilian Consul at Exaltacion, reported Assassinated·

From him I learned that Señor Barros Cardozo, Brazilian Consul in Exaltacion and the Mojos provinces, had been assassinated by one of his own Brazilian servants. I was pleased to find that this outrage had not been committed by Bolivian Indians, who are, as a rule, much quieter and more tractable than the mixed races of Brazil.

MISERICORDIA TO MADEIRA.

Proceeding onwards from Misericordia our canoes kept on the right bank, the country showing fairly level and with few igarapés (ravines) until about two miles above the fall, where the river bank appears to be rather low. Here it would be advisable to locate the line on the land side of a hollow that is to be found a few chains inland, especially as these rather low-lying lands extend nearly up to the cachuela of Madeira, which was reached on the evening of the 24th, after having passed four strong currents, and arrived at the principal fall, which was to be passed on the following morning with canoes half unloaded.

THE MADEIRA RAPID.

At the Rapid of Madeira there is a hill of perhaps 150 feet high on the railway side, but it is evidently an isolated bluff, and even if the line should not be taken inland, there is room at the foot of the hill on the river side, for the road. Arauz's road has been carried over this hill, but needlessly so.

JUNCTION OF THE RIVER BENI.

The whole of the 25th was occupied in passing the principal fall of the Madeira Rapids, and the currents above it. The junction of the River Beni occurs amongst the currents above the principal fall, and from this circumstance the fall has been named the "Cachuela do Madeira," as below the junction the Madeira river is said to commence. From the junction of the Beni to that of the "Itenez," the river is by some called the "Rio Grande," by others the "Itenez," and it is only above the latter junction that the Mamoré reigns. Above the Madeira Fall, and in the neighbourhood of the junction of the Beni, the river is much broken up by islands, and consequently the navigation of this stretch is almost, if not quite as bad as that of the "Rabo do Riberão." At the junction of the River Beni, the river appears to divide into two parts, and they appeared to me to be of almost equal width and volume of water.

THE LAYES RAPID.

Early morning of the 26th the expedition continued the ascent of the rapids of " Layes " formed by two small falls with a current below them. These were passed with canoes half-unloaded. Near the fall, but below it, is a wide stream on the right side of the river, which would probably require a span of hundred feet, while the earthworks hereabouts would be inconsiderable.

COCOA TREES VERY PLENTIFUL.

In this part of the river I noticed a great many wild cocoa trees, which, although growing almost universally on the banks of the Madeira and Mamoré, are hereabouts more thickly collected together. The fruit of these trees is of very superior quality, and with very little trouble an excellent plantation might be organized.

THE PÃO-GRANDE FALL.

The Fall of Pão-Grande we reached on the morning of the 27th, and had to unload all the cargo and carry it over the portage about a quarter mile in length. This fall is impassable by loaded canoes at any season, and only on the right bank can empty ones be hauled up the falls of a channel between an island and the mainland. This channel has two " saltos " or jumps each of about four feet in height, and it was wonderful to see how well the Bolivian Indians manage to make the canoes ascend these almost perpendicular falls. The right bank is very rocky, and considerable work in removing boulders may be met with here. Above the fall, there are several bluffs, and over one of these Arauz's Road has been taken. My assistant, Mr. Dalton, says of this part of the line, " Above Pão-Grande the road " is good for about five miles, when three or four very stiff rises are " met with, over these for some un-accountable reason the road is cut " whereas had it been carried along the river bank, an almost level " line would have been found." In this opinion I concur, and this fall will, I believe, be passed by the railway with a section somewhat similar to, but certainly not heavier than, that of the first three miles from San Antonio to Macacos.

PÃO-GRANDE TO BANANEIRA.

Between the falls of Pão-Grande and Bananeira the road is fairly level with no ravines that cannot be crossed by thirty-foot spans. In this stretch there is hollow land a short distance inland from the river bank. This low-lying land is probably inundated in the rainy season by drainage water that cannot get through into the river, but there is a width of high land close on the bank sufficient for the line, and probably the pools could be drained through this bank, or by deepening the existing igarapés.

The Rapids of Bananeiras.

The Rapids of Bananeiras were reached at early morning of the 29th, and ascended on the left bank. The principal fall had to be passed with empty canoes, the cargoes being carried over a very short portage across an island. The head currents of these rapids however give nearly as much trouble as those below Riberão, and are very dangerous. They occupied us for the whole of the 29th and the 30th.

Rubber gathering near Bananeiras.

At this fall, about eight or ten years ago, two Bolivians made a small settlement and collected rubber. One of them was returning to Bolivia with our expedition after having made a small competency by rubber-gathering on the Madeira River below San Antonio. He told me that he and his friends stopped at Bananeiras for nearly twelve months on the Bolivian or left side of the river. They found that the quantity of rubber given by the Seringa trees of the rapids was not so great as that given by those of the lower Madeira, but their health was uniformly good and they were not molested at all by savages.

Bananeiras to Guajará-Merim.

From Bananeiras up up to Guajará-Merim I was obliged to keep on the opposite side of the river to that on which the railway has to run, but the many glimpses I got of the railway side would lead one to believe that there do not exist any obstacles to construction. The surveys of the Public Works Company also warrant the same inference being drawn, the only noticeable feature on this length being that at about five miles below Guajará Guasu a marshy lake is said to exist, but the usual high bank appears to run along the river margin, with width sufficient for the location of the line.

Above Bananeiras and Guajará, the hills of the " Sierra da Paca Nova " show bold and high, being apparently some 30 to 40 miles inland.

PORT SELECTED BY ENGINEERS OF PUBLIC WORKS COMPANY BELOW GUAJARA-MERIM.

The rapids of Guajará, Guasu and Merim, offer no obstacle of any consequence to free navigation either in ascending or descending, the fall being but slight and distributed over nearly a mile of the course of the river. I am informed that one of the Engineers of the Public Works Company selected a landing place about a mile below Guajará-Merim, it being in his opinion suitable for a terminus, having good water access in the dry season and being apparently a healthy spot.

LAST OF THE RAPIDS PASSED.

The last of the rapids was passed on the morning of the 1st of June, after thirty-seven days' hard and constant labour in battling with the force of the many currents.

GENERAL REMARKS UPON THE COUNTRY THROUGH WHICH THE RAILWAY HAS TO PASS.

Before passing to the second division of my report I will shortly state my belief, that in the whole stretch of country from San Antonio to Guajará-Merim, there is no natural obstacle that should at all call for doubt as to the feasibility of the construction of the Madeira and Mamoré Railway. If we take the 240 miles of river bank into consideration, it may be broadly asserted that few if any rivers that are encumbered with rapids, have such almost uniformly level banks, as those of the Madeira. My assistant Mr. Grant Dalton, who had the opportunity of accompanying the Public Works Company's Engineers over the whole of Arauz's road from San Antonio up to Bananeiras, says in his report to the Company dated June 16th, 1874 ; —" There are no Engineering difficulties of any moment, tke road " will as a rule run through a level country, inequalities of ground " being rare between the Cachuelas. At the Cachuelas the ground is "invariably uneven, but will be easily got over with light banks and " cuttings in which not much rock will be found."

TIMBER FOR SLEEPERS, BRIDGES, &c.

Timber of good quality for sleepers and smaller bridges or culverts is found, throughout nearly the whole length of the line. The Public Works Company made a provisional contract with Señor

Ignacio Arauz for the supply of sleepers of good hard wood at three per milrei, or say eightpence a piece. There are many kinds of excellent hard woods, but it is difficult to get the proper names of them, as the workmen of Bolivia of course do not know the names used in Brazil. They are however universally good judges of timber and can select woods that will not perish when used in the ground. Pāo de Arco, Massuranduba, Laura, and Itauba, may be mentioned as being amongst the best kinds of timber for sleepers or bridging.

Mr. Caldwell's Contract.

A contract having been entered into by Mr. Josiah Caldwell, estimating is on my part unnecessary, but I may here say that in my opinion he has secured an excellent contract, and that I believe if he sends a few Engineers like those he sent out in the early part of last year and backs up their efforts by a judicious selection of imported labour, and a careful arrangement for supply of materials and stores, two and a half years counted from the commencement of the works should see the Madeira and Mamoré Railway an accomplished fact.

Climate at the Rapids.

The question of climate in the rapids is I think now proved to be, that although at San Antonio ague fever is prevalent, all the country above that spot is fairly healthy. In the clearing of Arauz's road, more than 150 people must have been engaged for upwards of twelve months and the deaths were I believe two or three only. The Engineers engaged in the survey of that road for the Public Works Company enjoyed good health when away from San Antonio, and out of the hundred individuals that formed the expedition that I accompanied up the rapids, only three lives were lost on the journey.

San Antonio or Theotonio as Head-quarters.

There appears to be great prejudice against San Antonio, and though I believe it to be unjust, still it would not be wise to shut one's eyes to its existence; it might therefore be prudent to remove the head-quarters of the construction staff somewhat higher up the country, say to Macacos or Theotonio.

Fish, Game, Plantains, &c.

Fish and game abound in the rapids, and plantations of yuca and plantains are easily and rapidly capable of being brought to form material helps to the provisioning of a body of labourers.

Rainfall.

The average rainfall at San Antonio for 1873 was 7·61 inches per month, making a total of 91·32 inches for the year, 17·55 inches of this fell during the six months of dry season from 1st of May to 31st of October, and the remainder 73·77 inches fell during the rainy season from 1st of November to 30th of April; but on 44 days out of the 181 in this period absolutely no rain fell at all, and only on 20 days during the same period did the rainfall of the 24 hours exceed a fall of one inch. In these 20 days 37·49 inches, or more than half the total fall of the six months' rainy season, fell.

Work may be carried on all the year round in Forests.

These figures prove, I think, that the works of the railway may be carried on all the year round, with intermissions of a few hours, or perhaps days, as the heavier storms of rain require. I believe that if the workmen are *well housed* they are as well off in the forest during the rainy season as they can be at San Antonio or any other head-quarters.

Huts—Palm Roofs preferable to Zinc.

For hut accommodation for the labourers, I am decidedly in favour of erecting good palm leaf huts; they last quite long enough for what is required of them, and make cool and well ventilated and water-proof huts when properly erected. Zinc roofing is suitable for stores or permanent buildings, but as it attracts the heat of the sun enormously, it requires a second roof of planking underneath it, and is therefore too costly to be used for hut accommodation for labourers.

General Information as to Operations of Contractors—Pará Agency.

To make complete the portion of my report that refers more directly to the railway, I cannot do better than introduce here a copy of a letter that I forwarded to the new contractors in January, 1874 in reply to a letter of theirs asking for general information.

My further experience and travel since that date has only served to confirm the views I then expressed, with exception of the advice given as to the employment of Messrs. Samuel G. Pond and Co., as agents in Pará. An inspection of their accounts rendered to the London office has convinced me that their charges are exorbitant.

LETTER TO DORSEY AND CALDWELL.

San Antonio, 2nd January, 1874.

Messrs. Dorsey and Caldwell.

Gentlemen,

According to promise in my letter of the 20th ultimo, which replied partly to yours of the 10th October, I now proceed to offer you a few further remarks upon the points raised in your letter referred to.

In order to reply somewhat categorically to your letter I have selected nine subjects therefrom. They are—

1. The river stages between Pará and San Antonio.
2. The most desirable mode of transporting materials.
3. Is it safe to send a steamer from Europe to San Antonio?
   3a. If so what is greatest draught allowable in the river Madeira.
   3b. From Pará to San Antonio in 10 days and from San Antonio to Pará in 5 days.
4. List of articles advisable for immediate dispatch here.
5. Advice and suggestions as to best manner of proceeding with the work.
6. How best to provide labour.
7. How best to house the labourers.
8. How best to feed and keep them.
9. Width and character of streams that railway is to cross.

RIVER STAGES BETWEEN PARÁ AND SAN ANTONIO.

1. The distances are reckoned as about 900 miles on the Amazon from Pará to Serpa and 600 on the Madeira from Serpa to San Antonio. Both rivers are fairly supplied with wooding stations.

WOOD OR COAL.

The price of wood is from 30 milreis to 40 milreis, say £3 to £4 per 1,000 sticks, which are here reckoned as being equal to a ton of coal, but if the coal is good I should say the ton would yield most work.

Stations on the Amazon.

On the Amazon the principal settlements are Breves, Gurupá, Mont Alegre, Santarem, Obidos, Villa Bella and Serpa, at most of which towns bullocks and general supplies can be obtained.

Stations on the Madeira.

On the Madeira, except Borba, a poor place, there are nothing but small stations of the "Seringueiros" or rubber gatherers. All these men keep wood on sale, but you cannot count upon supplies or provisions.

Obstacles to Navigation in the Madeira.

The navigation is free from all obstacles from Pará up the Amazon and Madeira until the Piedras de Urnas, on the latter river, are reached about three days' steam from Serpa. This is the first dangerous point, and for six months out of the year requires perfect knowledge of the channels. I believe there is plenty of water in these channels all the year round, but when the rocks are just covered by the water great care is necessary. The other dangerous points in this river above the last named are Marmelos, Das Abeillas, and Tamandoa; the first and last being sand banks, and the other a rocky channel.

Sand-bank, in Amazon near Villa Bella.

In saying that the Amazon is free from all obstacle, I should mention that there have been years when in the months of September or October a sand bank or bar, near Villa Bella, has only had eight feet of water on it, but I do not think this is well authenticated.

2. Transport of Materials up River.

In transport of materials we have to consider the ocean, the river and the road. My opinion is that at first the ocean and river traffic should be kept separate and distinct from each other, for the reasons that I have already given you in my letter of the 20th ultimo, namely scarcity of "practicos" (pilots), difficulties of clearing at Pará and opposition from the Amazonas Company, Limited.

Pará Agency.

I therefore recommend that the first shipments be made to Pará, and consigned to Messrs. Samuel G. Pond & Co., who will tranship them, and who have great facilities for getting goods through the customs. As they are our Company's agents, all matters in connection with the concession will be properly attended to.

OCEAN STEAMERS THROUGH TO SAN ANTONIO.

I could not advise that you try to send a steamer right through to this place this season, for you will hardly be ready to send a full cargo before March or April next, and the river begins to fall again in May. It appears to me that the heavy cargoes of rails, bridgework, &c., should be got ready for dispatch from Pará towards the end of next November ; by that time you would have been able to arrange for ocean-going steamers to run up here quickly in succession during the four to five months of full river, and you will have got good gangs of men here for unloading. But if you wish to run your ocean steamers up here before November, you must have a wooding station and tugs and barges ready below the Piedras de Uruas, so that you can come up from thence with a 2 feet 6 inches to 3 feet draught. But as you can make a very good start with the materials on the ground (of which more hereafter), I strongly advise you to come and see for yourselves, and perfect your arrangements after due consideration and consultation out here.

STEAMERS RUNNING FROM SAN ANTONIO TO PARÁ.

The Amazon Company, Limited, run a steamer up here regularly every month, starting on the night of 5th from Pará and arriving here on the 20th ; there is also another from Manaos, starting from thence on 27th, and arriving here on the 4th or 5th ; and I have no doubt but that you could get a special rate named by the Amazonas Company in London for the transport of your first shipments.

FREIGHTS BETWEEN PARÁ AND SAN ANTONIO.

£6 per ton is the lowest that has yet been obtained, but I think that to prevent opposition they would take £4 or£5.

TRANSPORT OF MATERIALS OVER LINE.

For transport of material over the road I strongly advise that as little as possible be sent ahead of the rails. As probably two-thirds of the line will be surface construction the rails should be laid down smartly, and if necessary temporary bridges set up at the ravines, so that an open road will always be kept behind the advanced parties. This is the American method of construction ; I need not therefore enlarge on the subject, and will content myself by saying that in my opinion this work will be best and most speedily carried out by the adoption of such a system.

3. SAFETY OF SENDING STEAMERS FROM EUROPE TO SAN ANTONIO.

As to the safety of sending steamers from Europe to this, as I have already stated, I see no reason why this should not be effected during the months of January to April. The total rise of the river here from its lowest is 48 feet 6 inches, but this height is only reached in March, but we may safely calculate on 25 feet of flood water from December to end of June at this place. The dangerous points on the river are, as already stated, four in number, and over these there is at lowest water always three feet ; therefore we ought to be able to calculate on 28 feet for seven months, and if the river were thoroughly surveyed I believe this would be found correct,

3A. GREATEST DRAUGHT PERMISSIBLE.

But the river is of various widths, and as we have no certain information as to rise and fall lower down, I think it well to say that steamers of fourteen feet to sixteen feet draught can come up from January 1st to April 30th.

3B. TIME OCCUPIED—COAL OR WOOD.

With reference to time occupied this depends on the fuel burnt ; if you send ocean steamers up they must burn coal all through their trip, and they ought to run up from Pará in thirteen or fourteen days at most. Certainly coal has gone up seriously in price lately, although in the States it is cheaper than at home, but even at £3 a ton here, if of good quality, I believe you will save expense by its use in general on these rivers. If you burn wood you must have large fire-grates and tubes, and these are very expensive with coal fuel. This, therefore, offers a subject for consideration in arranging the ocean and river traffic. To have supplies of coal to enable ocean steamers to run up the rivers, you must have deposits at Pará, and perhaps one station half way up, while, if you intend to burn wood on the rivers you must, I opine, change your steamer at Pará.

4. ARTICLES FOR IMMEDIATE DISPATCH HERE.

For a good start you will find everything here in the way of plant, for I have taken over all the late contractor's stores. Material for sixteen miles of permanent way is here, if we except "Ibbotson's Clips," which I much want to see sent to us, but I would advise you to lay in the road temporarily with the "fishes" we have here, get ahead thus over temporary bridges, and having created confidence relay again and put in the finished works.

MATERIALS REQUIRED FOR FIRST DISPATCH.

You will, however, have to arrange for sending a couple of locomotives with trucks, etc., before next wet season; also, you may think it convenient to send the remainder of the material required for the San Antonio Jetty, and bridgework for the first section of the line which is probably the roughest of the whole length. The only articles in plant that I can recollect as wanting at once are a good supply of rope falls of all diameters, also light chain, $\frac{1}{4}$ in. and $\frac{3}{8}$ in. in diameter. It will be well also to have a good stock of Piassava rope for canoe work, but this you will buy cheapest in Pará or Manaos.

5. SAN ANTONIO JETTY, &c.

One of the works that should be carried out during the coming dry season is the erection of the San Antonio Jetty; for which, send two vertical boilers, engines, jibs, crabs, etc., on travelling trollies; to lift to two or three tons, so that you may have full facilities for unloading your vessels to come up next fall. I have sent home my final suggestions for this work, so that Colonel Church will be able to settle the matter with you; and I need only say here that my plan No. 7, for an ⌐ shaped jetty is, I believe, the best and most economical that we can set up. This plan was sent to Mr. Hopkins with my No, 16. of 24th of June last, accompanied by a schedule showing the ironwork still required.

FAIRLIE LOCOMOTIVES.

For the rest of the work we have not yet sufficient data to decide on the details, but if we are supplied with the Fairlie Locomotive and the "Ibbotson Clip," we shall make a good job of the line.

Bridge Work.

For bridges I do not think you can do better than send out light iron pile shafts and girders for thirty and sixty feet spans, but pray take every care that all the parts are thoroughly interchangeable, otherwise great loss and delay will occur. There is no reason why the head of a pile shaft should not suit a cap or a coupling; those we have here are so arranged that they cannot be used indiscriminately.

Work from San Antonio end of Line.

As to the mode of doing the work, my opinion is that it must all be done from this end. The only works you could attack from the Guajará end would be clearing and earthwork, and by the time you got your rails and other iron up there from this end these works would probably have to be done over again. Push a road through from this end that you can get your trucks over and then finish up afterwards.

6. Labour.

The Labour question is the most difficult perhaps of all. The late contractors never attacked it with sufficient vigour. You must draw labour from every possible quarter and not depend upon any single one.

Bolivian Labourers—Coolies or Chinese.

Bolivia will undoubtedly send a valuable contingent, say 1,000 men, and these, if properly sought for, ought to be had for £3 to £4 per month and found. I do not think it will be much use sending into Bolivia from the Pacific side; wages are high there on account of the many roads going on in Peru and Chili; also there is great opposition to us there from the Peruvian interests. A commencement to recruit labourers should be made in the Mojos and Beni department, and if American or English gold for advances is sent up and the wages I have quoted are offered I believe the men will be found: then the search can be continued upwards towards Cochabamba from this side with greater prospect of success from the prestige it will have gained by partial success in Mojos and Beni. You must always have a good number of Bolivian Indians for canoe work, clearing and other timber work, but your earthwork and other construction ought to be done with imported labour. On this point you will not need advice from me as to the sources from which it can be drawn;

I will therefore only say that I believe the East Indian coolie or the Chinese would do well here ; a good many of them would have ague fever, but they would recover as the work of clearing went on.

7.—Huts.

Decidedly the best houses for this country are made with palm boards and palm leaf roofs, and up the line the palm trees are very abundant ; near a station such as this they get scarce, and it is not desirable to build houses close together with palm leaf roofs as fire then spreads rapidly.

Zinc Roofs required at Stations.

At stations therefore zinc roofs boarded underneath will be found the best and most economical. Asphalted felt has been used with success on many works, but I am not much in favour of this article ; it requires good boarding underneath, and even then is apt to leak at the nail holes. Here the alternations of sun and rain are very trying, and "felt" would I think deteriorate rapidly. Zinc sheeting is put together rapidly and lasts well ; any rough boarding underneath answers its purpose of stopping the radiation of heat.

8.—Food Supplies.

Considering that you have perfectly free navigation from San Antonio to Pará and other ports of Brazil where supplies are obtainable in large quantities, I think that bullocks, farinha, charqui, coffee, sugar, and rice should be expected from Brazil for some time to come. Once we have a temporary road through, then good supplies will undoubtedly come down from Bolivia , and by that time producers there will have become convinced of the reality of the undertaking and will have raised crops for our supply and for export. At present I do not see how we can expect Bolivia to yield large quantities of cereals, for as there has hitherto been no means of exportation, only sufficient for home consumption has been raised.

General Store.

I would impress upon you the vital necessity of at once establishing a good store here with food, clothing, etc., both for Europeans and Indians. By this you will be able to keep your men well contented, and while charging prices far below those current on the river, you will be able to make such profits as will materially reduce the price of labour. It would be impossible to tell you in writing the kind of stores required. I would therefore suggest that you purchase a first stock in Pará, as unless you have the proper things in calicoes, dungarees, shirtings, etc., losses will occur.

Cost of Keep of Labourers.

The average calculation of this river for the keep of an Indian is a milrei or two shillings per day, but I have kept them well for much less, and my practice is to estimate at two shillings per day and save 30 to 50 per cent.

Medical Requirements.

You will doubtless send a good medical staff with officers accustomed to treat tropical maladies such as fever, ague, dysentery, etc., and pray by no means forget to let the doctors bring out a fresh stock of vaccine matter, as few, if any, of the Bolivian Indians are vaccinated.

9.—Streams to be crossed.

Detailed information as to width and character of streams is not to hand yet. There are two large rivers to cross, the Yaci-Paraná and Tres-Irmãos ; the former of these spreads itself at its junction with the Madeira, but in neither case will it be necessary to have openings of greater individual width than 60 feet. With a succession of these we can bridge these rivers rapidly and well, for as there is no great current down either of them there is no necessity to give them greater waterways. For all other streams 30 and 60 feet spans will be ample, placing one or more in each case as circumstances may require.

I think I have now replied to the various queries in your letter, and I assure you that both myself and my assistant, Mr. Dalton, who has been here since the works started, will most gladly render you any assistance or co-operation in our power.

I am, Gentlemen,
Yours very faithfully,
(*Signed*) E. D. Mathews,
Resident Engineer.

II.—GUAJARÁ MERIM TO EXALTACION.

PORT ABOVE THE FALLS.

The river above the falls narrows considerably, its general breadth being about half a mile, with deep water giving every facility for free navigation. Above the Guajará Merim there is very good quiet water on the railway side, where a good harbour can be made, should further experience prove that the fall cannot be easily passed by steamers.

ISLANDS OF CAVALHO MARINHO.

The Islands of Cavalho Marinho, about ten miles above Guajará Merim, are several in number, but there is plenty of water in the channels, so that they offer no obstruction to navigation. Canoes are obliged to keep crossing from side to side of the river, in order to avoid the strong currents usual at the bends, but a steamer would be able to keep mid-channel, as the ordinary current of the river does not I think run more than two miles per hour.

THE HILLS OF PACA NOVA.

The range of hills called " La Sierra da Paca Nova " by Messrs. Keller are very plainly seen from the islands of " Cavalho Marinho." Messrs. Keller depict them as trending out to the " Cachuela do Pão Grande, but this I do not think is correct. They seemed to me to finish off a good distance from the river, and the hills of " Pão Grande" are I believe quite separated from those of Paca Nova.

THE RIVER TO JUNCTION OF ITENEZ.

There is nothing special to report of the river from " Las Ilhas do Cavalho Marinho " to the junction of the river Itenez, which we reached on the 7th of June, after seven days' hard paddling, and I might say seven nights as well, for the practice is to paddle as much as possible during the night so as to avoid the chance of attack from the savages : thus very long hours of work are made during this portion of the journey, from 14 to 18 hours being the usual allowance of paddling given to the men.

JUNCTION OF ITENEZ WITH THE MAMORÉ.

The river Itenez, whose waters are clean and dark coloured, is wider than the Mamoré at the Junction. The waters of the Mamoré are probably greater in volume, as they are deeper and give their whitish clayey colour to the united waters below. Both these rivers are exceedingly handsome at the junction, and so fine a "meeting of the waters" it would be difficult to match.

THE RIVER MATOCARI.

On the 12th of June we passed the river Matocari on the right bank. This stream is said to be navigable to the villages of San Joaquin and San Ramon, and might doubtless be utilised during the rainy season, instead of taking goods for those villages round by the river Itenez and the Fort of Principe da Beira.

GRAZING LANDS BELOW EXALTACION.—PROPOSED TRANSPORT OF CATTLE TO SAN ANTONIO.

Above this river we commence to pass many open lands or "pampas" on either side of the river, with excellent grass for cattle feeding grounds. The largest of these cattle feeding establishments is called "La Estancia de Santiago" and was the property of the late Don Barros Cardoza, Brazilian Consul in Bolivia for some years. He is reported to have had nearly 8,000 head of cattle, and I am credibly informed that, had he not lost his life in the early part of 1874, he had intended to drive a large number of his cattle by land from Guajará Merim to San Antonio down the road cut by Señor Ignacio Arauz. From his estancia to the first cachuela, the cattle would have been taken in canoes, or on rafts, a comparatively easy work, as the navigation is entirely free from other obstacle than a few "playas" or banks of sand which stretch out into the river at low water, leaving, however, in every case a channel deep enough and wide enough for craft drawing not more than 7 or 8 feet. Pasturage for the cattle would be easily found at night along the river banks, which are covered with "capim," a rough wild grass, or "chuchia," a kind of wild cane, the succulent points of which are greedily eaten by the cattle.

PRICE OF OXEN AT THE ESTANCIAS AND AT SAN ANTONIO.

Oxen in the estancias of Mojos are worth from 15 to 20 pesos faibles, say about £2. 10s. to £3. 10s. a head, while at San Antonio and on the higher Madeira they are worth from 80 to 100 milreis or £8 to £10.

COCOA PLANTATIONS OF EXALTACION.

On the 15th the canoes arrived at the "chocolotales" of Exaltacion, which are very large plantations of cocoa trees belonging to the government. These plantations are farmed out by the authorities of Exaltacion and Trinidad to speculators who make good profits as there is no labour, or very little, expended in clearing. At the proper season a party of Indians descend the river from Exaltacion, collect the fruit, clear away a little of the rubbish from beneath the trees, and then the "chocolotales" are abandoned to the savages and wild animals till the annual time of collecting again comes round. Doubtless these "chocolotales" were planted at the time when the Indians were more populous in the fifteen missions of the depart ment of the Beni.

EL CERRITO : THE CO'S ESTATE NEAR EXALTACION.

" El Cerrito," the National Bolivian Navigation Company's property, was reached on the 16th. This place is named from the only small hill that is to be found on this part of the river for many miles, and is said to be the only spot that, at exceptional high floods, is not inundated. There are four houses all belonging to the Company, well built with palm leaf roofs and wattle and dab walls. Here I found a Bolivian named " Melchior Cruz," in charge with five peons, acting under the orders of Señor Francisco Ceballos, who lives in Exaltacion and has an " Ingenio de caña ," or sugar mill there. Señor Ceballos had been left in charge by Dr. Velarde. The men are employed in a " chaco," or plantation, which is certainly very well kept. The ground cleared may be about ten acres, and is well stocked with yucas (mandioca) plantains and sugar cane. The products are sold to passers by when possible, and the crops of yuca made into " harina de yuca," or mandioca flour, while the cane is sent to Sr. Ceballo's mill for grinding. In consequence of the small demand in comparison with the large supply of these articles in Exaltacion, it is probable that the plantation does not yield at present more than sufficient to cover the expenses.

THE ESTATE OF GREAT SERVICE DURING CONSTRUCTION OF THE RAILWAY.

To the enterprise, however, the plantation will be of great service when the works are in progress. A steamer with canoes in tow could take large supplies of the products of the estate to Guajará Merim, and thus the enterprise would be sure of obtaining the necessary vegetable food for the labourers at reasonable prices. The Bolivian agriculturists of the Beni are all most patriotically disposed in favour of the railway, but I fear their patriotism would not prevent their putting exceptionally high prices upon their productions if they saw that the enterprise was dependent upon them. In my ascent of the rapids, I had to pay a Bolivian, who was descending to the River Madeira, 12 milreis, or £1. 4s. 0d., for an arroba of 25 pounds weight of mandioca flour, worth about a peso or a peso and a-half in Exaltacion, say 3s. 2d. to 4s. 10d. It will be well, therefore, that the Company should have means of its own of supplying fruit and breadstuffs at reasonable prices.

THE S.S. EXPLORADOR.

The small steamer the "Explorador" is here, and in safety. As far as one could tell, the hull seems tight and good, as she did not make any water to speak of. The boiler is said to be useless, having burst twice on the trip up the rapids. The wood fittings are in fair condition; the paint, gilding, and varnish only being damaged. Whether the whole of the machinery was in working order I could not say, but my cursory inspection showed it to be not much damaged by exposure.

EXALTACION.

On the 17th I arrived at Exaltacion, which is the first village in Bolivia that a traveller up the Mamoré meets with. Here I had to stay eleven days, it being necessary to find a new crew for the further journey to Trinidad, the men that I had brought from San Antonio all belonging to Exaltacion, or to the villages on the Maddalena and Baures rivers. I was exceedingly well received by Sr. Francisco Ceballos, and by the majority of the inhabitants, but the Corregidor, one Faustino Varoma, a nomination of the party in power under the "Frias" regime, did not seem inclined to lend assistance to anyone belonging to the enterprise.

In Exaltacion there are about a dozen Bolivians of some sort of position and education, a "cura," or parish priest, and the before-mentioned Corregidor. All these men, except the "cura," keep small stores, and exchange their goods with the Cayubaba Indians for chocolate, mandioca, sugar, or rice, all of which articles are sent to Trinidad or Santa Cruz for sale to merchants established there.

POPULATION OF EXALTACION DECREASING — KELLER'S OPINION OF
THE CLIMATE OF EXALTACION DISPUTED.

The Cayubaba Indians are, I fear, decreasing rapidly in numbers,
the deserted houses and the lines of the old streets giving a sad and
desolate look to the place. The present population cannot be more
than 1,500, and I should judge that less than fifty years ago there
must have been nearly 4,000 Indians in the old mission of "La
Exaltacion de la Santa Cruz." Señor Keller has given a graphic
description of the town, it is, therefore, unnecessary for me to occupy
space and time on the subject; I observe, however, that "amongst
the causes which tend to contribute to the decadence of so flourishing
a town," he cites the fevers, which he says of "late years appear to
have taken an endemic character." To this I would remark that in
exceptionally high floods, which appear with a rather remarkably
regularity to occur about once in seven years, the lands of Exaltacion
are flooded to a depth of perhaps six inches, and after the retirement
of the waters, ague fever may be epidemic or prevalent, but I could
not agree that fever in Exaltacion is "endemic," or peculiar to the
country—on the contrary, I should say that "El Cerrito" and
"Exaltacion" are generally very healthy places.

CAUSE OF THE DECLINE OF THE INDIAN POPULATION OF THE BENI.

The reason for the decline of the Indian population is to be found,
without doubt, in the baneful effects to Bolivia of the rubber-col-
lecting trade of the Madeira and Purus rivers. This trade is the real
cause that is rapidly depopulating not only Exaltacion, but all the
towns of the Department of the Beni.

To take the year 1873 as an example of the working of the emigration from Bolivia to the rubber districts of Northern Brazil. In that
year 43 canoes descended the rapids from Bolivia, with merchants on
their way to Europe with ventures of "cascarilla" (cinchona bark), or
with speculators in the rubber "estradas" of the Madeira river;
while in the same year 13 canoes only ascended to Bolivia. We may
average the Indians that leave Bolivia with these canoes at 10 per
canoe, and thus we have an exodus of 430 Indians from their country
in twelve months, while only 130 return in the same period; we thus
have 300 Indians lost to Bolivia in 1873, and as the rubber collecting
fever has been decidedly on the decrease for the last four or five
years, the year 1873 does not give a fifth of the number of Indians

that have left in previous years. We may, I venture to think, estimate the drain to human life that the Department of the Beni has suffered from the Northern Brazilian rubber trade at 1,000 men per annum during the decade of 1862 to 1872. The worst feature of this emigration is, perhaps, the fact that rubber speculators and merchants descending the rapids will not allow the Indians to take any of the females of their families with them. This is done on account of avarice in some cases and necessity in others, which prompt the "patron" or owner of the descending craft to load his canoes as fully as he can with his merchandise, reserving as small a place as possible for provisions, which, on account of the quantity of farinha consumed, occupy so much space that every mouth that requires to be filled, without its owner being able to assist in the propulsion of the craft, becomes a very serious consideration. Thus it arises that in every town of the Beni the females are in a majority of perhaps five to one over the males, and the populations are decreasing.

Proposal for Re-populating the Beni.

Here I may mention that in Trinidad I had a conversation with Don José Manuel Suarez, the prefect of the Beni, upon the subject of the serious depopulation of the Department and that at his request I addressed some remarks upon the subject to Señor Don Mariano Baptista, the Minister of Home Government and Foreign Affairs in Sucre. Sr. Suarez proposes that the Government of Bolivia should send a special mission to Brazil, to ask to be allowed to interview all the Bolivian peons that are in the province of the Amazon, whether engaged with rubber collectors or others, engage them if possible for work on the Madeira and Mamoré Railway, remitting through Government agency a portion of their earnings to their destitute female relatives in Bolivia, and at the termination of their agreements sending them back to Bolivia. These Bolivian peons are held by the "Seringueiros," or rubber collectors, and in this respect there is nothing to choose between Brazilians or Bolivians, in a perfect state of slavery, by means of debt and drink. At most of the Barracas on the Madeira river, where the Seringueiros live, the Sundays are passed in perfect orgies of drunkenness. On that day it is that the peon delivers over to the "patron" the rubber that he has collected during the week; he is then treated liberally to white rum (called "cachaça" on the river), and when under the influence of this liquor he is induced to buy trinkets, calicoes, ribbons, and other articles that he could very well do without. These are charged to him at enormous prices, whilst his rubber is credited to him at inversely corresponding low ones, and he is thus kept under a heavy load of debt, and cannot, under the Brazilian laws, leave his "patron" until it is worked off, which happy event the patron takes care shall not happen. If a Bolivian authority, aided by the Brazilian officials, were to visit these unhappy exiles, and to settle between "patron" and "peon' the just state of the accounts, paying the amount fixed upon to the patron, I believe that a thousand Bolivian peons could with ease be gathered together on the banks of the Madeira river. They could work on the railway for two or three years with advantage to themselves and their country, and return to their native villages at the expiration of their agreements with a small fund in hand. This would be the only and the most expeditious method of re-populating the now half deserted villages of the Department of the Beni. The Minister at Sucre heard my arguments and acknowledged them to be true, but as at home it is said to be "a far cry to Loch Awe,' so in Bolivia it is "a far cry from Sucre to the Beni," and the Government were too much occupied in devising means for securing their own retention of power in the capital to be able to devote time or thought to places so distant as Trinidad or Exaltacion.

THE PORT OF EXALTACION. —"TIERRAS DISBARANCANDAS" FALLING BANKS.

The port of the town of Exaltacion is situated at the apex of an immense bend in the river Mamoré, each arm being at least a league in length. The wind therefore exerts a great force on the craft made fast at the foot of the bank, which rises more than 50 feet above low water level. On this part of the river boats are also much exposed to danger from the falling banks, called "Tierras Disbarancandas." The Mamoré, and indeed all the rivers of the Beni valley, are for ever shifting their courses in many parts of the forests through which they flow. They undermine the banks on one side, which falling away form the numerous curves, on the convex shore of which the mud and sand brought down by the current is deposited, and playas and banks termed "igapo" lands are formed. On these a forest grows in course of time and "vargem" lands arise.

CHANGES OF THE COURSES OF THE RIVERS.

The river on the concave side of the curve is continually causing the trees of the terra firma to fall and obstruct its course, a barricade or " palisada " is formed, the river then returns in exceptionally high floods to its old course, bursting through the vargem and igapo lands. and so the ever recurring changes of the river course continue. In illustration of this I saw on the river Chapari a place where the current was breaking down a bank that was apparently " terra firma," and had trees growing on it that were of great age. At the foot of this bank and under some 15 feet of earth was a deposit of timber blackened and in fact almost carbonised by time and pressure of the super-incumbent earth. From the manner in which these logs of timber were deposited, one above the other, it was evident that it formed part of a huge collection of drift-wood such as may often be seen collected together in many parts of the rivers. In all works on the river side, great care must be taken to see that the bank is not a " tierra-disbarancanda," or of slipping nature. In the Cachuelas, however, this feature of the river does not appear to exist, but on the Mamoré all " chacos," " barracas " and " pueblos " are placed some distance from the river, generally from ½ a mile to a mile, so that they may not be exposed to this danger.

LAND-SLIP AT THE PORT OF EXALTACION.

Whilst I was at Exaltacion an enormous mass of the bank at the port gave way and fell into the river, causing the loss of one man and a large canoe. This landslip measured more than 100 feet in length, the breadth of earth that fell being more than 30 feet at top, which was upwards of 40 feet above the then water level. It is therefore evident that a more secure situation must be sought for the port of Exaltacion when any navigation of the Mamoré commences. The port which is used at high water, about a couple of miles higher up the river than the one used for the greater part of the year, is, I think, always preferable notwithstanding its greater distance from the village when the water is low.

PROVISIONS FROM EXALTACION.

Provisions are at present rather difficult to obtain in any quantity at Exaltacion, from the fact that no stocks are kept, the demand being very limited. There is, however, no doubt but that when the owners of plantations and feeding grounds in or near Exaltacion become convinced that the railway will be constructed, they will at once prepare large quantities of various articles suitable for provisioning the workmen on the railroad, or for sale on the Madeira and Amazon rivers. The following are some of the prices current when I was in Exaltacion.

PRICES OF PROVISIONS AT EXALTACION.

" Farinha de Yuca " or mandioca flour, 12 reales (4s. 10d.) per Bolivian arroba of 25 lbs. Rice, 6 reales the arroba in the husk, this only produces about 15lbs. when husked, thus the dressed rice may be put at 2d. per lb. Sugar, brown in cakes called " empanisadas " $\frac{1}{2}$ a reale, say 2$\frac{1}{2}$d. per lb. Ordinary aguadiente, 8 reales (3s. 2$\frac{1}{2}$d.) per " frasqueira " of 3 bottles, a stronger and better sort of spirit fetches 8 reales per bottle. Meat, fresh, sells at 1 peso of 8 reales (3s. 2$\frac{1}{2}$d.) per arroba of 25lbs., when preserved by being salted and dried in the sun it is called " charqui," and sells at 3 pesos (9s. 7$\frac{1}{2}$d.) per arroba, say 4$\frac{3}{8}$d. per lb. Wheaten bread is very scarce, when obtainable, costs $\frac{1}{2}$ a reale, say 2$\frac{1}{2}$d. for a small loaf that may perhaps weigh a couple of ounces. The flour comes from Cochabamba.

LABOURERS FROM EXALTACION.

Very few men for the works could be counted upon as obtainable at Exaltacion, as I do not think there were more than 100 men in the town at the time I was there, and the few "patrons" of the place appeared very anxious to engage the men that returned from the river Madeira to Bolivia. Exaltacion, however, will probably always be of importance to the enterprise, from its being the best point at which to unite labourers from the other "pueblos" previous to their descending the river to the railway, and also its being the market at which to obtain the above mentioned supplies for the provisioning of the labourers when at work. I would, however, recommend that the Company's possession at "El Cerrito" should be utilised as much as possible in these ways, as there the Indians would not be exposed to be tempted away from their contract by the native patrons of Exaltacion and its neighbourhood.

WOOD FUEL FOR STEAMERS.

Before concluding this chapter of my report, I would advert to the difficulty that to me appears to exist in supplying the wood necessary for the fuel of the steamers which will ply from the upper terminus of the railway to Exaltacion. This portion of the Mamoré is much infested by savages, principally " Chacobos," and there is no doubt but that they would prove troublesome for a considerable time to isolated settlers. Messrs. Keller give the distance from Guajará Merim to Exaltacion at 72$\frac{1}{2}$ leagues say 200 miles. On account of the nume-

rous " playas " or sand banks it would not be possible to navigate by
night except when perhaps the moon was very clear, and allowing for
the strength of the current and the smallness of the steamers that will
probably for some time be used, it would be necessary to have wooding stations say about 50 miles apart from each other. Three stations
would therefore be required and I would suggest that the Brazilian
and Bolivian Governments be requested to make these stations military
outposts, each to be garrisoned by about 40 to 50 men who could also
be employed in cutting the wood for the steamers. The price agreed
upon by the navigation company and the governments would defray
the expenses and a great step would be taken towards repressing the
savages in their attacks on the passing craft.

III.—EXALTACION TO TRINIDAD.

START FROM EXALTACION.

Having secured a crew of eleven men in Exaltacion after considerable difficulty and some opposition from the Corregidor, I started on
the 28th of June for Trinidad.

CROPS OF GRAIN GROWN ON THE SAND BANKS.

The river on this section preserves a very uniform width of from
800 to 1,200 yards and runs through level plains of very rich alluvial
soil on which the Cayubaba Indians of Exaltacion and the Mobimas
of Santa Ana have numerous plantations, some of the playas or sand
banks are of enormous size, and on them the Mobimas during the dry
season plant several kinds of " frijoles " or beans, and also a sort of
pulse called " mani " from which they make their favourite beverage
of " chicha," and which they also use as a substitute for rice ; there
are also many pampas for cattle grazing.

LANDS VERY SUITABLE FOR SETTLEMENT.

The lands on this portion of the river would I think be very
suitable for emigration, as the climate is good, the land of excellent
quality for the production of crops of sugar-cane, rice, maize
plantains and every other description of tropical produce, or for
cattle rearing. Also there are no savages in this district, the
country between Exaltacion and Trinidad being traversable freely on

either side of the river. The river abounds in fish, and wild ducks, herons, storks, and other waterfowl are plentiful on its banks.

RIVER YACUMA AND VILLAGE OF STA ANA.—TRADE TO REYES, &c.

On the 30th, we passed the mouth of the river Yacuma, a day's paddle up which river is the village of Santa Ana the home of the Mobima Indians.

It is by this river that traders take goods for the pueblos of Reyes, San Pablo, San Borja and Santa Cruz all of which are peopled by the Maropa Indians, and from which villages a trade is carried to the towns of Apolobamba in Bolivian territory and Sandia in Peruvian. The river Yacuma is said to be navigable all the year round for large canoes, and is free from savages, therefore, a small steamer may be advantageously employed here after the construction of the railway of the cachuelas.

STEAMERS TO TRADE ON THE AFFLUENTS OF THE MAMORÉ AND ITENEZ.

One of the merchants who accompanied me up the rapids, sold goods to the value of £3,400 in Exaltacion to a trader who sends canoes up the affluents of the Mamoré or the Itenez to the various pueblos of the Department of the Beni, such as San Joaquin, San Ramon, San Nicolas and San Pedro on the Machupa river, Maddalena on the Itonama, Concepcion de Baures and El Carmen on the Baure river sometimes called "Rio Blanco," the above mentioned town of Reyes and others on the Yacuma, and San Ignacio on the Jamucheo. Considerable trade will doubtless be opened up with these towns and villages, and work will be found for two small steamers, one to run on the affluents of the Mamoré and the other on those of the Itenez the head-quarters of both being at Exaltacion or "El Cerrito."

RIVER JAMUCHEO AND VILLAGE OF SAN IGNACIO.

On the 1st of July we passed the mouth of the river Apiri on which there are no villages, and on the 3rd and 4th we got to districts where the Canichana Indians have their chacos on the Mamoré. At night of the 4th we stopped opposite the mouth of the river Jamuchéo. When this river is full the pueblo of San Ignacio, on its banks can be reached in six or seven days, but when dry three weeks are often required as the canoes have to be dragged over the many shallow parts of the river bed.

PORT OF SAN PEDRO.

Half a day's journey from the River Jamucheo is the port of San Pedro the pueblo of the Canichana Indians. The village is about two leagues from its port on the Mamoré, and is situated on the source of the Machupa river which is as aforesaid an affluent of the Itenez.

THE CANICHANA INDIANS OF SAN PEDRO.

At the port of San Pedro on the Mamoré there is a large and well built shed for the sentinel who takes charge of the canoes belonging to the villagers, and from the style of the work, the quality of the timber and the tidiness of the place, my previous favourable opinion, obtained by the employment of a few Canichanas in San Antonio, was confirmed to the effect that these Indians are the most desirable of any of the various tribes of the Beni. They are excellent workmen with the axe and are I think less addicted to the use of ardent spirits than the Cayubabas or the Trinitarios.

ROAD FROM THE MAMORÉ TO SAN PEDRO REQUIRED.

Traders going to San Pedro use the port on the Mamoré, instead of making the round by the river Itenez. Their canoes are hauled up on land and dragged by oxen across the two leagues of pampa between the Mamoré and the Machupa. The Navigation Company will do well to make a corduroy road over this track so that the town of San Pedro may be accessible during all seasons from the Mamoré.

SAN XAVIER.

At night of the 6th we were opposite to the port of San Xavier a village of small importance situated on a creek running into the right bank of the Mamoré.

THE RIVER YBARI.

On the 8th we entered the Ybari, on which river the town of Trinidad, the capital of the Department of the Beni is situated, and on the 9th after eleven days good work we arrived at its port, the town being distant about two leagues from the river.

TRINIDAD.—PRINCIPAL TRADE.—SEÑOR IGNACIO BELLO.—JOURNEY TO COCHABAMBA INSTEAD OF SANTA CRUZ DECIDED UPON.

In Trinidad I was exceedingly well received by all the inhabitants, who are enthusiastic in favour of the railway, which affords them the only hope of arresting the decay of the department arising from the emigration before referred to of the Indians to the Madeira river. There are here a few merchants of considerable position and resources whose principal trade appears to be the export of cocoa to Cochabamba and Santa Cruz, receiving in return flour from Cochabamba, and dry goods from Santa Cruz, brought from the town of Curumbá on the river Paraguay. The principal of these merchants, Señor Ignacio Bello, has always been anxious to do everything in his power to assist in the construction of the railway, and received me with great kindness; and as I found that he was on the point of making a journey to Cochabamba, I decided to proceed in his company, and give up any idea of visiting Santa Cruz, more especially as I was informed that the river Piray was very dry, and the savage Sirionos who dwell on its banks were very active, having attacked several canoes during the months immediately prior to my arrival in Trinidad.

COUNTRY ROUND TRINIDAD.—PERIODICAL INUNDATIONS.—SPLENDID CATTLE.

The country round Trinidad is a flat pampa, with a rough kind of tall grass, which requires burning frequently. These pampas are almost annually flooded, and are, I think, more subject to these inundations than those of Exaltacion. The inundations at times rise up to the town itself, there being only one street that is said to be left dry on these occasions. The grazing lands, however, generally have some slight eminences upon them, where the cattle find refuge during these floods, and as, upon the retirement of the waters, the grass springs up with renewed vigour after the rubbish has been burnt away, the cattle thrive excellently, and are really handsome animals, being nearly twice as large as the Brazilian oxen. They would, indeed compare very favourably with our ordinary English cattle.

PRICES OF PROVISIONS.—PRICES OF IMPORTED GOODS.

Prices of provisions are much the same as those current in Exaltacion. I observe, however, that Manchester goods, such as calicoes, longcloths, ribbons, &c., are brought to Trinidad from Curumbá *via* Santa Cruz, at prices far below those at which they can be brought at present from Pará *via* the Cachuelas, and it is evident that when the trade in these goods is carried up the Amazon and over the railway, the merchants of Pará must be contented with smaller profits than those they now obtain. Pará, on the Amazon, and Curumbá, on the

Paraguay, are both Brazilian ports, and I presume that the same tariff of customs rules alike at both places, nevertheless calicoes, bought in Pará, that cannot be sold in the river Madeira for less than 250 or 300 reis—say 1s. to 1s. 2½d. per yard, can be bought in Trinidad at 2 reales, or 9½d. ; also, longcloths, on the Madeira, sell at 200 to 300 reis (9½d. to 1s. 2½d.) per yard, and are only worth 1½ to 2½ reales, say 7¼d. to 1s. per yard in Trinidad. It must, however, be noted that only very low quality goods are brought from Curumbá, and that the secret of business in Trinidad seems to be to sell at a low price without regard to quality.

Articles most suitable for Importation to Trinidad.

The articles that leave the best profit when taken up the Cachuelas are iron pots, enamelled saucepans and other general ironware for house use, also claret of a low class, for anything called "wine," and sold in bottles with pretty "etiquettes" fetches 8 reales (3s. 2½d.) per bottle, and no Bolivian in the Beni would pay more even for "Chateau Margaux" or "Chambertin."

Small-pox at Trinidad.

The town of Trinidad did not display a very animated appearance at the time of my visit, but that was not to be wondered at, as small-pox was very prevalent.

Population Decreasing.

The people also complained bitterly of the great emigration of the Indians during past years to the Madeira river, and as in Exaltacion, the remains of streets that were well populated but a few years ago, tell a sad tale of the results to Bolivia of the rubber fever of the Amazon valley.

Madeira and Mamoré Railway the only Salvation for the Beni.

The construction of the Madeira and Mamoré Railway is the only event that can save the once flourishing department of the Beni from becoming again the hunting grounds of the savage Siriono and the wild beasts of the forest. The few merchants of the town are so convinced of this that they are merely living on in the hope of being able to wait for the realization of the undertaking, and Sr. Ignacio

Bello, the wealthiest of them all, in money or in cattle, made the journey in my company to Cochabamba, with the avowed intention of purchasing a property there and leaving the Beni. He gave as a reason the pressing need of education for his children, but if that were all, he could send his children to Cochabamba and remain himself in Mojos. I prefer to think that the true reason for his leaving is the increasing poverty of the Department. The construction of the railway would, however, arrest entirely the decay by affording a ready means of transit to a good market for the chocolate, sugar, oxen, hides, skins, and other produce of which the inhabitants are now only able to realize but a small amount in value compared with what they will be able to, when the line is open, and some of the exiled Bolivian peons have been brought back to their homes.

RATE OF WAGES FOR PEONS.

In consequence of the scarcity of hands, the peons now get 30 to 40 per cent. more for their journeys than they did a couple of years ago ; thus, from Trinidad to Coni, they now get eight pesos for the up-river voyage, and two for bringing back the canoes, while formerly the price was six to seven pesos for the round trip. The monthly rate of pay does not, however, seem to have altered much, as it is still about five pesos (16s.) per month, so that the price established in San Antonio of "ten pesos fuertes," or £2 per month, should be sufficient to induce patrons and peons to go to the works.

LABOUR FROM THE BENI FOR THE RAILWAY.

Being desirous of testing the possibility of obtaining labour from the Beni, for the railway, I invited some of the principal residents to express to me in writing their willingness to assist in the construction of the line when resumed, explaining to them the great assistance they would be rendering to the contractors by taking their labourers to the works, and undertaking the entire care of their men in manner similar to that done by Señor Ignacio Arauz, in the cutting of the track from San Antonio to Guajará Merim. My ideas were responded to by the principal men of the place, namely, Sr. José Suarez, the Prefect ; Sr. Ignacio Bello, his son-in-law ; Sr. Miguel Maria Cuellas, Sr. Manuel Becerra, Sr. Jesus Arauz,, brother of Sr. Ignacio Arauz, and Sr. Francisco Ceballos, of Exaltacion. By means of these "patrones" I believe that 5 to 600 men could yet be raised, for the works of the railway, in the various pueblos of the Beni ; and, as I maintain, that whoever carries out these works must always have a certain number of Benianos for timber work, and for navigation purposes, amongst the rapids, I am of opinion that no better plan can be adopted than that of entering into contracts with some of the above-named parties for the supply of this labour. All responsibility as to housing and feeding these peons would thereby be taken off the principal contractor's employés.

IV.—TRINIDAD TO CONI.

Start from Trinidad.

On the 19th of July I started with Don Ignacio Bello and other traders, there being in all eleven canoes, with an aggregate number of a hundred Indians in the crews that then left Trinidad for the port of Cochabamba, on the River Coni, an affluent of the River Chapari, which itself is an affluent of the Mamoré. All these canoes, except my own, were laden with cocoa in the bean, or "pepita," as it is there called, a few tiger skins and tamarinds being the only other articles that were taken up for sale in the interior of Bolivia.

Outlet from Trinidad to Mamoré.

Canoes ascending the river from Trinidad do not have to return by the rivers Ybari to the Mamoré, as there is a lagoon about a league from the town, from which a creek or "curiche" gives egress to the principal river. This creek is not, however, large enough for even small steamers, such as the "Explorador," which would always have to descend the river Ybari to its mouth, before they could ascend the Mamoré, which above Trinidad, still preserves a bold and wide course, with free facilities for navigation, and has many plantations and sugar estates on its banks.

The River Securé.

On the third day out from Trinidad, we passed the entrance to a large lagoon situated on the right bank of the river, and then on the left bank we passed the mouth of the River Securé, which has its rise in the mountains of the northern part of the province of Cochabamba. This river though broad and wide for a great portion of its course is very shallow, and entirely unnavigable from the driftwood and timber collected therein. An expedition sent up by the Prefect of

Trinidad shortly before my arrival, returned with the only result of the impracticability of the river for any kind of navigation.

BELLA VISTA, THE PORT OF LORETO.

The same day we got to Bella Vista, the port of Loreto, a small village inhabited by Indians of the Trinitario family. Here is a "trapiche" or sugar mill and many large and well-kept cane fields and planta-tions, but the land is much cut up by old river courses called "madres," and must be entirely inundated in exceptionally high floods.

SIRIONO SAVAGES.—COCOA PLANTATIONS.

Above Bella Vista the right bank is subject to visitation by the "Si-riono" savages, and it is advisable therefore for canoes to keep on the left bank, on which, a short distance in the interior, are plantations of cocoa (chocolotales) belonging to the Government.

JUNCTION OF THE RIVER GRANDE.

On the 24th, the sixth day after leaving Trinidad, we got to the " Junta de los Rios," being the junction of the Rio Grande with the Mamoré.

THE RIVER PIRAY.

Of the navigation of the Piray or Sará, which is an affluent of the Rio Grande flowing from the district of Santa Cruz, I can only speak from hearsay. It is said to be free from all obstacle to navigation for at least eight months of the year, but for the re-maining four I should think it would be closed to steamers, as even canoes have to be unloaded and dragged over the shallows, which I am informed are of frequent occurrence. The Rio Grande at its mouth appeared to be very dry, while the Mamoré had plenty of water and was about 500 yards in width.

THE RIVER CHIMORÉ.

The next day we entered the River Chapari, leaving the Mamoré, or as it is there called the Chimoré, on our left. I was only able to ascend the Chimoré a short distance, but that was sufficient to con-vince me that it is a far superior river to the Chapari in volume of water : and I am informed by Dr. Velarde, who has ascended it to the port of Chimoré, distant only about two leagues from Coni, the port of Cochabamba, that it is far preferable to the Chapari for navigation, having no great amount of driftwood and no rapids as far up as he penetrated.

THE RIVER CHAPARI.

The River Chapari, though about 200 yards in width at its junction with the Mamoré, is in my opinion entirely useless for trading purposes. It is a very serpentine river, and the bends are so continually changing their courses that the whole river is one succession of sand banks and stockades of dead wood, through which even canoes are with difficulty manœuvred.

THE CHAPARI USELESS FOR NAVIGATION.

The ascent of the Chapari occupied our party eleven days of constant struggling with the stockades of dead wood and the rapids of the River Coni on which the port is situated, and I think the conclusion may at once be arrived at, that a better channel for navigation must be found, and this is, I believe, afforded by the River Chimoré.

CONI, THE PORT FOR COCHABAMBA.

Coni is distant about 45 leagues from Cochabamba, and is a small clearing with a few huts where the mule drivers and traders from Cochabamba remain while waiting in the dry season for the canoes from Trinidad. As the mule drivers only come to Coni when they expect to get a freight, intending travellers must make arrangements beforehand to have animals ready for their arrival, otherwise they will have to foot it over very bad roads.

THE YURACARÉ INDIANS.

The district is the home of the Yuracaré Indians, who are called savages, but are very friendly and well disposed. They are nomads in so far as that they only live in one clearing for perhaps two or three years, until they are tired of the spot or fancy that the "chaco" does not yield so well as it did at first. They then make a move to another part of their district, which ranges from the higher waters of the Chapari to the foot of the hills of San Antonio and Espiritu Santo on the road to Cochabamba.

TRADE VIÂ CONI.—NEW PORT FOR THE NAVIGATION COMPANY.

At this place I found a Corregidor and one or two traders, who had come down from Cochabamba, either to receive the departmental tolls on the traffic, or to receive their cargoes of cocoa, and ship the return loads of wheaten flour or salt. There were about 200 mules in all waiting at Coni, the cargo for each one being 8 Bolivian arrobas, or about two cwt. During the five or six months that this trade is open, more than 1000 cargoes, of 8 arrobas each, are received from Cochabamba in salt or flour, while a similar quantity of cocoa or leather and tiger skins is returned, and this trade is carried on under every possible difficulty of miserable roads, and defective means of navigation.

Coni is about 950 feet above sea level, and has a delightful climate ; while the vegetation is not of that dense and rank nature found on the main portions of the Amazon and Madeira rivers, so that ague and fever is very little if at all known. The Navigation Company will have to set up a port and clearing at the small Indian village of Chimoré on the river of the same name, and then the present port of Coni will be abandoned.

V.—BOLIVIA—CONI TO COCHABAMBA—COCHABAMBA TO SUCRE.

Start from Coni—Utilization of the Yuracaré Indians.

On the 9th of August I left Coni with five mules, an arriero, or mule driver, and also taking with me the Cacique, or headman and five boys of the Yuracaré Indians, who were useful on the road in carrying small packages, and with their bows and arrows shooting fish in the river, or game in the forest, so as to provide a repast at the halts at night. These Indians always volunteer to accompany a traveller of any note, and as it is well that they should always be disposed to especially assist anyone connected with the navigation or railway enterprises, I did all I could to keep them friendly, and believe that when operations are resumed, should it be intended to take men down to the works from Cochabamba, these Indians can be made very useful in clearing ground at the Chimoré port, and in taking the peons down to Trinidad, or even to Guajará Merim. They could also be utilized as guards against the few hostile tribes of the Mamoré, such as the Sirionos and the Chacobos, while, if properly managed, it might be possible to induce a hundred or two of them to locate themselves at the wooding stations necessary on the Mamoré, and referred to in the second chapter of this report. As these Indians are more expert with their bows and arrows than the tribes of the lower Mamoré, their presence would effectually check the incursions of those savages that refuse to approach civilization in the slightest degree.

PACHIMOCO.

The first night's halt was made at one of their chacos, called Pachimoco, a single house where lives the Intendente of the tribe, the next in authority to the Cacique. Here I was well treated, the supplies of the chaco, such as yucas, maize, and plantains, being freely offered to me, while the centre of the hut was given up to me for my camp bed.

THE RIVER SAN ANTONIO—THE YUNGAS OF ESPIRITU SANTO.

On the second day we came to the River San Antonio, an affluent of the San Mateo, which is an affluent of the Chapari. This river, where we first crossed it, is about a mile in width, forming at this season an immense playa of water-worn stones, through which the river finds its way in shallow channels, whilst during the rains the whole is covered with a very turbulent and rapid stream. Having crossed the river, the track leaves the plains of the Mamoré and the Chapari and commences to enter upon the mountainous districts called the Yungas of Espiritu Santo. The track so far is but a path cut through the forest, and is in wet weather quite impassable, from the depth of the mud and the numerous holes in which mule and rider may easily come to grief; but over the hilly lands it has in former days been roughly paved after the fashion of old Spanish roads.

THE MOUNTAIN ROAD VERY BAD.

The third day we followed up the course of the San Antonio, descending into it several times and riding up and down the "cuestas," which are, perhaps, the roughest mountain paths I have yet ridden over, the zig-zags being of the shortest and steepest possible. These paths are, as is usual in mountain districts, much worn by the trampling of the mules, and the drainage of the rains which hollows them out, so that at times one appears to be riding between two walls of earth.

CRISTAL MAIO.

The third halt was made at a place called Cristal Maio, at an elevation of 1,920 feet above sea level. Here we came upon the first settlements of Bolivians of the type found generally in the interior, such as Cochabambinos, Pazeños, and others. Planting "coca" is the only occupation of the settlers in this district, who are continually extending their clearings down the slopes of the hills towards the plains below.

THE COCA PLANT.

This plant is a small tree, allowed to grow to four or five feet in height, and planted in rows kept in excellent order. The leaves only are valuable, being collected carefully, dried in the sun, and then when pressed into seroons they fetch at Totora, the great coca depôt of the district, from 11 to 16 pesos the "sesta" of 22 lbs., or say 1s. 7d. to 2s. 4d. per lb.

USE OF THE COCA.

Without this stimulant, used largely in the interior of Bolivia and Peru, but unknown by, and apparently useless to, the Indian of the plains of the Mamoré, the Quichuan or Aymará Indian appears unable to exist, for it serves him on his long journeys with his mules or llamas in place of food or drink.

BEST DISTRICTS FOR THE GROWTH OF COCA.

The best districts for its production appear to be the eastern slopes of the northern hills of Bolivia, and a very large and valuable trade is carried on at most of the towns of the republic in this article.

CONTINUOUS RAINS.—ZINC ROOFED HOUSES.

There is said to be much rain in this district, the months of August and September alone being blessed with a few fine days. The houses, in view of this constant rain, are built with roofs of very steep pitch, at an angle of 60 degrees perhaps, whilst the settlers who can afford it have covered their houses with sheets of tin or zinc, brought at great expense from Cochabamba and the Pacific coast. The timbers used are extraordinarily heavy, in order that the frequent violent wind storms may not overturn the houses. A foundation wall of dry stone, raised 18 inches from the ground, is first placed, and on that a bed plate of hard timber dressed with the axe to 12 or 14 inches square, then uprights and wall plates to match complete the solid framing.

The fourth day's ride continued to be over very rough ground, the whole of the distance done being over a succession of "cuestas," which in places were actual staircases. At one moment one would ascend to 3,000 feet elevation, and then quickly down to 2,500, and even 2,000, the road being dug out of the side of the rock, just wide enough to allow of the passage of one mule with its burden.

Gold at Minas Maio.

At a place called Minas Maio, there is a tradition that gold has been found in the sands of the ravine, and at Espiritu Santo the river bed was full of quartz stones, from which I selected a few that have since been declared by a geological authority to have come from gold bearing reefs.

Metalliferous Character of the District.

There is also much ironstone of a shaley character in some of the ravines; copper pyrites also abound in some of the quartz stones in the rivers, and there is no doubt that the region is highly metalliferous, and will well repay explorers when the improved navigation of the Mamoré shall have caused the amelioration of the present bad roads of the district.

"El Chaco."

The fourth night was passed at a house called "El Chaco," 3,250 feet above sea level. The settlers appear only to produce provisions sufficient for their own use, and I found it impossible to purchase even a fowl or a few eggs at any price, so that had I not taken a small supply of "charqui" and tinned meats with me, I should have passed a bad time on this part of the road.

Forage of the District.

A grass called "saracachi" is grown for the mules, and is, I believe, peculiar to the district; it grows in bunches, and is cut off at the ground, leaving the roots, which soon afford another growth.

Coffee Grown.

Coffee of a very superior class grows here, but does not appear to be cultivated largely, owing, doubtless, to the greater profits yielded by the coca.

Wages of Peons.

The labourers or peons are all of the Quichuan race, and earn 4 reales, say 1s. 7d., per day and their rations.

The mountains at "El Chaco" rise on either side of the river to a height of probably 6,000 or 7,000 feet, their tops, generally in the clouds, being clothed with forest.

Landslip at Lina Tambo.

The fifth days' ride was from "El Chaco" over the cuesta of Lina Tambo, the summit of which was 6,150 feet above sea level. There had been a tremendous landslip, the path having been covered up for more than a mile in length, a distance that had to be traversed in drenching rain over the debris of the mountain side. After passing the summit we crossed a river called the San Jacinto (another affluent of the Chapari and Chimore system) over a bridge of rough timbers spanning perhaps 60 feet, with dry stone abutments and timber longitudinals. The fifth night was passed at a " pascana" or resting place by the road side, we having failed to reach any habitation.

Toll at Los Jocotales.

On the sixth day we passed a few houses called "Los Jocotales," where a toll of 2 reales, 9½ pence is collected upon all cargoes of merchandize carried by mules or donkeys, but the traders complain bitterly of having to pay tolls on a road that is so little cared for as this is.

Inca Corral.

We then passed over a very high ridge, on the summit of which the Aneroid marked nearly 8,000 feet above sea level, and the night was passed in a small hut at Inca Corral, a wide valley, at an elevation of 7,715 feet, and running nearly N. and S., down which the wind blew with searching force, the thermometer at night sinking to 39½° Fahrenheit. The country at this height is open, the few trees being quite stunted and covered with mosses and lichens. Maize of a very large size and very sweet is grown, also barley and potatoes, while a root called "Yacunes," somewhat like a small yellow carrot, is eaten raw and esteemed a great delicacy by the few cotters who live here to tend the cattle which are in fair numbers.

Frost at Night.—Cuesta de Malaga, the highest Pass.

Rising early in the seventh morning, the grass was covered with hoar frost and the cold extreme. Proceeding up the valley or raised plateau of Inca Corral we soon began the ascent of the Cuesta de Malaga, the highest and last before reaching Cochabamba, and really on the dividing ridge of the watersheds of the Mamoré and Rio Grande systems. On this ascent all trees stop at about 10,800 feet elevation, while the still higher tops of the mountains had snow in their crevices. I was told that in 1873 a snow-storm in August filled up the pass, and that an arriero and his "recua," or drove of mules, perished in the drift, a cross on the pathway attesting the fact. The summit of the pass I made to be 12,550 feet, and soon after passing it we came to cultivated lands and some isolated huts.

Barley and Potatoes.

Barley and potatoes are the only products, the soil being very barren and full of stones. No trees are visible, and only a rough long grass from which a few sheep and oxen find sustenance.

Total Change in the appearance of Nature.

From this point onwards into Bolivia the face of nature seems entirely changed, the luxurious tropical vegetation of the plains being left behind, and the stony barren hills of the Andes, with their occasional irrigated valleys, entered upon.

Cochi-Janchi.

About 5 p.m. of the seventh day we got to the village of Cochi-Janchi, at 10,950 feet above sea level, a village of mud or adobe houses, which, from the absence of any trees, look very dreary. The people, however, must be very laborious as the hills are much cultivated, some up to their summits of probably 12,500 or 13,000 feet elevation.

Potatoes sent to Cochabamba.

The village has a church and probably fifty or sixty scattered farms, and supplies Cochabamba with potatoes, in which a large trade appears to be done, both in fresh fruit and preserved, called "chuño," which is really nothing more than a frozen potatoe and a most horrid substitute for the real article. In this state, however, it is said to keep for any length of time, and forms the staple article of food with the Quichuan Indian.

Road to Coni by Bandiola.

From the village there is another track that leads to Coni by passing through a district called Bandiola to the east of Espiritu Santu, but that track also has to pass the ridge of Malaga, and I was credibly informed that the cuestas in that direction are far more severe than those over which I travelled, and as my Yuracaré Indians refused to return to their homes by the Bandiola road, I think it may be agreed that it is not a practicable one. The Malaga cuesta, although rising to a great height was not very steep, the ascent and descent being so gradual that I did not once have to dismount from my mule in going over it.

On the eighth day, leaving Cochi-Janchi in the morning, we passed another cuesta, whose summit was only 600 feet lower than that of Malaga, and passing several villages came about midday to the pueblo of Sacaba, a considerable town about four leagues from Cochabamba.

Sacaba.

From Sacaba the shortest road to Cochabamba is over a ridge of hills that stretch out into the plain where the city stands, but the animals being very tired and footsore we made the detour round the hills arriving at the town at dusk.

Cochabamba.

Cochabamba, probably the most important town of the Republic of Bolivia, is situated in a plain 8,450 feet above sea level, overlooked by the snow-clad heights of Tunari and Larati. The town is well built, with regular streets, a handsome Cathedral and other public buildings.

Agricultural Products.

The population probably is about 50,000; the greater portion of whom are engaged in agriculture, for Cochabamba may certainly claim to be the agricultural capital of Bolivia, whilst La Paz may be termed the mineral capital. Large stores of flour, maize, corn, barley, potatoes, fruits of all kinds such as apples, pears, apricots and strawberries are forwarded from Cochabamba to the central towns of the Republic.

COMMERCIAL FIRMS.

Here there are three German houses of considerable wealth and position, and the American firm of Haviland and Keay who are contractors for public works, and for the coaches that run to Arani through the valley of Cliza.

MINERAL WEALTH OF THE DISTRICT.

Very little if anything has been done in mining in the neighbourhood of Cochabamba, owing to its still greater distance from the coast than Potosí, Oruro, or many of the other mineral towns of Bolivia but I am sure that there is immense wealth in minerals in all the hills encircling the plain in which the city stands, and samples of manganese, silver and lead ore that I have brought from the outcrop of lodes in the district attest the fact.

CLIMATE.

The climate of Cochabamba may I think be classed amongst the finest of the world, it enjoying an almost perpetual summer, whilst the nights are pleasantly cool and strengthening to constitutions depressed by the humid heat of the Madeira and Amazon valleys.

FEELING OF PUBLIC OPINION ON THE ENTERPRISE.

From my reception in Cochabamba it is evident that all classes and shades of political opinion are, notwithstanding the disappointment caused by the abandonment by the Public Works Company of its contract, resolved to do all possible to assist the enterprise to completion. Cochabamba sees in it, its only hope of emancipating itself from the heavy costs and charges levied upon all its European necessities by the merchants of La Paz and the Pacific Coast, and also its only hope of securing an outlet for those abundant agricultural products that now are lost for want of customers.

INTERVIEW WITH THE MUNICIPALITY.

The Municipal Council invited me to a special session and requested me to give them information upon certain points that were in doubt with reference to the future prospects of the enterprise, and also as to the supposed difficulties of construction of the railway and the unhealthiness of the climate.

Address sent to Congress by the Municipality.

The remarks that I addressed to the Municipality being satisfactory it was decided that a note should be sent to the Secretaries of the Congress sitting in Sucre, praying the Congress to take the question of continuance of support to the enterprise into early consideration and praying that a favourable solution might be given to the matter, whilst all the Deputies from the Cochabamba district, some twelve in number, received special instructions to vote in favour of and to protect the enterprise.

Why Popular Meetings were not held.

Public meetings in favour of the enterprise would have been organised during my stay, had it not been feared that they might give rise to some political manifestations of disapproval of the acts of the Government in their interpretation of the decree of November, 1873, and as party spirit was very much excited amongst Government supporters or " Rojos," Quevedistas and Corralistas, it was thought advisable not to make the Company's affairs a vehicle for political disturbances.

Labour obtainable.

With reference to the prospect of obtaining labourers from this district I think that at least a thousand good workmen might be obtained from the valleys of Cochabamba, Cliza and Arani ; and nothwithstanding the distrust left by the acts of the Public Works Company in returning peons without due payment for their loss of time suffered in a fruitless journey from Cochabamba to the Beni, or from Chiquitos to Santa Cruz, I had many applications from peons desirous of proceeding to the railway, while two well known contractors offered to enter into engagements for either subcontracts or for supplying labour for the works.

Totora Route to Sucre determined upon.

I determined to make my journey to Sucre by way of Arani and Totora, in order to see the new cart road in course of construction between these places, and also at Totora, to obtain all the information possible with reference to the advisability of a road from thence to the proposed new port on the Chimoré.

Tarata and Cliza.

Leaving Cochabamba on the 31st of August, in one of Messrs. Haviland and Keay's coaches, we were soon in the pampas, and passed the towns of Tarata and Cliza, both populous and flourishing places. August being one of the dry months, the pampas were bare and dusty, but evidently in the spring and showery seasons large crops of barley, wheat, and maize are raised. The pampas are dotted over with the dome-shaped huts and houses of the Quichuan Indians, and from their being built in mud and stones, the country presents features similar to those of many of the plains of Central India.

Punata.

The first night was passed at Punata, a town with about 16,000 inhabitants, and a place of considerable trade in cereals, felt hats, and ponchos made from vicunha and other wools. Here I was introduced to a Señor Manuel Arauco, who is the principal man of the place, and was most hospitably received by him. He has a most interesting museum of products of the province of which he is a native, and showed me some excellently tanned specimens of leather; also various dyestuffs and drugs collected in the forests of the hills which border on the plains of the Beni.

Exportable Products of the Province.

Amongst other articles I noticed and brought away small samples of brasiletto wood, campeachy, a root which might be utilised as a purple dye instead of Orchella weed also turmeric collected near Santa Cruz, and called " coorcama " in the district. As this article commands a large sale in Europe, at about £30 a ton, it might with great advantage be made an article of export, viâ the railway of the rapids. Señor Arauco, who is a most talented man, spoke with great enthusiasm of the benefits to accrue to the eastern provinces of Bolivia from the completion of the railway and navigation schemes.

Arani.—Lagoons near Vacas.—Irrigation Works.

Leaving Punata, the road passes through Arani, a small town at the end of the pampa, and then commences to ascend the hills leading to Vacas and Totora. Near Vacas are the lagoons from which it is proposed to take water for the irrigation of the pampas of the Cliza valley. The lagoons are three in number, the largest being about 1½ leagues in length by half a league in breadth. The works, which have been carried out by Messrs. Haviland and Keay, of Cochabamba, for Mr. Meiggs, of Lima, are now in abeyance, and it was supposed that they would be abandoned, the reason being that it was surmised, that if the channel cut from the lagoons to the pampas were opened, the lagoons would drain dry in about four years, and that no return for the capital spent (about £50,000) would be obtained. I think it may be considered that the lagoons are only drainage deposits from the surrounding hills, which attain altitudes of 14 or 15,000 feet, the lakes themselves being about 9,500 feet above sea level, while the pampas to be irrigated are at an altitude of nearly 9,000 feet. There are no rivers to empty themselves into the lagoons, and there is only the drainage of the hills during the rainy season to depend upon. Some authorities aver that the lagoons are supplied by springs, while others think that they are decreasing yearly in size ; and it is probable that they are parallels on a small scale of Lake Titicaca, in the north-western corner of Bolivia, or of the Lake of Valencia, in Venezuela —lakes that are known to be decreasing yearly, from extended agriculture, and in the case of Lake Valencia denudation of forests. If irrigation could be taken to the pampa lands of the Cliza valley, they would, perhaps, become the richest agricultural plains in the world, as their climate, owing to the considerable elevation, is suitable for the production of almost any cereal.

Path from Vacas to the River Chimoré.

Vacas is a small Indian village of no other interest than that it is said, that from there exists a path that leads to the Chimoré and Coni rivers, and its position on the map would lead to the belief that it is favourably situated for explorations to these rivers.

New Road between Arani and Totora.—Pay of Labourers.

The road from Arani to Totora has been made without any engineering help, and consequently the grades are very uncertain, and the routes taken might in many cases have been improved upon. Notably one part of the road, about two leagues before arriving at Pocona, has been taken over a ridge, the descent from which is accomplished by a zig-zag of 3 inclines of possibly 1 in 6 at least, while the road might have been taken up a quebrada, and the abrupt descent avoided by a continuous grade of perhaps 1 in 25. However, considerable work has been undertaken and some of the cuts are of great depth, one point of rock and earth being cut to 50 feet depth ; on the whole the work reflects great credit on the contractor Señor Demetrio Jordan, of Cochabamba, and is the first road undertaken in the republic by a Bolivian contractor. The tools in use by the peons were of the most miserable description, and considering this, it is clear that the Quichuan Indians may be made very fair navvies; their daily wage on the road was 4 reals, or say 1s. 7d., without provisions.

POCONA AND TOTORA.—PROPOSED ROAD FROM THE RIVER CHIMORÉ TO TOTORA.—INTERNAL SYSTEM OF RAILWAYS FOR BOLIVIA INITIATED.

Pocona is but a small village, but the next town arrived at, Totora, has about 15,000 inhabitants, and possesses considerable trade principally in coca, it being the chief emporium in eastern Bolivia for that article. Coffee, flour, sugar and potatoes are also articles of export, while foreign merchandize finds its way here from Sucre and Santa Cruz, numerous droves or "recuas" of mules and donkeys being met with between Totora and Sucre. I took a good deal of information in Totora with reference to roads to the head waters of the Chimoré, and was introduced to two of the principal men of the place, Don Eugenio Soriano and Don Saturnino Vela. These gentlemen are owners of cocales or plantations of "coca" in the hills which form the watersheds of the affluents of the Mamoré and of those of the Rio Grande. Señor Soriano has made a track from Totora to Arepucho where his cocales are situated and is now cutting a further track from Arepucho to the Chimoré ; he is confident a much better road is found *via* Arepucho than by Espiritu Santo and the Yungas of San Antonio, and gives me the distances as follows : Totora to Arepucho, 14 leagues, and Arepucho to the Chimoré 12 leagues, total 26. The road from Cochabamba to Coni *via* Espiritu Santo is 44 leagues. If a road by Arepucho and Totora is made the 26 leagues can be easily done in three days, and from Totora to Cochabamba, the road being good, can be done in two days, making five days in all, the last day of which can be done in coach from Arani. The Espiritu Santo route cannot be got over in less than seven days, and the road is not susceptible of much improvement, while the wide crossing of the River San Antonio renders the route almost impracticable for general traffic. I therefore conclude that our efforts should be directed to making a road from the Chimoré *via* Arepucho to Bacas or Totora, this latter place having the advantage of being a good starting point for a road to Sucre as well as to Cochabamba and Oruro. Thus a future would be prepared for an internal system of railways for the eastern part of the republic, in communication with the Madeira and Mamoré Railway.

From Totora to Misque—Misque.—Aiquile.

The height of Totora is about 10,000 feet, and the road rises slightly on leaving the town, then crosses a large plain about 500 feet below the Totora hills, then rises very sharply to 11,500 feet and descending a very steep and bad "cuesta" gets to a quebrada, up which the road for Misque is taken. Misque is an old cathedral town, once of considerable importance, but of late years much abandoned on account of the ague fever that is said to be prevalent there. It is in a beautiful plain 7,000 feet above sea level, and as irrigation is carried on to a great extent, the fields and feeding grounds were in very fine order. From Misque to Aiquile, the road taken up a wide and almost dry river bed is fairly level, the greatest altitude passed over this section being about 9,000 feet. Aiquile, the next town of importance, is a thriving place of about 4,000 inhabitants, from this place a road branches off for Santa Cruz. Aiquile seems a very healthy town, and many people have of late years moved to it from Misque ; its elevation above sea level is 7,850 feet ; its trade seems to be entirely in agricultural products.

Sugar Cane at Quiroga.

Chinguri, Quiroga, and Palca are the next villages passed through ; they are unimportant and call for no remark, other than at Quiroga, about 7,000 feet above sea level, there are several very large and fine cañaverales, or fields of sugar cane, small in height and of slender growth, but said to yield good produce. Cultivation was well carried on there, considerable work in aqueducts, channels, and other irrigation requirements being carried out.

The "Rio Grande."

About six leagues from Quiroga we entered on the "Rio Grande," flowing between two ranges of hills, on either side of which a road might, with considerable ease, be made. The track now crosses the river about seven times, and then comes to a bridge now broken down entirely ; the stone abutments of very inferior masonry are, however, still standing. The roadway was, I am told, a suspension affair, which, being hung with too great a sag, was soon washed away. The river offers great facilities at the site for a span of a hundred feet, as two points of rock jut out from the main hills on either side, giving good foundations for the abutments. The Rio Grande, therefore, does not offer any very great obstacle to the construction of a good road to Sucre. In the rainy season, when the river is full, a "balsa," or raft, lower down the river than the old bridge, forms at present the means of communication.

FLOUR MILL AT CANTO MOLINO.

After leaving Palca, the short but sharp "cuesta" of the "Jaboncillo," so called from the greasy nature of the earth, a kind of clay slate, has to be surmounted. The next, called "Masa-cruz," rises to an elevation of 8,550 feet above sea level, and on the Sucre side falls 1,350 feet in about a couple of miles. At the foot of this hill, at the junction point of three large ravines, is a flour-mill, called "Canto Molino," where most of the wheat and maize grown in the Cochabamba district is sent for grinding. The power used is water, the machinery an old-fashioned vertical wheel, while grind-stones of fair diameter and hard grit are found in the neighbouring hills.

THERMAL SPRING AT HUATA.—FIRST VIEW OF SUCRE.

From Canto Molino the road continues up a ravine of from two to three hundred yards in width, with hilly country on either side, which offers good sidelong ground for a road, but here, as is usual throughout Bolivia, the road is taken up the bed of the river, which, from being dry for the greater part of the year, offers a ready-made road good enough for animal traffic. At Huata, in this ravine, there is a bathing establishment at a thermal spring, much used by the residents of Sucre. Beyond this village there is a very sharp cuesta, which rises to 10,000 feet above sea level, and shortly after surmounting the crest of the hill, one comes in sight of the capital of Bolivia, "Sucre," or "Chuquisaca," as it is called in the Quichuan tongue. The first appearance of the town is very pleasing, from the number of churches, convents, and other large public buildings, but the country round is very bare and dull looking, water being very scarce and only to be met with in the bottom of the deep ravines.

VI.—SUCRE.

ARRIVAL AT SUCRE.

On the eighth day from Cochabamba I arrived at Sucre, the time occupied usually in travelling between the two places being five to six days, the detour that I made by Totora being two days' journey longer than the route by the valley of San Pedro.

SUCRE.---PUBLIC BUILDINGS.—BAD MUNICIPAL ARRANGEMENTS.— SMALL POX.

Sucre, although nominally the capital of Bolivia, is not the largest or most influential town of the republic, being surpassed in population, and consequently in commerce, both by Cochabamba and La Paz, while Potosí, once more populous than any of the three, has still nearly as many inhabitants as Sucre, which may be said to have a population of about 25,000. There are no manufactures and but little agriculture, so that the town owes its importance to its being the constitutional seat of government and the site of the principal universities and religious establishments of the country. The principal cathedral is dedicated to "Nuestra Señora de Guadalupe," and the Archbishopric of La Plata, with the see of Chuquisaca for head-quarters, once had jurisdiction over all the ecclesiastical dignities of Southern South America. There is a palace for the President, a Hall of Congress, and some imposing-looking structures amongst the colleges, convents, and other public buildings; whilst the streets and squares being broad and fairly well paved, the town has altogether a very decent appearance, although it is to be regretted that the sanitary arrangements of the Municipality should in this city, as well as in all the other principal towns of Bolivia, be conspicuous for, and remarkable solely from, their absence. For this reason it is that small pox hangs for such long periods of time about these cities, and kills yearly large numbers of the Indian population, who, averse to vaccination in ignorance of its benefits, fall easy victims to this terrible scourge of South American cities. Whilst I was in Sucre this plague was rife, my own servant boy, a Cruzeño, falling a victim.

DR. VELARDE—REPORT PRESENTED TO CONGRESS—ARTICLES PUBLISHED IN NEWSPAPERS.

Here I met, for the first time, Dr. Juan Francisco Velarde, who had temporarily resigned his position as agent of the Navigation

Company in Bolivia, in order to represent the town of Santa Cruz in the Congress ; he received me most enthusiastically and kindly. Dr. Velarde seemed to look upon my unadvised coming to Sucre as something almost providential, as he had begun to have serious misgivings as to the treatment the affairs of the enterprise would receive at the hands of Congress, owing to the active hostility of its enemies and the lukewarmness of its friends ; the advent of a coadjutor was to him, therefore, an event of unparalleled importance. Finding that the question of the " Empresa Church," as the affairs of the Navigation and Railway Companies are universally spoken of throughout Bolivia, could not come to discussion in the Congress for some weeks, I applied myself to sounding the views of the deputies with reference to the enterprise, and informing the public generally of its views and aims. I also prepared a report upon the events that had transpired in connection with the railway since the letting of the contract in 1872, and had it printed for and delivered to the deputies in Congress. Dr. Velarde had, in the month of July previous to my arrival, prepared and published a most lucid and painstaking pamphlet concerning the affairs of the enterprise and the causes that had delayed its realization, but nevertheless the enemies of the railway continued to be very active ; and as certain schemes which had for their object the carrying of Bolivian commerce by the waters of the River Paraguay to the Atlantic at Monte Video and Buenos Ayres, were being brought before the Congress by competing concessionaries, Dr. Velarde and I considered it necessary to keep the public instructed, by means of newspaper articles, in the condition of and the aims and objects of the " Empresa Church ;" these articles we brought out in " La Cronica," an independent journal published in Sucre during the sittings of Congress.

New Concessions being applied for in Congress—Caracoles and Mejillones Railway—La Paz and Lake Titicaca Railway.

The schemes for which new concessions or renewal of old ones were being sought in the Congress of 1874, were mostly connected with the opening up of the eastern side of Bolivia. With regard to the Pacific side, the Concessionaire of the Caracoles and Mejillones Railway, Sr. Braun, asked for the payment of interest due on the bonds issued by the Bolivian Government for the construction of that railway, but failed to get any satisfactory resolution. Sr. Zara, the Concessionaire of the La Paz and Yungas Railway, of two feet guage, in the department of La Paz, sought for and I believe obtained from the Government a concession for an extension of that railway from La Paz to the Lake of Titicaca; this concession did not, however, come before the Congress, the Government appearing to have power to grant it without the necessity of an application to the Assembly.

SCHEMES FOR ROADS TO THE RIVER PARAGUAY—SENOR PARADIZ'S SCHEME.

The applications for concessions on the eastern side were four in number, all having for their object the construction of roads across the unknown territory which separates Bolivia from the River Paraguay. One scheme taken up in Sucre by a Señor Paradiz, obtained the recognition of concessions granted in 1853, for the construction of a road from Santa Cruz *viâ* Chiquitos to a port on the Paraguay River, to be called Port Vargas, situated about 180 miles below the Brazilian port of Curumbá, in the disputed territory of the Gran Chaco. The projector of this enterprise estimates that with about £60,000 he can complete his cart track, establish a rural colony at this port, and place two steamers, a schooner, and sundry lighters upon the river, and as he has secured the right to all duties that may be levied at the port for a period of eight years, he calculates that a profit of over 70 per cent. will be earned. His estimate of cost is, I think, far too low, and his chances of success are but small, as he has not been able to form his Company or obtain assistance in Buenos Ayres, for which town he started in December last from Sucre.

SEÑOR M. S. ARANA'S SCHEME.

The second scheme is propounded by Señor Miguel Suarez Arana, a Bolivian gentleman of good family. He proposes to construct two cart roads, one from Santa Cruz to an undefined port on the right bank of the River Paraguay, and the other from a town called Lagunillas, in the Cordillera, to the same port. The concessionaire asks for two-thirds of the duties to be created at the proposed port for a period of 40 years and seeks for tolls, premiums and other advantages. This proposal, however, did not come before the Congress, but was referred for settlement to the Government under special powers granted shortly before the Congress closed.

CAPTAIN GREENLEAF CILLEY'S SCHEME.

The third scheme is brought forward by Captain Greenleaf Cilley, a retired commander of the United States Navy, who proposes to construct a railway from Santa Cruz to the territory of Otuquis on the upper waters of the river of the same name, an affluent of the River Paraguay. The length of this line would probably be not less than 300 miles, and Captain Greenleaf Cilley, estimating the cost of construction at £8,000 per mile, asked for a guarantee of 7 per cent on the expenditure, and for two leagues of land on either side of the line. But Bolivia is not reckless enough to promise a guarantee on such an enormous capital, and Captain Cilley had to strike the guarantee clauses out of his application. At the time I left Sucre the concession had not been granted even in this mutilated state, but was waiting for a report by the Council of State. Captain Cilley is the representative of the Vernet family of Buenos Ayres, who are descendants of the " Oliden " who received a concession of lands from Bolivia nearly fifty years ago. These lands are high and well suited for cultivation of coffee or cocoa, and Captain Cilley hopes to be able to attract emigration to them, if he can obtain the concession for the railway and funds wherewith to construct it.

SEÑOR REYES CARDONA'S SCHEME.

Sr. Reyes Cardona, for some time Minister to the Brazilian Court at Rio Janeiro, proposes a colossal scheme of railroads commencing at Bahia Negra on the Paraguay, crossing the deserts of Izozo to Santa Cruz, and from thence by Sucre on to La Paz. This vague scheme seemed to possess only one fixed idea, which was to seize the £600,000 deposited in the Bank of England for the construction of the Madeira and Mamoré Railway. In what part of the grand scheme of internal railways for Bolivia the fund was to be spent did not appear to be of much consequence, so that it was handed over to Sr Reyes Cardona, to be dealt with as his self-acknowledged " honour, talent and patriotism " should direct.

EFFECT OF NEW PROPOSALS UPON PUBLIC OPINION.

These various schemes tended to weaken public interest in the opening up of the trade route for Bolivia viâ the rapids of the Madeira and the River Amazon, and though the route of the Paraguay could never be at all antagonistic to that of the Amazon, some sections of the Bolivian public were glad to accept the delay in the construction of the Madeira and Mamoré Railway as an excuse for listening to new schemes and new interests.

5

THE BRAZILIAN MINISTER AT SUCRE ASSISTS THE ENTERPRISE.

The "Empresa Church" was, however, very fortunate in finding a staunch advocate and firm friend at Sucre, in the person of Sr. Lionel d'Alencar, the Minister of Brazil to the Republic of Bolivia. This gentleman whom I had the honour of knowing in 1868 at Caracas, he being then Chargé d'Affaires to the Republic of Venezuela, took a lively interest in forwarding the prospects of the enterprise by sending friendly communications to the Minister, Señor Mariano Baptista, expressing the great interest with which Brazil looked upon the Madeira and Mamoré Railway, and assuring the Minister of the desire of the Brazilian Government to assist Bolivia by offering the guarantee upon the extra capital, which it was known had been asked for at Rio Janeiro by the Company's agent there. Sr. d'Alencar also gave his opinion to the effect that Brazil would not look with any favour upon or assist schemes having for their object the carrying of Bolivian traffic by the River Paraguay to the Atlantic.

POLITICAL PARTIES.

Political parties in Bolivia spring up, change or die out according as some leading spirit rises up for a time and soon gives place to another and newer man, for politics in Bolivia cannot be defined in any other way than as purely personal.

QUEVEDISTAS.

The party once led by General Melgarejo, formerly President of the Republic, has now for its chief General Quintin Quevedo, whose followers are termed Quevedistas. To General Quevedo belongs the honour of having not only been the first to demonstrate practically the existence of a trade route for Bolivia by the Madeira and Amazon River, but also of having been the introducer of the grand idea to American enterprise.

THE POLITICAL CHARACTER GIVEN TO THE ENTERPRISE.

The "Empresa Church," therefore, at present, doubtless, has to struggle against the dislike inherent to it in the minds of supporters of other political parties in Bolivia, it being generally considered throughout Bolivia as a Quevedistic creation; and here I must do the party the honour due to it of saying that it has throughout the struggle loyally supported the enterprise.

CORRALISTAS.

Another party in Bolivia is termed that of the Corralistas, from its component followers acknowledging as their head Dr. Casimiro Corral, Minister of Home Government and Foreign Affairs in the Republic, under the presidency of General Morales. This party is a large and influential one, and its members, with the exception of some resident in La Paz, have always supported the various decrees that have been passed in favour of the construction of the Madeira and Mamoré Railway.

ROJOS OR BALLIVIANISTAS.

The third party in Bolivia is the one at present in power, and is headed now by Don Tomas Frias, the actual President of Bolivia. The terms "Rojos," or Reds, and "El partido Ballivian," are some-what indiscriminately applied to this section of public opinion, which includes many independent members, as well as many of the supporters, and co-political religionists of the lately deceased President Don Adolfo Ballivian.

It is well known that Señor Ballivian returned to Bolivia from England in the year 1872, with a fixed resolve to crush the Madeira and Mamoré Railway and the National Bolivian Naviga-tion Company; and, notwithstanding the failure of his project for a new loan proposed by him to the Congress of that year, the declared object of the new loan being principally the buying up of the old one, and the cancellation of the subventions to the enterprise, his more declared disciples in the "Rojo" party still perpetuate his idea, and endeavour in every way to throw obstacles in the path of the enterprise in its efforts to re-habilitate itself. The more indepen-dent members of this party do not, however, join in this blind policy of the Ballivianistas, and have afforded very valuable aid to the enterprise during the late discussions in Bolivia.

THE OPPONENTS OF THE ENTERPRISE DEFINED.

From this short review of the politics of Bolivia, it will be seen that the only opponents of the "Empresa Church" form but a small minority of the inhabitants of the Republic, and may be said to be extreme Ballivianistas, including such men as the Señores Aramayos, father and son, of whom more hereafter, and Señor Nicolas Acosta, at one time Secretary to General Narciso Campero in the Legation to London. A few of the leading men of La Paz, whose hard-headed, restrictive notions still lead them to believe that the act of giving freedom and life to the eastern provinces of Bolivia will inevitably bring death to the western ones, have, from the inception of the enterprise, steadily opposed it, and, still consistent to their retrograde principles, continue to do so upon every possible opportunity.

INTERVIEWS WITH PRESIDENT FRIAS AND THE MINISTERS.

Shortly after my arrival interviews were granted me by the President, Sr. Frias, and the Ministers of Government and of Finance, Señores Baptista and Dalence, the result of these interviews being to convince me that the enterprise need not look for any active support from the present Government. In fact, I soon discovered that the policy of the Government of Bolivia was one of masterly inactivity, the Government being so fully occupied in maintaining its own power that it feared to bring on any question in Congress that might at all provoke discussion or create criticism of any of its acts. However, Señor Baptista, the most prominent member of the Government, assured me that the enterprise need not fear any opposition from his Government which would submit itself entirely to any instructions that might be furnished it by the Congress then sitting.

PROJECT OF A DECREE SUBMITTED TO CONGRESS.

The next step, therefore, was to formulate a project of law having reference to the affairs of the enterprise, and get it submitted to the approval of Congress. Señor Velarde, therefore, proposed the form of decree which was submitted to Congress on the 21st of September, having obtained the signatures of six other deputies in its favour. This project stipulated that the remainder of the 83 per cent. of the product of the loan of 1872 should continue to be applied to the construction of the Madeira and Mamoré Railway, that the Companies should give guarantees for the completion of the line, that the Government should consider the time for the completion of the road as extended to two years from the recommencement of the works, and should enter into negotiations with Brazil for the obtaining of the necessary favours with regard to the free navigation of the River Madeira and the freedom of the traffic from all taxation, as well as to solicit from Brazil its aid in the providing of further funds; that a permanent legation should be established in London with powers to see to the proper inversion of the deposited fund and to put an end to all litigation, and that speedy provision should be made for a commencement of the collection of duties on all traffic coming into or going out of Bolivia by way of the Amazon. The project was passed to the Commissions of Finance and Industry for their report, but, owing to press of general business and the many sittings that the Commissions held before coming to a resolution on the subject, a whole month passed away, the reports on the project not being returned to Congress until the 19th of October.

CAUSES OF DELAY IN OBTAINING A HEARING IN CONGRESS.—OCTOBER ELECTION OF OFFICERS OF CONGRESS FAVOURABLE.

Questions as to the powers to be given to municipal authorities throughout the Republic, and disturbances in La Paz occupied the attention of Congress about this time so much that I feared the subject ɔ the " Empresa Church" might get shelved altogether, but on the 9th of October new officers of Congress were elected, and a staunch supporter of the enterprise, Dr. Martin Lanza, of Cochabamba, was named President, whilst Señor Velarde was appointed as one of the Secretaries, while, to balance this favourable election, Señor Nicolas Acosta, the most active of our declared enemies, was named as co-secretary. This election of officers for the "mesa" or table of Congress served to prove that the friends of the enterprise were in a majority of at least two to one in the Assembly, and enabled Dr. Velarde to make sure of bringing the question of the railway to a hearing in Congress.

INTERVIEW WITH THE MINISTER OF FINANCE.

About the second week in October it became evident that the Commissions would report favourably on Dr. Velarde's project of law regarding the enterprise, and I called on the 13th on Dr. Dalence, the Minister of Finance, suggesting to him that, in view of the evident desire of the majority of the Bolivian public that the enterprise should be carried to completion, it would be advisable to forward instructions to the Financial Commissioners in London to cease obstructing the Company in its efforts to prosecute the work, and at least order them to delay all legal proceedings on their part until the wishes of Congress should be made known.

ORDERS SENT TO LONDON TO SUSPEND LEGAL PROCEEDINGS.

The Minister informed me that he would consult the President and the Cabinet as to the desirability of adopting my suggestions; and at a subsequent interview assured me that a letter had been sent to Messrs. Aramayo and Terrazas, ordering them to suspend all legal action. The letter of the Cabinet in Sucre, dated October 17th, 1874, and published in "The Official South American Gazette" for 2nd of February, is probably a correct copy of the instructions then sent to the Commissioners.

REPORT OF THE MAJORITY OF THE COMMISSIONS.—REPORT OF THE MINORITY.

The United Commissions of Industry and Finance comprised fourteen Members of Congress, and of these eleven signed a report favourable to the continuation of the enterprise. This report was most carefully drawn up, and successfully combating all the arguments of the opposition, offered for the acceptance of Congress a decree nearly similar to the project proposed by Dr. Velarde. The minority of three recapitulated all the accusations that have been brought against Colonel Church by the opponents of his enterprise, and proposed a short decree to the effect that as Colonel Church had not complied with the law of the 5th of November, the Government should, without loss of time, proceed to rescind the contracts, in conformity with the said law. This proposition was signed by the Deputy Nicolas Acosta before referred to, and by Deputies Luis F. Lanza and B. Villalva, of La Paz.

ADVERSE REPORTS BY THE BOLIVIAN CONSUL AT PARA, &C.— ADVERSE REPORTS REPLIED TO.

The Government, on the 23rd of October, published in their Official Gazette, "El Rejimen Legal," a letter received from a certain Dr. Benjamin Lens, who had in April, 1874, been sent down, *via* the Cachuelas, as Consul to Pará. This gentleman had probably left Bolivia with the idea of undertaking sub-contracts on the railway—a work for which he doubtless thought his position as Consul would highly recommend him, as he might, with his consular powers, be able to collect many of the Bolivian Indians on the Madeira River. He was, therefore, much disgusted at finding the works at a standstill, and addressed a communication to his Government, full of exaggerations and mis-statements, as to the condition of San Antonio, the materials and buildings there existent, and the conduct of the employés of the Railway Company. Dr. Lens, however, while villifying all those connected with the enterprise, could not help admitting the usefulness of the railway to his country, and stated to his Government his conviction that "it would be absurd to assert the uselessness of the scheme, because the probable growth of Bolivia, if the nineteen Cachuelas were overcome, is incontestible." He also stated that "the towns of the Beni, Santa Cruz, and Cochabamba call for the execution of the great work;" that "the track which had been made on one of the margins of the Madeira, proved that the railway could be completed in two or three years," and that "its incontestible benefits would amply remunerate Bolivia, by the encouragement of its commerce and the augmentation of its finances, without taking into account the increase of its industry, principally that of agriculture, its chief element of life." The Prefect of Trinidad also sent up a communication by one Señor Juan de Dios Molina, regarding the climate of San Antonio, and stating that on account of the death of the Chief Engineer sent there by the new contractors, the whole of the staff had been returned to London, and the place abandoned; while from Trinidad came another assault upon the enterprise, in the form of a petition by one Nazario Buitrago, for permission to seize the Company's property at the "Cerrito," near Exaltacion, declaring that the same was left without any care on the part of the Company's employés. To the first two of these attacks I replied in a report to the Government, dated the 28th of October, and Señor Velarde replying to the third one about the same time, the Government published our replies in the same official Gazette, thus removing in great measure the bad effects caused by the hostile communications.

TREATY WITH CHILI.

The settlement of boundaries with Chili came on for discussion about the middle of the month of October, and occupied Congress until the 7th November, being argued with much warmth by the opponents and supporters of the treaty concluded by the Government with the Chilian Minister Sr. Carlos Walker-Martinez. The final modifications of the Government treaty by the Assembly defined the northern Chilian boundary as the 24th degree of latitude, from the Pacific Coast to the highest points of the Andean range (excepting towns already under Bolivian Government, such as Antofogasta), and therefore reduces the Bolivian coast line to a still smaller extent than that afforded by the miserably small slip given to Bolivia at the time of the partition of the Spanish Empire of South America into Republies.

THE MINISTRY ACCEPT THE REPORT OF THE MAJORITY OF THE COMMISSIONS.—DR. QUIJARRO PROPOSED AS NEW ENVOY TO LONDON.

Finding that our question now stood some chance of coming on for discussion, and seeing the evident probability that the decree proposed by the majority of the Commissions of Industry and Finance, would be accepted by a majority in Congress, Dr. Velarde and I thought it advisable to seek an interview with the Government, and again endeavour to induce the Ministers to declare openly in favour of the enterprise. On the 3rd November we had a satisfactory interview with Sr. Baptista, Minister for Home and Foreign Affairs, who assured us that the Government accepted the report and proposal of the majority of the Commissions, and would support the same in Congress. The Minister also said that the present Financial Commissioners, Messrs. Aramayo and Terrazas, would have to be recalled as the Government was convinced that the Companies could never work in harmony with those gentlemen. To that proposition Dr. Velarde and I cordially agreed, and took the first occasion to place the name of Dr. Antonio Quijarro before the Minister, as one most suitable in every way to bring the various matters in litigation between the Government and the Companies to an amicable adjustment. Dr. Quijarro was deputy for Cobija, and a staunch supporter of the policy of the Cabinet of Don Tomas Frias, but at same time a warm friend to the enterprise, and heartily desirous of aiding in the opening up of the commercial route for Bolivia, of the Amazon and Madeira Rivers.

DEBATE ON THE ENTERPRISE COMMENCED.—THE DEBATE ADJOURNED AND RESUMED.—SPEECH OF THE MINISTER OF FINANCE.

On the 7th of November, a Saturday, the Chilian question was finally disposed of, and the reports of the majority and minority of the Commissions of Finance and Industry regarding the "Empresa Church" were read to Congress, but no discussion took place until the following Monday, when, after the election of the President and Secretaries of Congress, who have to be re-elected monthly, Dr. Velarde opened the debate with great moderation and lucidity, shortly stating the necessity that had arisen that Congress should express its views as to the necessity of removing the obstacles to progress that the Financial Commissioners had raised in London, and impressing upon the deputies, by the reading of copious extracts from scientific authorities, the advantages that would accrue to Bolivia from the establishment of communication between Bolivia and Europe *via* the Atlantic route. On the 10th, Dr. Velarde concluded his opening speech, and the opposition was commenced by Sr. Nicholas Acosta, who, making a violent and untruthful attack upon Colonel Church and all connected with the enterprise, begged the Deputies to decide that the enterprise should be put an end to at once and for ever. Dr. Miguel Rivas, Deputy for the Department of the Beni, and a supporter of Government, carried the debate over the 11th with an eloquent speech in favour of the project of the majority of the Commissions, Dr. Martin Lanza and Señor Enrique Borda, both of Cochabamba, also warmly supporting the enterprise. The debate was then adjourned to the 17th to allow for the election of a new Council of State, an operation that occupied the Congress for an entire week, owing to the excessively close balance of parties amongst the Deputies. On the 17th, General Quintin Quevedo ably defended Colonel Church against the accusations of Sr. Acosta, and Señores Sanjines and Navarro spoke in favour of the cancellation of the concessions to the enterprise, grounding their opposition principally upon the failure of the Public Works Company to fulfil their contract. Dr. Velarde replied to the arguments of the opponents, and the debate was again adjourned to the 19th, when Dr. Dalence, the Minister of Finance, being requested to attend and state to Congress the views of the Cabinet, said that the Government had thought it its duty to carry out the law of the 5th of November, 1873, by the nomination of Messrs. Aramayo and Terrazas as Financial Commissioners, and to support their action in London in consequence of their unfavourable reports as to the organization of the Companies, and the probability of the construction of the line, but that the Government were extremely desirous of seeing the Madeira and Mamoré Railway constructed, and was determined that not one penny of the deposited funds should be spent in any other object. The Government consequently considered that the project of law submitted by the majority of the Commissions should be adopted by the Congress. This speech of the Minister of Finance did the enterprise great good by determining the minds of many Deputies of the Ministerialist party, who had hitherto been in doubt as to the course that would be taken by the Government.

DR. QUIJARRO'S SPEECH—FIRST VOTING ON THE BILL—SECOND AND
THIRD READINGS.

The same day Dr. Antonio Quijarro, Deputy for Cobija, and one
of the newly-elected Councillors of State, delivered an eloquent
oration, embracing a complete *resumé* of the history of the enterprise
which he had supported from its inception, and while deploring the
delay and discredit that had occurred by reason of the default of the
Public Works Company, Dr. Quijarro earnestly recommended the
Assembly to vote for the decree presented by the majority of the
Commissions, and prove thus to the world that Bolivia really was most
desirous of emerging from her isolated position, and intended to
uphold the proper fulfilment of the contracts she had made. The
friends of the enterprise being determined that the project should be
fully voted before the close of Congress, which was to take place on
the 23rd of the month, called for a permanent session, and at 10 p.m.
of the 19th the first voting was taken, only about eleven Deputies
out of fifty-two present supporting the Opposition. On the 21st the
Commissions of Finance and Industry met at 8 a.m. prior to the
opening of Congress for the day, and agreed upon the final law that
should be submitted for the second and third readings. These readings
were quickly passed, a very feeble opposition being kept up by Señores
Acosta, Sanjines and Mas, while Dr. Belisario Salinas, Deputy for
La Paz and the Government candidate for the next Presidential term,
spoke very favourably of the enterprise, and eulogized warmly the
steadfastness displayed by Colonel Church in defending it against
attacks from all sides.

BILL PASSED.

The proposal of the majority of the Commissions, which again
declared the unaltered desire of the great majority of Bolivians to
have the Madeira and Mamoré Railway constructed, then became law,
except that it required sanction by the President of the Republic,
who affixed his signature and promulgated the decree in due form on
the 25th of November.

PUBLICATION IN SUCRE OF ADVERSE PAMPHLETS —OPPOSITION TO THE
ENTERPRISE BY F. AVELINO ARAMAYO.

During the days on which the debate was in progress the enemies of
the enterprise were in full activity, and several adverse pamphlets
appeared in Sucre, some employing fair argument as their weapon,

and some violent abuse and scurrility. One of the latter class, published by Señor F. Avelino Aramayo, son of Señor Aramayo, one of the Financial Commissioners in London, is entitled, "Bolivia y su Crédito Victimas de la Especulacion Church," or "Bolivia and her Credit Victims of the Church Speculation." This scurrilous sheet attacked Colonel Church in a most shameless manner, and contained all the erroneous arguments of the opponents of the enterprise, such as that the Public Works Company, Limited, was formed solely for the purpose of bringing out the loan of 1872, and other statements equally false. I replied to Señor Aramayo in the newspaper "La Cronica," already referred to, and exposed the fallacy of his arguments, while he, replying to me in another pamphlet, entitled, "Church y los suyos," or "Church and his Followers," boasts of the fact that the whole of the opposition to the enterprise had been created by himself when secretary to the deceased President Ballivian, who in 1873 was occupied for some time in London as Financial Commissioner of Bolivia. The younger Aramayo also acknowledges, in the pamphlet, that certain letters that appeared anonymously in the "Reforma," a newspaper of La Paz, in 1873, were written by him from London to President Ballivian, and states that he, by his representations of the mismanagement and "inconveniences" of the enterprise, induced his father (one of the present Financial Commissioners), General Campero (for some time Bolivian Minister to London), and Mr. Sampson, late city editor of the "Times," to believe that the "Empresa del Oriente," or enterprise of the East, was premature, and that the undertaking it would bring disastrous results to Bolivia. This Aramayo also carried his hostility to the extent of placing small handbills in the seats of the Deputies in Congress, insinuating that employés and others directly interested in the Company should not be allowed to vote in the project then before Congress with reference to the enterprise. This shaft had direct reference to Dr. Velarde, Deputy for Santa Cruz, and previously agent for the Company, who had been specially sent to Congress by the Cruzeños to protect and further the interests of the undertaking, but had also indirect reference to General Quintin Quevedo, who, Sr. Aramayo affirmed, was a director and shareholder, and to other Deputies who Sr. Aramayo ventured to hint had been bought over in the Company's interests. This uncalled for active opposition from the Aramayos, and the extremely vulgar style in which it was carried on, instead of damaging the Companies, had, I believe, an effect the reverse of the intended one, for it opened people's eyes to the fact that the desire of the Aramayo family was to crush the Madeira and Mamoré Railway, in direct opposition to the wishes of the majority of the Bolivian nation, for the sole purposes of obtaining the manipulation of the fund deposited in the Bank of England, and maintaining the isolation of Bolivia from the contact of the civilizing power of steam navigation on its eastern rivers.

FURTHER REMARKS UPON THE OPPOSITION TO THE ENTERPRISE.

The result of the contest in the Congress was to convince me, after much investigation, that the opposition to the Madeira and Mamoré Railway is kept up by but a small minority of interested parties in Bolivia, and by some of the principal merchants of Tacna and La Paz, who see in the success of the railway a break up of the monopoly of trade with Bolivia that they have so long enjoyed, and a consequent probable diminution of their profits. The oppositionists in Bolivia may be divided into Aramayistas and Paceños, of which the first are clients of, or related to, the numerous family of Aramayos, who are to be found in almost all the western towns of Bolivia, but whose reputation, if one may credit the pamphlets published in 1866 and 1867 by the creditors of the bankrupt mining firm of Aramayo Brothers, is not of the most immaculate.

LA PAZ V. COCHABAMBA.

The Paceños are, as a rule, very jealous of the growth of Cochabamba, and of the importance that will accrue to that town from the opening of the railway, and some of the Deputies from the La Paz provinces having at the inception of the undertaking opposed it, think it necessary to be consistent to the last, and register annually their feeble protest against the inevitable opening up of the eastern plains and rivers of their country; but deputies Dr. Belisario Salinas, of La Paz, and Señores Roman and Merisalde, of the Yungas of La Paz, deserve to be mentioned as having emancipated themselves from these narrow ideas, preferring to assist in the general development and welfare of their country rather than restrain their efforts to the benefit only of their own immediate provinces.

SPEECHES AT CLOSE OF CONGRESS.

On the 25th of November the Congress was closed with speeches from the President of the Republic, Señor Don Tomas Frias, and the President of Congress, Dr. Serapio Reyes Ortiz, the latter of whom congratulated the Assembly of 1874 upon "having dedicated the national resources, obtained upon credit, to place Bolivia in contact with the new world that is called the rich and exuberant valley of the Amazon." He also stated that "it was a matter of high consideration and satisfaction for the Assembly to hear from the speech of the President of the Republic, that that official, assuming rather the character of Prime Minister, had been, and would always be, anxious to show respect and homage to the majority of the Assembly, even at the sacrifice, should it be necessary, of his own opinion."

DEPARTURE FROM SUCRE DELAYED. FINAL INTERVIEW WITH THE MINISTRY. CAUSES OF DELAY IN THE MISSION OF DR. QUIJARRO. EXTRACT FROM LETTER RECEIVED FROM DR. QUIJARRO.

After the closing of Congress, I immediately determined to return home, taking with me an official copy of the decree passed, but found it impossible to obtain animals for the journey which I at first wished to make by way of La Paz, in order to afford the fullest possible information with reference to the position and prospects of the enterprise, to the newly-arrived Minister of the United States. The Ministry announced the intention of moving the seat of government temporarily to La Paz, in order to carry out some new arrangements proposed for the service of the custom houses of the coast; disorder in the battalion stationed at Cochabamba also took place, and these events rendered it almost impossible to hire or purchase animals for a journey. I was therefore detained until the 23rd of December when I was fortunate enough to get mules as far as Potosí, where I hoped to find postal mules for the prosecution of my journey to La Paz. Previous to my departure I had interviews with the principal Minister, Señor Baptista, who delivered me an official copy of the decree to be handed to Colonel Church, together with his assurances that Dr. Antonio Quijarro had been named the new Diplomatic Representative and Financial Commissioner in London, it being hoped that the Government would be able to dispatch him without loss of time. I was also informed by the Minister that it was the intention of the Government to seek a temporary loan from the merchants of Sucre and Potosí for the purpose of dispatching Dr. Quijarro on his mission, and several of these merchants assured me that they would be willing to respond to the request. The revolution now in progress broke out early in January, and doubtless has prevented the Government from carrying out their promises, for Dr.

78

Quijarro writes me from Sucre with date of the 20th of February last, saying that "the Minister Baptista had written to him from La Paz on the 12th of that month, to the effect that as soon as the pacification of Cochabamba was effected the Government would occupy itself in forwarding matters in London; and that he hoped that by one of the mails of March he would be able to send a final and satisfactory notice, for that it was clear, that the Government, whose prestige had increased by a victory over the united parties of Quevedistas and Corralistas, was firmly resolved to fulfil the wishes of the Assembly." These advices Dr. Quijarro requested me to participate to Colonel Church with his remembrances, and I am therefore convinced that Dr. Quijarro at his arrival will be found most anxious to co-operate cordially with the Companies in the prosecution of the railway and navigation enterprises.

SHORT REMARKS UPON THE FINANCES OF BOLIVIA.

A concluding paragraph of this division of my report may, perhaps, be advantageously devoted to a few remarks upon the financial position of Bolivia. Colonel Church, in his various papers notably that entitled "Bolivia and Brazil in the Amazon Valley," contributed to the "Fortnightly Review" of November, 1870, has so clearly and truthfully shown the vast undeveloped resources of the country that it would be superfluous for me to enlarge upon the subject of the "riches" of Bolivia, I would therefore limit myself to a consideration of its "poverty" from a financial point of view, the first and principal reason for which is without doubt its isolated position, which cannot be better described than by the words of, perhaps, the most talented Minister that Bolivia ever possessed, namely, Don Rafael Bustillo, who, to quote from Colonel Church's paper above referred to, wrote to the Brazilian Minister, in 1863, as follows:—"Bolivia, "as your Excellency knows, occupies a territory entirely central in "this vast continent. It has but five degrees of latitude on the "Pacific coast, and even this is disputed in part by the Republic of "Chili.* "Bolivia is seated upon the masses of silver of the "double range of the Andes. She has a territory fertile beyond "measure, where the treasures of the most opposite climates are "grouped together. With all this Bolivia perishes from con- "sumption for want of methods of communication which may "carry to the markets of the world her valuable productions, and

* This five degrees has now been reduced to two and a-half, by the treaty offered to Chili by the late Congress of Bolivia.

"stimulate her sons to labour and industry." The only certain means of ameliorating the poverty of Bolivia by the providing of a mode of transport for its at present useless riches, lie, in my opinion, on the eastern side of the Republic, for nature has declared that the route to Europe shall not be a western one : and the enterprise of the navigation of the Madeira and Amazon Rivers with the railway of the rapids, offers a more speedy and economic transit than can be afforded by any scheme having the River Paraguay for its basis. In regard to time the Madeira and Mamoré Railway can with ease dispatch its freights from the port of San Antonio to Europe in 28 days or possibly less, while the Paraguayan route from Bahia Negra to Europe will occupy at least 40. In reference to cost the Madeira and Mamoré Railway offers to carry a ton of freight from the centres of Bolivia to its market in Europe for £15, whilst the lowest estimate that I have seen for the Paraguay route was that of £26 per ton proposed in 1858 for the navigation of the Vermejo.

But another reason for the present financial poverty of the country is the absolute non-existence of any financial talent or even ordinary account keeping knowledge amongst the Ministers and officials in power of late years. The absence of any proper knowledge of national account keeping (for if the shortcomings of the officials of the Finance Department of Bolivia are not to be set down to want of knowledge, they must be charged to want of candour or straight-forwardness) has been thoroughly exposed in the matter of the loan of 1872. In the financial accounts for 1873 the debt appears as 8,500,000 Bolivian dollars or £1,700,000, the correct nominal amount of the loan, and although no notice is taken of the operation of the sinking fund which by the end of 1873 had paid off a first drawing of £34,000, there is in the estimated outgoings of the Treasury a credit taken for the service of a loan of £2,000,000. The Government also allowed Sr. Aramayo to charge Colonel Church with having received £10,000 of the loan and devoted the sum to his own purposes, whilst it must have known that the whole amount was fully and faithfully accounted for. In the published national accounts for the year 1871 the government acknowledges the receipt of the amount, and the same Aramayo's name appears as a witness to President Ballivien's receipt for £5,000 of the amount, whilst the remaining £5,000 was paid directly to the Government by Messrs. Erlanger and Company, but no account is published of the mode in which this £10,000 was expended by the Government, who allow the idea to prevail in Bolivia that the sum was given to Colonel

Church by the Government of General Melgarejo. Further the officials of the Treasury persistently deny that they have received any accounts from the firm of Messrs. Lumb, Wanklyn and Company in respect to the disposal of the loan, whilst in the report of Señor Pedro Garcia, Financial Agent of the Republic in 1872, to the Minister of Finance and Industry, published in "La Paz" in September of the same year, appears an account of the firm showing the state of the accounts up to the 9th of May of that year, and it is well known that several copies of the final accounts have been duly forwarded to the Bolivian Government.

The national receipts that even at the time the country gained its independence amounted to $2\frac{1}{2}$ millions of hard dollars, in 1873 only reached the poor total of 2,566,034 Bolivian dollars, or say £513,207, whilst the Minister of Finance declares a requirement of 3,660,679 dollars, or say £732,135, thus showing a deficit of £218,928 without offering to Congress any plan for equalizing the national account, and it is a fact that in the Congress of 1874 not one proposal whether financial or political, except the treaty with Chili, emanated from the Ministry. In the Ministerial statement of the National Finances for 1874 one sees at a glance that there is no effort made to equalize income and expenditure ; for not much more than £20,000 per annum is got from the Bolivian people by any kind of direct taxation. Customs rentals cannot be expected to increase until the Madeira and Mamoré Railway commences to create new entries on the eastern side, for Peru only can benefit by any growth of commerce on the western side, having stipulated with Bolivia that it shall only pay her £81,000 per annum out of the receipts of the port of Arica. It is certain, therefore, that direct taxation must be resorted to, and as there is no individual poverty visible in Bolivia, I cannot see any reason why the Government should not be able to show easily a fair balance-sheet which should meet the current wants of the nation, and provide honourably for the service of the public debt. A writer in " La Cronica " of Sucre, of October 23rd, 1874, after proposing the imposition of an Income Tax, a revised stamp act and a new mining code adds, " That such are the methods that " in his judgment should be taken in the then sitting Congress. The " country would not refuse its aid to avoid the national bankruptcy " imminent. In Europe confidence would be re-established, and re- " sources which the country requires for the regulation of its finances " would be obtained for the payment of the most onerous debts, the " recal of the bad money circulating and the opening of good roads

" between the principal cities, the effect of these measures being to
" raise Bolivia to the level of the civilization at its doors, and placing
" it in that high position which its natural riches and its central posi-
" tion amongst the nations of South America entitle it to."

VII.—SUCRE TO LONDON.

Start from Sucre.—Potosí.—Mines of Potosí.—Mint at Potosí.

Starting from Sucre on the 23rd of December, I arrived at Potosi
on the evening of the 24th, and remained there until the 26th, partly
on account of a slight attack of ague fever that I suffered from
there, and partly waiting for postal mules to take me on the first
stage of the road to Oruro. Potosí, once the largest and most populated town in Bolivia, is situated at about 13,200 feet above the level
of the sea, and owes its importance to the famous "Cerro" or Hill
which rises 2,300 feet above the town. This famous hill still gives
occupation to a population of about 25,000, while the ruined portion
of the city shows that but a few decades since a population of three
or four times that number must have found a livelihood from the
mining operations so easily carried on whilst the lodes could be worked
without any special appliances. The deposit of silver ore in this hill
may be said to be practically inexhaustible, for as soon as internal
means of transport in Bolivia shall be sufficiently advanced to allow
machinery to be brought to Potosí, the old levels and shafts can be
again worked, as the entry of water to these workings has been the
sole cause of the decline of the mining trade. At Potosí is the only
steam engine and machinery to be found in Bolivia, a machine for
coining having been at great expense taken there during the presidency of General Melgarejo. This machinery is made by Messrs.
Morgan, Orr and Co., of Philadelphia, and is managed by American
engineers and employés.

Potosí to Oruro.

From Potosí to Oruro was a journey of seven days, performed by
the aid of posting mules, the ordinary stages between each posting
house being about five or six leagues. The postal service is fair,
good baggage mules being kept, but a saddle mule is not to be

obtained at any price, an intending traveller must therefore provide an animal for his own especial use. This part of the road is excessively mountainous and barren, the only inhabitants being a few Quichuan Indians, who at each of the post stations grow a little barley and potatoes for their own wants, and for sale to travellers, asking exorbitant prices for their products, as much as twenty shillings being paid at times for a quintal, or 100 lbs. of barley in the straw. The town of Poopo, a day's ride prior to arriving at Oruro, is situated in the midst of several flourishing tin mines, and the whole district of Oruro is noted for its richness in this respect.

ORURO.—MINING OPERATIONS.—FREIGHT OF ORE FROM ORURO TO TACNA.—ORE TO BE CARRIED OVER THE MADEIRA AND MAMORÉ RAILWAY.—GOLD DISTRICTS NORTH OF LA PAZ AND COCHABAMBA.

Oruro, with a population of about 8,000, is especially famous for the mines of tin, copper and silver in its immediate neighbourhood. There are here several Englishmen and Americans engaged in mining, with as good results as can be expected to be obtained in the absence of other means of transport for the ore than llamas and donkeys. I brought home with me a small collection of minerals from this district, many of them containing as much as 10 per cent. of silver and from 40 to 50 per cent. of tin. The practice here is to pick from the stuff sent up from the shaft the best looking pieces only, the selecting being carried out by Indians who soon get very expert in the work. The selected mass is then broken up by hand into small chips, washed in order to rid it of as much of the earthy impurities as possible, and packed into small bags of 50 lbs. each, two of which form the burden of a llama. The freight paid for carriage to Tacna by llamas, the cheapest mode of transport that exists, is from 2 to 3 pesos per quintal, say £7. 5s. to £10. 15s. per ton. This rate is not very high, but the time occupied on the journey is very great, the principal house in Oruro assuring me that it often took twelve months before they could obtain a return from their investments in tin ore, or "barrilla," as it is there called. When the Madeira and Mamoré Railway is completed, and a fair road made from Totora to the port on the Chimoré, the existing road between Oruro and Cochabamba will be improved, and the whole of the mineral traffic of the central Andean Valley of Bolivia will doubtless find its way to European markets over the Amazon route, even before the finances of the country shall have sufficiently improved to enable an interior system of railways to be commenced. There is no great difference in distance between Oruro and Tacna, and Oruro and the port on the Chimoré, but a journey with animals by the latter route offers far less risk, as pasturage is plentiful everywhere, and there is no danger of the animals dying from the effects of the "soroche," so fatal on the pass of Tacora, the highest point of the Andes, passed on the road from Oruro to the Pacific Coast. One of the English miners, a Mr. Penny, who works a flourishing mine about a mile from the city of Oruro, and who has been resident in Bolivia for more than twenty-five years having travelled over most parts of the Republic, informed me that he had paid particular attention to the country north of La Paz and Cochabamba, and that he considered that the richest gold quartz in Bolivia, if not in the world, would be found in the mountains of that district, for he was certain that all the rivers that discharge into the Beni or the Mamoré flow over gold-bearing quartz.

The Revolution.

Arriving at Oruro I found that the revolution which commenced early in December by a mutiny of the 3rd Battalion at Tarata, near Cochabamba, had extended to La Paz, where the 2nd Battalion had declared General Quintin Quevedo President of the Republic.

Route by La Paz Abandoned.

The postal service between Oruro and La Paz was entirely disorganised, while the Government forces under General Hilarion Daza were on the road between the two towns, it was therefore impossible to prosecute my journey to La Paz, and I determined, after being detained at Oruro for a week purchasing animals, to take the straight road to Tacna by way of Tacora. During the week a night attack on the Government house and barracks of Oruro was made by a party in favour of the revolution, but after a smart fusilade, in which two of the soldiers defending the barracks were killed, the attacking force had to retire without having gained any advantage.

The Desaguadero.

I left Oruro on the 9th of January, the first day's journey being ended at "La Barca," on the banks of the "Desaguadero," or riverine canal which unites the lakes of Titicaca and Poopo or Choro. This canal has to be crossed on a pontoon or raft, kept in its course by a hide rope stretched from bank to bank, a distance of about 300 yards, the tolls being 1 reale, say 4¾d. per mule, ½ reale for a donkey, and five llamas are passed for a reale. As the traffic is considerable the ferry must be a fortune to the proprietor.

LLOLLIA.

A day was lost at Llollia, a small Indian settlement 24 leagues from Oruro, the mules having during the night escaped from the mud-walled enclosure in which they had been placed, and on the 13th we arrived at the village of Curahuara de Carangas, situated in a plain 12,890 feet above sea level.

CURAHUARA DE CARANGAS.

This is probably one of the oldest Spanish settlements of the Andes, the church being said to be more than 200 years old. Near by are mines of silver and tin, the ores being of rich quality, but they are not worked now, although the inhabitants of the place believe that the district will some day turn out to be a second Caracoles. Shortly after leaving Curahuara the track has to ascend a very steep and stony ascent leading to a vast plateau, 800 feet higher than that on which the village is situated, and from this plain rise the snow-covered peaks of Sahama and Las Tetillas, the former of which is said to rise to an elevation of 21,500 feet. Passing the postal stations of Chocos, Sepulturas, and Cosepilla, where provisions for travellers and forage for animals are scarce and excessively dear, barley in the straw costing £2 per quintal of 100 lbs., being rations for four animals for one night, we arrived at Tacora on the 16th, this being the first station in Peruvian territory.

PERU ENTERED AT TACORA.

At Tacora post-house the track for La Paz meets the one from Oruro, and a few leagues beyond takes over the pass of Tacora, probably 14,800 feet above sea level, and the highest elevation travelled over in crossing the Andes between Peru and Bolivia. At this pass the soroche is very fatal to animals, the path being bordered with bones and carcases of the mules or donkeys that have succumbed to its fatal effects.

Projected Railway between Tacna and La Paz.

Descending the mountain on the western side, I observed the marks left by the surveyors of the Tacna and La Paz Railway, an undertaking that appeared to me to be an almost impossible one, not so much from the height ascended to, as from the enormous amount of earthwork that must be encountered in following the curves of the ravines, up which the line is to be taken. In Tacna I was informed that the surveys demonstrated the practicability of the line, and that the ascent would be made with gradients not exceeding 3 per cent., but estimates of cost of construction do not appear to be made public.

Tacora to Tacna.

From the Tacora Pass to Tacna is a ride of a day and a half, principally down a narrow ravine, which would be barren were it not for the irrigation carried on at every available spot from the small supply of water at the bottom of the gulch.

New Road from Tacna to La Paz.

A short distance from Palca, a postal station nine leagues from Tacna, we came upon the new road that is constructing to La Paz; this will be a great improvement upon the old track, and proves that Peru is making every possible effort to keep her trade with Bolivia.

Tacna.

Tacna, where I arrived on the 18th of January, is situated in a broad and arid plain, to which water has been brought at great expense by an aqueduct, which, passing the hill of Tacora, leads a supply from the head waters of the river Maury, which flows from the eastern side of the Andes into the river Desaguadero of the central plain of Bolivia. This work has been carried out at very great expense, and does not yield a sufficient supply for the wants of Tacna and its vicinity. The commerce of Tacna is undoubtedly of very important character, but one sees at a glance that the only " raison d'être " of the city is owing to its being situated on the one exit practicable at present for Bolivia to the Pacific Coast. There are no Peruvian towns in communication with Tacna, and the long lines of mules and donkeys seen continually entering into or leaving the town, are all either destined for, or coming from, the neighbouring Republic of Bolivia.

Arica, the port of Tacna, is united to it by a railroad fairly con-structed and decently maintained, the works of the line having been but slight. The country between the termini of the line is a barren desert. Arica suffered greatly from the earthquake and tidal wave of 1868, the disastrous effects of which are visible in the ruined portions of the town, and the hulls of vessels washed far inland; but the custom house and pier have been rebuilt, and the place is slowly re-covering, although it is probable that it will never again enjoy its former importance, as the merchants of Tacna have reduced their establishments at Arica to the lowest possible point.

MESSRS. J. CAMPBELL & Co.

In passing Tacna and Arica it was impossible not to observe the hostility evinced by the only English firm of merchants there esta-blished, to the enterprise of the Madeira and Mamoré Railway. This firm, Messrs. John Campbell and Co., is in conjunction with and practically the same as Messrs. Hainsworth and Co., and Mr. John Hegan, of London. These firms own nearly all the stock of the Tacna and Arica railway, and in order to increase the value of that property have laid out a considerable amount of capital in projecting a railway from Tacna to La Paz, doubtless rightly judging that, as they hold the key to the coast in the Tacna and Arica railroad, they will be for years to come the masters of the position. The hostility of the firm of Messrs J. Campbell and Co. is thus probably accounted for, and I would do the general body of merchants of Tacna, comprising French and German establishments of considerable eminence, the justice to believe that they do not share the narrow views of the English firm, for some of the principals of the foreign houses alluded to evinced to me great interest in the progress and welfare of Colonel Church's enterprise, stating their opinion that as its realization would immensely assist the development of the internal resources of Bolivia, it must indirectly benefit instead of damage the Pacific trade and credit; for Bolivia offers a field sufficiently large for the introduction of capi-tal and commercial intelligence on both sides of her at present isolated position: and from one of the German houses of Tacna I received every possible courtesy whilst I was denied any attention at all by the English firm, simply, I believe, because I belonged to Colonel Church's enterprise, for my letter of introduction from Sucre ought to have entitled me to a different reception than that I met with at the hands of the said firm.

Leaving Tacna on the 23rd of January, I embarked the same day on the Pacific Steam Navigation Company's steamship the "Lima," and the rest of my journey viâ Callao, Panamá, Aspinwall, Jamaica, and St. Thomas' requires no comment, it being terminated by arrival at Plymouth on the 1st of March.

VIII.—RESULTS TO BE EXPECTED FROM THE ENTERPRISE.

GENERAL ANTICIPATIONS OF RESULTS FROM ENTERPRISE.

It is not perhaps too much to say that the realization of the joint enterprises of the Madeira and Mamoré Railway, and the National Bolivian Navigation Company, will change the entire character of not only the eastern provinces of Bolivia, but also of the Republic itself, whilst at the same time the Brazilian provinces of Mato Grosso and the Amazonas will be most materially benefited.

I believe that I am correct in stating that a maxim of political economy is to measure the value of a nation, not so much by the amount of its mineral wealth, as by the development given to its agricultural resources, for the former source of national greatness must sooner or later decay, if due attention is not paid to the creation or fostering of the latter. It may, in the case of Bolivia, seem absurd to say that its mineral wealth can ever appreciably decrease, and certainly such an assertion or fear must to any one that has passed over the higly metalliferous districts of Potosi, Oruro, and the whole central plain of the country, appear entirely groundless, but the examples of California and Australia teach us that though mineral discoveries are the first cause of the creation, and the settlements of new countries, it is their agricultural development that causes them to take rank amongst the nations of the world; and this it is that Colonel Church's enterprises will do for Bolivia, for there can be no doubt but that their realization will place Bolivia in the foremost rank of the Republics of South America. No scheme that has for its object the opening up of the country on any other sides than its northern and eastern can effect this result, for there alone exist immense plains and tracts of country suitable for any kind of agriculture or cattle rearing. On the western side of the Republic the barren and inaccessible heights of the Andes forbid any attempts at settling, while the southern and eastern territory of the Gran Chaco is a cheerless swamp, never capable of affording a home to other than the irreclaimable savage, or the wild animals of the fast decreasing forests of the continent.

Few, perhaps, are the enterprises that can offer to create and unfold such vast industries as those found in the districts, to be benefited by the opening of the Amazonian route to the interior of the continent ; for as we descend in an eastward journey from the barren summits of the Andean Passes, we find that the railway will prove the outlet not only for the mineral riches of Bolivia, her wools, hides, and other animal products, the drugs, dyes, and other commercial values of her unexplored forests, but will also develop agricultures that already exist in considerable scale on the descending plateaux of her eastern plains. Commencing at altitudes of 12,000 feet to 9,000, barley and potatoes are grown ; from 9,000 to 6,000, corn, potatoes, apples, pears, and all kinds of fruit ; from 6,000 to 2,000, coffee, coca, cocoa, and plantains ; while from 2,000 down to the plains of the Madeira and Amazon Valleys, cocoa, plantains, sugar-cane, maize, mandioca and other yams, tobacco, and other tropical products grow luxuriantly.

Politically considered, the enterprise will be of vast benefit to Bolivia, for her population, naturally a laborious one, will find employment in the impetus given to commerce, and will consequently become less turbulent, the need of, or occasion for, revolutions decreasing commensurately with the interest that each one will find in their increasing prosperity. A bond of unity will also be created for Bolivia, with her powerful neighbour Brazil, she being thereby rendered more secure from the encroachments of the Republics of the Pacific seaboard.

The results, financially considered, will be that probably a trade equal, if not superior to that now carried on through the Peruvian towns of Tacna and Arica will be created on the eastern side of the Republic ; and from the customs' receipts of this trade Bolivia would have far more than sufficient to keep up the service of, and rapidly pay off both her internal and external debts. Taking the year 1873 as a guide, we find that the imports through the port of Arica amounted to £1,422,369, and the exports to £860,607. Of these figures three-fourths of the imports, or £1,066,776, and £812,345 of the exports fairly belong to Bolivia, making a total of £1,909,121

Bolivian commerce that passes through Peru, whose duties may reasonably be averaged at 20 per cent. on the value of the trade, so that Bolivia annually affords Peru a rental of more than £381,000, out of which she magnanimously grants Bolivia a subsidy of £81,000. That Bolivian commerce is not decreasing is proved by the fact that the exports of Arica for 1874 exceeded by nearly a million hard dollars those of 1873.

Bolivia is generally supposed to have rather more than 2,750,000 inhabitants. The present Arica statistics, therefore, give something less than 14s. per head per annum, and it is not unreasonable to suppose that a similar amount of trade will soon be carried over the Madeira and Mamoré Railway. The loan of 1872 demands an annual service of £136,000, and figures that I have put together for paying off the other debts of the country would serve to show that another £120,000 per annum would release Bolivia from all its financial embarrassments. The total requirement of £256,000 would in a very few years be provided by the customs duties collected on the eastern route, for the amount is not equal to three-fourths of the duties now shown to be received by Peru from Bolivian commerce. Bolivia would still have the Peruvian subsidy, the profits received from sale of guano on the Pacific Coast, and her departmental rentals, for the general government expenses and the improvement of her internal means of communication. It is not, therefore, too much to say that the realization of the Madeira and Mamoré Railway may be made the means of materially changing and improving the present deplorable financial and political appearance of the Republic.

With regard to the profits to be realized by the railway from the traffic to be carried over it, I am prepared, after considerable investigation, to accept the figures given by Colonel Church in his " Preliminary Report made upon the Railway " in 1870, as sufficiently correct for a preliminary estimate, indeed there are many items of Bolivian produce omitted, which will doubtless find their way to the Amazon Valley, such as flour, barley and potatoes from Cochabamba, and wines and liquors from Cinti. In Colonel Church's estimate I would reduce the number of passengers looked for, but at same time the cost to the enterprise of their transportation is placed at such a high figure that I do not think the net returns will be at all decreased by the alteration. Also the cost of freight for a ton of goods over the railway is set at £1 which is at least 10/- more than it should ever amount to. Colonel Church's estimate of a net return from passenger and

7

freight traffic amounting to £108,040 may therefore be taken as the result that is likely to be arrived at within a short time after the opening of the line, and is sufficient to provide for a dividend of 10 per cent. upon the possible cost of the line : this dividend being as Colonel Church points out entirely apart from all traffic that will arise from the settlement of the lands under the grants of Brazil of 100 square leagues or 576,000 acres ; 23,040 of which have been allotted at San Antonio to the Company by the commission appointed by the Imperial Government for the demarcation of lands in the Madeira Valley.

In the hope that may desire to afford the fullest possible information upon every point that has come under my notice, has not led me to incur the fault of having been too prolix, I have the honour to remain,

Gentlemen,

Your most obedient Servant,

E. D. MATHEWS,

Resident Engineer.

Assoc. Inst. C. E.

London, 19, Great Winchester Street,
May 31st, 1875.

www.ingramcontent.com/pod-product-compliance
Lightning Source LLC
Chambersburg PA
CBHW020908210326
41598CB00018B/1806